The Devil's Cormorant
A Natural History

Richard J. King

UNCORRECTED PROOF

To be published by University of New Hampshire Press/
University Press of New England
on October 1, 2013

Price/ISBN: $29.95 (cloth) 978-1-61168-225-0
$24.99 (ebook) / 978-1-61168-474-2
Pages/Trim: 360 pp., 21 illus., 2 maps, 6 x 9"

For more information please contact:
Sherri.L.Strickland@Dartmouth.edu
University Press of New England
Or Call: 800-639-6102 ext. 238

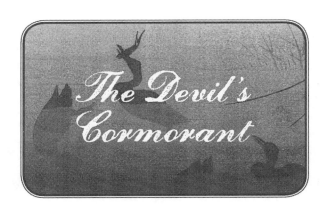

UNIVERSITY OF NEW HAMPSHIRE PRESS

DURHAM, NEW HAMPSHIRE

The Devil's Cormorant

A Natural History

RICHARD J. KING

University of New Hampshire Press

An imprint of University Press of New England

www.upne.com

© 2013 Richard J. King

All rights reserved

Manufactured in the United States of America

Designed by Eric M. Brooks

Typeset in Adobe Jenson, Calluna Sans, and

Magesta Script by Passumpsic Publishing

University Press of New England is a member of the
Green Press Initiative. The paper used in this book meets
their minimum requirement for recycled paper.

For permission to reproduce any of the material in this
book, contact Permissions, University Press of New England,
One Court Street, Suite 250, Lebanon NH 03766;
or visit www.upne.com

Library of Congress Cataloging-in-Publication Data
[to come]

5 4 3 2 1

For my parents

"Your father's right," she said. "Mockingbirds don't do one thing but make music for us to enjoy. They don't eat up people's gardens, don't nest in corncribs, they don't do one thing but sing their hearts out for us. That's why it's a sin to kill a mockingbird."

HARPER LEE,
To Kill a Mockingbird, 1960

Several men were shooting at black cormorants; and it developed that everyone in Cape San Lucas hates cormorants. They are the flies in a perfect ecological ointment. . . . They dive and catch fish, but also they drive the schools away from the pier out of easy reach of the baitmen. They are considered interlopers, radicals, subversive forces against the perfect and God-set balance on Cape San Lucas. And they are rightly slaughtered, as all radicals should be.

JOHN STEINBECK & ED RICKETTS,
Sea of Cortez, 1941

Contents

Illustrations appear after page 000.

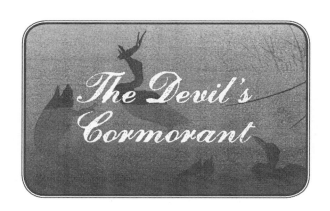

The Devil's Cormorant

March

A thin black line wavers across the sky. Perhaps they are a flock of undisciplined geese, but even from this distance they appear to be something else. They seem smaller than geese, more slender, their wing-beats quicker. As they gradually descend from their high flight, several of the birds in the line pause their flapping for several seconds. They soar, then resume beating their wings. These brief glides, in small groups, continue smoothly along the flock so that the string of birds seems from afar to pulse within its fluid formation.

The birds now fly close overhead. They begin to pull up vertically and alight on the island with their wide, webbed feet. Aside from the soft *shush-shush* of their wings, these animals have been completely silent: not a single honk or caw or whistle announces their arrival.

They land on a clump of rocks called Gates Island, just by the mouth of the Mystic River at the far eastern edge of Long Island Sound. Few people notice. From the town on the opposite side of the river, on a clear day, Gates Island is barely a gray-blue bulge on the horizon. That town is a summer tourist village that once built ships and still clings to a small commercial fishing fleet. This morning the mouth of the river is empty save for a couple of bald boats that were never hauled out for the winter, a few navigational buoys, and the hundreds of floats that mark the places of summer yacht moorings. It is quiet enough to hear the breeze over the river's surface, the plunk of oars into the water, and the occasional chime of a halyard clinking against a metal mast.

It is a cold morning when the first cormorants arrive, only a couple hours after dawn. The water directly beneath the sun is sparkly and bangled. Gates Island has neither trees nor any real foliage, only the spongy brown dirt and sand that have settled between piles of granite rocks and boulders. The cormorants find that heavy winter weather has blown away nearly all of the nests built last year.

In their current form, cormorants have migrated along the eastern coast of North America for millions of years. The earliest progenitors of this species lived perhaps sixty million years ago. Evolution has crafted an animal that can migrate the length of a continent, dive and hunt in the pitch dark beneath the water's surface, perch comfortably on a branch or wire, walk on land, climb up cliff faces, feed on thousands of different prey species, and live beside both fresh and salt water in a vast global range of temperatures and altitudes, often in close vicinity to man.

Six species of cormorants live in North America. In all there are about forty species of these birds, often known as shags, and they roost beside nearly every major body of fresh and salt water except the central Pacific Ocean. Cormorants live in Siberia and Tierra del Fuego, in Israel and India and Indianapolis, beside the Amazon, the Nile, and the Yangtze. Cormorants soar in the high altitudes of Nepal and Peru and dive to bone-chilling depths in both the Arctic and Antarctic seas.

This March morning the cormorants return to Gates Island and have a look around. They are hungry for fish. They shit sometimes.

This ability to catch fish and the nature of their guano have brought cormorants an extraordinary amount of human attention.

Gifu City

JAPAN

The cormorant is the bridge between God and man.

JUNJI YAMASHITA,
master cormorant fisherman, 2001

I'm sitting on a veranda that opens onto a well-manicured garden. From a perch in a small tree that extends over a pond, a single cormorant presides like a dark prince. The bird's gray feet are wrapped around the highest limb. A few more cormorants roost in a tall cage at the far side of the garden, while others waddle around freely or stand on the edge of half an old boat that rests beside the water. Occasionally one of the birds croaks, and once in a while I can hear the clinking of ceramic from the adjoining coffee shop. The name of this café is the Japanese character for cormorant, pronounced simply as *oo*.

One of the cormorants stands almost at my feet, by the step to the veranda. He is thirty-five years old. Apparently this bird has been blind for a few years. His eyes are glazed over with cataracts. The animal is so old and weary that he can't lift his neck. He sits perpetually curled over, with one side of his face on the stone.

Our host, the owner of the garden and the coffee shop, is Mr. Junji Yamashita.[1] He is a master cormorant fisherman, an *usho*, on the payroll of the emperor. He is one of only six usho in Gifu City, the most esteemed area for cormorant fishing in Japan. His official title is Cormorant Fishing Master of the Board of Ceremonies of the Imperial Household Agency. It is a hereditary position. Yamashita's extended family of cormorant fishermen can be traced back seventeen generations.

3

I've spent over a dozen years studying cormorants, observing a local colony of these birds, and traveling to places such as Gifu City to learn more about cormorants and our relationships with these seabirds. I first stepped onto this admittedly quirky path because I was looking for an interdisciplinary project for my master's thesis. I wanted to study one subject in a way that would require the gobbing together of literature, history, science, and current management policy.

One morning I read a front-page article in the newspaper about fishermen from Henderson Harbor, New York, going out on an island and slaughtering thousands of cormorants. I read this at the kitchen table. My housemate at the time, Munro, told me about how during his years in Japan he had watched men using these birds to catch fish. These men considered cormorants to be sacred animals. While eating a bowl of Cheerios, reading the article, and speaking with Munro, I remembered once learning in a literature class the contrast between a lovely poem by Amy Clampitt, titled "The Cormorant in Its Element," in which she extols the mysterious grace of how these birds can both fly and swim, against the epic poem *Paradise Lost* by John Milton, in which he used a cormorant perched in the tree to evoke Satan himself. I realized that I also knew of a small colony of cormorants at the mouth of my local Mystic River, on Long Island Sound.

Here was the project I'd been searching for.

So I began exploring more about cormorants. I soon found that this family of birds, for all their physical abilities and vast global range, has been reviled, gunned down, and demonized for at least five hundred years, particularly in Europe and North America. Yet in other parts of the world, either for their abilities to help men catch fish or the immense value of their guano for fertilizer, these cormorants have been valued, honored, and protected for more than fourteen hundred years.

I finished my thesis and went on to work on other projects, but I kept studying and thinking and observing cormorants while getting my PhD and then between the duties of my current day job as a literature professor. Over the years, I've kept gathering a guano-load of bizarre and compelling historic and contemporary stories about our relationship to these particular birds. All the while I've been trying to figure out why and how and where and when it was exactly that we either love or hate these animals. And in doing so, in studying the cormorant in literature and human history and science and in current management policy, I've arrived at a few larger lessons about our ever-shifting relationship to animals and the broader natural environment.

Studying the cormorant properly requires a circumnavigation of the earth, which we'll do in this book in huge flights across time and geography. Most of these places I have visited myself over the years, such as islands in Antarctica and the Galápagos. I traveled to touch the corpse of an extinct cormorant preserved at the research collections of the British Museum. I flew to Peru to put my feet on the most famous guano hills in the world. Closer to home, I visited Henderson Harbor, the Mississippi Delta, and the mouth of the Columbia River. Other spots I have visited vicariously through historic journals, correspondence, interviews, literature, and fine art. These are Cape Town, Bering Island, and the sheer cliffs of the Aran Islands off the Irish coast.

This book explores all these places and a range of different people and their varied relationships to cormorants. The people appear in this order: an usho, a charter fishermen, a writer of short stories, a 1940s naturalist-explorer, a future wildlife manager, a curator of bird skins, a sculptor and her favorite 1740s naturalist-explorer, a novelist, an aquaculture researcher, two guards of a guano island, and a veterinarian-bird rescuer. And then me: the first wing-bat, as far as I'm aware, to embrace this cormorant in a full-length cultural and natural history that migrates across the globe.

The trip to Japan was my first big expedition. I wanted to start with the oldest of stories and one of the strongest of connections to these birds. I was able to persuade my housemate Munro, fluent in Japanese, to come along. We recruited our other housemate, a biologist nicknamed Hoss.

From mid-May to mid-October each year, Junji Yamashita and his five fellow usho fish most nights, as long as it is not a full moon, an exceptionally high tide, or the water is not too cloudy with rain or sediment. The usho fish the Nagara River, which runs through the middle of Gifu City in a valley winding between steep hills. Gifu Castle, a modern reconstruction of a fortress that dates back to 1201, overlooks this portion of the river from the tip of Mount Kinka.

The Japanese word for cormorant fishing is *ukai*. It is a performance. Each night hundreds of spectators — mostly older-generation Japanese — come in buses and pile onto the dock, where one of the fishing masters puts on a microphone.

The usho explains his clothing and the history and process of ukai. He grabs a cormorant out of a basket and shows, with the help of an assistant, how he ties a string—a collar—around the base of the bird's neck. He feeds fish to the cormorant to demonstrate how its long neck expands with the food in its gullet. To the gasps and shutter-clicks of the spectators, the usho then makes the bird regurgitate four long fish out again—all at once onto a tray.

After the introduction, the usho, his assistants, and the cormorants—who are carried four birds to a basket—return upriver as if departing backstage. They go to join the others and prepare for the evening's fishing.

Some spectators get ready to watch ukai from the rocks along the shore, but the more luxurious way is by dinner boat, which is what we do tonight. We take off our shoes and find a seat. The crew poles the boat upriver and under the bridge. We watch the sun go down behind the mountains. Under paper lanterns, we eat and drink, anticipating the fishing. The boats play recorded traditional Japanese music, and at one point we watch a boat float by that is filled with entertainers in traditional Japanese dress. Some passengers on our boat light sparklers. By chance, this was one of the special evenings when there are also fireworks from the shore. The city had closed the roads beside the river to remove any noise and light from cars.

The first thing we see of ukai itself is a glow behind the bend of the river. The light is from the fires. Soon the six boats appear around this bend. Everything goes silent. The river is transformed. We can hear the oars through the water and the splashing of the birds. The crews of the dinner boats maneuver their vessels in a choreographed dance around the fishermen, staying out of the way while still providing every passenger a close-up view.

Ukai is spellbinding. This first time I see it I'm embarrassed to admit that all I can think is, how is it possible that this has not been made into a scene in a Hollywood movie?

Mr. Yamashita, like the other five usho, commands his own boat: a long narrow craft with a thin wood extension off the front, a sort of bowsprit. Yamashita stands up at the bow to tend his fire. This blazes forward in an iron basket, hanging from a wood pole fit into a slot at the base of the sprit. Yamashita has wrapped the base of this pole with leaves, providing lubrication as he rotates it from side to side, depending on where he wants to cast the light. He fuels the flames with slats of pine. The fire reflects off the water, attracting the fish. Embers pop and spark, floating down and sizzling into the water and dusting back toward Yamashita with the forward glide of the boat.

To protect himself from the fire, Yamashita wears the usho uniform. He looks like a sorcerer. He has wrapped a dark linen cloth around his head to shield his hair. He wears a full-length black or dark-blue cotton kimono that he has folded up under his straw skirt. This skirt extends below the knee and repels water, insulates, and protects his legs from the beak and feet of the birds. Despite this gear, Yamashita still has small burn holes in his kimono and in his *muneate*, an additional layer of cloth worn over his chest. On his feet, he wears special straw sandals with only half a sole, so that his heels meet the deck to keep from slipping.

While tending the fire, Yamashita manages a flock of twelve cormorants swimming in the river. He holds their leashes, the *tanawa*, in one palm while with his other hand he adjusts each string—plucking, overlapping, tugging, slacking—as the birds dive, fish, and return to the surface with their gullets full. While still managing the others, Yamashita pulls individual cormorants back to the boat, picks them up, squeezes the fish out of their necks into a basket, then returns each bird to the water. The fire lights up the cormorants' wet heads and the patch of white and yellow on their faces. When not diving, the birds swim ahead of the boat and under the flame, as if pulling the craft forward, as if the cormorants are dignified little black foxes guiding a dark Merlin.

The leashes fan out from Yamashita's hand. He has tied each line loosely enough around the bird's neck to not restrict movement, but firmly enough to stop the passage of larger fish down into its stomach. Yamashita explains to us later that the cormorants swallow small fish while they are diving. He says they could even eat the larger fish if they really wanted to, but there is an "understanding." The birds drink the oils from the fish, which is their favorite part, Yamashita says. Then the birds allow him to take their catch, since the cormorants know they'll get whole fish later.

The material and lay of the leash resists tangling and fraying. If Yamashita pulls the tanawa hard it won't break, but if he twists it sharply against the lay of the line, it will snap. In this way he can free a bird if its leash gets caught under a rock. Between the leash and the collar around the neck is a thin stick of stiff plastic, which hangs parallel to the cormorant's back. This helps keep the line out of the bird's way and helps the usho handle the cormorant. This stick used to be made of whale baleen.[2]

In each boat are two or three assistants. At the stern, steering with either an oar or a long bamboo pole, is the *tomonori*. At the waist is the *nakanori*, who helps to navigate with a smaller oar. This man also helps manage the incoming fish as well as the cormorants and their baskets. Yamashita fishes with an optional fourth man, a *nakauzukai*. This fisherman sends six additional cormorants into the river and manages them from the waist of the boat.

The demonstration of ukai lasts less than one hour. The climax is the *sougarami*, where all six boats form a line across the river and drive the fish. The usho shouts a guttural "ho-ho" while a crew member in each boat bangs his oar against the hull. Aside from a few camera flashes, the entire performance feels ancient.

After the sougarami, the fishermen drift their boats over to the beach in order to let their cormorants onto the shore. The birds croak and walk around, some standing with their wings spread. The usho removes the leashes and collars. Without irony, Yamashita and the other usho reward these birds with fish.

As the cormorants eat, the dinner boats return under the bridge and to the dock. Like after a sad movie, most of the spectators appear subdued and pensive. To increase the odd impression, the Gifu City tourism council plays through giant speakers a Japanese version of "Auld Lang Syne." The song echoes across the river. It is surprisingly moving.

If you think I'm exaggerating, that this melancholy sounds like nonsense, consider that people have been feeling this for hundreds of years. In the seventeenth century, the Japanese poet Matsuo Bashō wrote of ukai: "The cormorant fishing of Nagara River at Shō in Gifu is famous throughout the country and it is as fascinating as the accounts. Without wisdom and talent, I cannot possibly exhaust the scene in words, but I long to show it to those whose heart understands."[3]

Bashō wrote a haiku about his experience with ukai. Zempei Yamashita, Junji's second cousin, translated the verse this way:

> After the brightest sight
> Of the cormorant-fishing
> There remains a loneliness alone,
> The gaiety diminishing.[4]

In today's Japan, ukai is a cultural display. It is a form of entertainment and a way to preserve historical skills and practices. In this way it is somewhat like a falconry show in England or certain types of American rodeos. The cormorant fishing masters are in the same class in Japan as traditional sake brewers. Over one hundred tools of ukai are listed as "Important Tangible Folk Custom Cultural Assets of the Nation." The sound of the usho knocking the boat and his "ho-ho" shout is officially recognized in Japan as among the country's "top one hundred soundscapes."[5] To the emperor himself are presented the first fish caught in the cormorants' gullets each season, delivered officially from the governor of Gifu Prefecture and the mayor of Gifu City. When members of the royal family and their guests come to Gifu, they view ukai in a private sanctuary along the river.

People have been using cormorants to help them fish for at least fourteen hundred years. People have fished with cormorants in China, Japan, India, Korea, and possibly even in ancient Egypt.[6] Pictures and textiles left by the Chimu culture in Peru suggest they, too, domesticated cormorants to fish.[7]

It is not clear who exactly originated the practice. Fishing with cormorants in China has been conducted since at least the Sung dynasty (960–1279). Some archaeologists believe they have evidence of the practice in China as far back as 317 BCE.[8] Eventually fishing with cormorants became widespread and commercial over a large part of China, but now it is conducted predominantly in a few areas in the southern provinces, often as a tourist display, but also still as a local artisanal fishery.[9] The birds have been known colloquially in Chinese as "black-headed nets," "fish catching gentlemen," and "black devils."[10]

The first definitive written record of cormorant fishing in Asia or elsewhere, however, is in Japan. It seems to have come from a Chinese envoy visiting Japan in 607 CE. This is in the Chinese *Sui shu:* "In Japan they suspend small rings from the necks of cormorants, and have them dive into the water to catch fish, and that they can catch over a hundred a day."[11]

On the Nagara River near Gifu City, census records as early as 702 suggest there were cormorant fishermen living and working by the river.[12] In the twelfth century, twenty-one usho houses stood in seven villages along the Nagara.

In the early sixteenth century a female usho fished on the river. Her name was Ako. She apparently was not only skilled, but also fished with twelve cormorants, perhaps one of the first to use this many at one time. One Japanese scholar wrote: "This lady of old is regarded as the mother of ukai."[13]

In 1564 Nobunaga Oda, a warlord and major figure in the history of Japan, helped further ukai on the Nagara River. He treated cormorant fishermen with respect, declaring the position of usho as honorable.[14] Oda granted ukai and the fish caught by the cormorants a special status at Gifu Castle. He gave the usho money for rice each month and a new boat every four or five years. Soon Oda invited the servant of another lord in the region to come see ukai as a spectacle. Historians mark this as likely the first instance of ukai perceived as entertainment. In the following years, cormorant fishing started to become a court event, viewed by the nobility.[15]

On the Nagara, cormorants catch for the usho a fish named *ayu*. During the time of the Tokugawa shogunate, ayu became a popular sushi fish for the military. Thus ukai came under direct control of a feudal clan. The usho received money for rice in exchange for ayu sushi packed in barrels and sent to the shogun's castle in Tokyo. The shoguns outlawed dams, weirs, and net fishing on the Nagara River anywhere near where cormorant fishing took place.[16]

Ukai thrived under feudal protection for centuries.[17] In addition to the practice of using cormorants for fishing, cormorant feathers were gathered as roof materials for sacred birthing huts. This was presumably derived from ancient beliefs — both in Japan and China — that if a woman in labor had a cormorant in her arms or at least held a cormorant feather, her delivery would be faster and easier.[18]

The Imperial Restoration of 1868 pushed out the Tokugawa shogunate, which meant a temporary end to the stewardship of ukai. Yet soon the new emperor came to see ukai in Gifu for himself. Then he came back a couple of years later to watch it again. Before the end of the nineteenth century, he had issued a decree declaring that cormorant fishing in Gifu Prefecture was now protected by the Imperial Household Department.[19]

The city of Gifu gradually began to center its cultural and economic identity on this history and practice of cormorant fishing.

The first European description of cormorant fishing in Japan seems to be from the diary of English merchant Richard Cocks in 1617, which describes a type of ukai conducted from the shore. He wrote: "Soyemon Dono made a fishing over against English howse with cormorants made fast to long cordes behind their winges, and bridles from thence before their neckes to keepe the fish from entring their bodies, so that when they took it they could take yt out of their throtes againe."[20]

By the late nineteenth century a few more accounts had been published, including an 1889 article in the *Times* in Britain, written by Major General Henry Palmer of the Royal Engineers. Palmer's description of the practice on the Nagara River is strikingly similar to how ukai is performed there today. Palmer went on an evening dinner boat. The spectators drank hot tea and ate fruits, sweetmeats, and eel. He described first seeing the glow of the fires and then the sounds of the cormorant fishermen banging the hulls and shouting to urge on the birds.

Palmer wrote: "Next appear the forms of the boats and the swarthy figures of men, thrown up with weird, Rembrandt-like effects against the inky blackness of the night; and in the water round about the boats are numbers of cormorants, behaving to all appearance in the maddest fashion."[21]

Palmer described seven boats with four men in each. The birds had metal rings around their necks:

> The master is now the busiest of men. He must handle his twelve strings so deftly that, let the birds dash hither and thither as they will, there shall be no impediment or fouling. He must have his eyes everywhere and his hands following his eyes. Specially must he watch for the moment when any of his flock is gorged—a fact generally made known by the bird itself, which then swims about in a foolish, helpless way, with its head and swollen neck erect. Thereupon the master, shortening in on that bird, lifts it aboard, forces its bill open with his left hand, which still holds the rest of the lines, squeezes out the fish with his right, and starts the creature off on a fresh foray— all this with such admirable dexterity and quickness that the eleven birds still bustling about have scarce time to get things into a tangle, and in another moment the whole team is again perfectly in hand.[22]

Historians believe a few such Western accounts encouraged Europeans to take up the practice. Over the centuries it became an aristocratic sport. The earliest known record of semidomesticated cormorants in Europe is in Italy, revealed in a painting by Vittore Carpaccio, called *Hunting on the Lagoon*, painted about 1495.[23] What is happening in this painting has been much debated. It depicts several men in shallow boats aiming stretched bows toward the water in which cormorants are swimming. Some have interpreted the scene as the hunting of cormorants or just persecution of these birds for sport. A few details, such as cormorants standing serenely on several of the boats, reveal that what these men seem to be doing is actually shooting clay balls at fish, using their trained cormorants to scare these fish up to the surface.[24]

In sixteenth-century England, Elizabeth I placed a bounty on cormorants and other birds, declaring them "pests of the crown." Her successors James I and Charles I, however, maintained a "Master of the Royal Cormorants." This man was charged with keeping and training the birds to fish for the pleasure of the king.[25] In 1618 James I was so infatuated with watching his cormorants that he built a house and ponds for these birds, as well as for his trained ospreys and otters. The site was where the current Houses of Parliament stand.[26]

In the nineteenth century the sport of fishing with cormorants was revived by a few men in England, most famously Captain F. H. Salvin, who fished with the birds for at least thirty-five years. He had experience training hawks. The first cormorant that he trained in 1847 he named Isaac Walton. (After this prize bird died, they learned by dissection that Isaac was actually a female.)[27]

Across the Channel in France, the sport goes back to at least 1609. King Louis XIII owned trained cormorants. They were a gift from King James, arriving with an English keeper.[28] On record at Fontainebleau Palace is a position from at least 1698 through 1736 called *garde de cormoranes*.[29] In the nineteenth century M. Le Comte Le Couteulx de Canteleu owned cormorants at his castle and even wrote a short book on the subject. His *La Pêche au Cormoran* (1870) opens with an etching of a young Frenchman holding a cormorant named Tobie.[30] (See figure 3.) The young man, wearing breeches and a beret, holds his cormorant like a falcon. This was often the way cormorant fishing was performed in Europe, walking down to the water with your prize bird on your arm. Some even put little hoodwinks on the cormorants.[31] In the late nineteenth century the French painter Henri de Toulouse-Lautrec fished with a trained cormorant. The artist would promenade "Tom" on a leash in the evenings. He brought his cormorant into cafés, where it supposedly drank absinthe. (To Toulouse-Lautrec's horror, Tom died when a man shot him mistakenly while hunting.)[32]

In 1903, a French fisherman named Pierre-Amédée Pichot found the taming of the birds fairly easy. He wrote of the cormorant: "His heart is very near his stomach, and one may be reached by way of the other." Pichot continued: "If you feed him out of your hand, you will have trouble to prevent him from following you everywhere, ascending the stairs behind you, perching on your furniture, and leaving on all pieces incontestable traces of his rapid and abundant digestion."[33]

Even today in small surprising places outside of Asia, such as in Macedonia, locals still fish with cormorants, whether for food or catering to tourists. American author and sportsman Daniel Mannix fished with cormorants with some success in Pennsylvania in the 1960s.[34] A small group of indigenous people in Peru keeps domesticated cormorants on Lake Titicaca, the highest navigable lake in the world.[35]

Ayu (*Plecoglossus altivelis*), the fish that the cormorants hunt on the Nagara River, translates to "sweetfish." It is closely related to salmon and found only in the streams, lakes, and coastal waters of Asia. Ayu can grow up to one foot (30 cm) long. Toward the mouth of the river, the fish eat algae, plankton, and crustaceans, but up in the Nagara the ayu are mostly vegetarian, scraping their food off rocks and leaves with specially adapted teeth.

The ayu has a varied role in Japanese culture. Scientists like to count ayu, using them as an indicator species to gauge the health of a river. Fish farmers raise ayu in Japan. Recreational anglers vie for them, too. One sporting method is to put a live ayu on a hook. Since ayu are aggressively territorial, a fish will come out to attack another and get snagged.

Maritime anthropologist Tomoya Akimichi wrote that the ayu is probably the most popular freshwater food fish in Japan.[36] Tourism brochures claim that ayu smells like watermelon. The fish is served smoked, preserved in salt, grilled, fried with tempura batter, served raw as sashimi, and fire-dried for making broth. Ayu is most popular as *shioyaki*, where it is skewered, rolled in salt, then roasted.

During our visit to Gifu, Munro, Hoss, and I tried them all. We even bought ayu-flavored snack crisps and gift-wrapped ayu pastry. Ayu does indeed taste good. To me it has a mackerel flavor. (It does not smell like watermelon.)

I first tasted *shioyaki ayu* when I was so kindly treated to dinner by Mr. Katuzo Hino, one of the volunteer tour guides in Gifu. In her purse, his wife smuggled into the restaurant a sample of her own specially prepared *shioyaki ayu*. She felt that I should try it. I smiled nervously. I didn't know how to eat it. Mr. Hino had some English, but not a great deal, and Munro and Hoss were off singing karaoke. The grilled ayu did not seem a chopstick food. I kept smiling and took it up with my fingers. I bit into the middle of the fish as if it were corn on the cob.

The Hino family started giggling.

I apologized deeply.

Mr. Hino said: "No, no. Good. You eat it like we do."

Yamashita and his fellow usho use Japanese cormorants (*Phalacrocorax capillatus*).[37] Japanese cormorants are most commonly found on the rocky coasts of central and northern Japan. They also breed on the coasts of nearby islands and the facing shores of Siberia, China, and Korea. The feathers of the adults, like those of the majority of cormorants around the world when breeding, have iridescent greens, bronzes, and purples within their glossy black. Adult Japanese cormorants also have thin white feathers, called filoplumes, that extend out from their head and neck. Individuals can have so much white feather fluff that it looks like they have a white beard. Hence the Latin species name, *capillatus*, which means "hairy." Japanese cormorants have a white-speckled band around each bright emerald eye. As with every cormorant species except the Galápagos flightless cormorant, the females tend to be just slightly smaller. Female Japanese cormorants, as with all cormorants, have the same plumage as the males. It is difficult to distinguish gender in cormorants. Scientists who absolutely need to get it right use DNA samples.

There is also a freshwater cormorant in Japan, the great cormorant (*P. carbo*). This bird looks similar to the Japanese cormorant but is also native to the interior rivers and lakes. The great cormorant is the most widespread of all cormorant species, ranging from Tasmania to Ethiopia to Nova Scotia. Though it might seem more logical for usho to use this great cormorant, since they are already adapted to fresh water, Yamashita and his fellow fishermen of the Nagara River have long preferred the Japanese species. They believe this cormorant to be heartier and easier to train.

Japanese cormorants are also a bit larger than the greats in this part of the world, so they are capable of catching larger fish.

Scientists and fishermen have found cormorants to be pretty intelligent as bird brains go. No other birds anywhere in the world have been trained to catch fish for man in any substantial way. Cormorants are certainly highly trainable. British and Israeli biologists who have raised wild cormorants in research aviaries have been able in a matter of weeks to devise intricate experiments by positively reinforcing the birds by feeding them fish, evocative of dolphin training.[38] English researchers in the 1970s observed a set of domestic cormorants in China that could count to at least seven, since the birds would cease their work until rewarded for catching that many fish.[39]

Throughout the world cormorants in the wild work collaboratively to catch fish, and numerous accounts from both China and Japan describe trained cormorants fishing together to help haul an especially large fish within the grasp of their master.[40]

In China, cormorant fishermen often work their birds without a leash or neck collar of any kind. The cormorants respond entirely to the commands of the owner, even when fishing among large groups of other fishermen and their own birds. This is documented in a short film by Frédéric Fougea titled *He Dances for His Cormorants* (1993).[41] Fougea crafts a story around a cormorant fishing family and their close relationship with these birds. Their cormorants do not have anything around their necks. Zong Man, the patriarch of this Chinese family, walks around with cormorants on his shoulder, beside him while he eats, and with several of them on a pole while he is driven in a little car to an alternative fishing site. When a chick is born, Zong Man stays with the newborn for several days in isolation so that he may feed the chick in private and the bird will imprint on him.

In the documentary, Zong Man has named his birds much as Santa has done with his team of reindeer. The cormorants are Escaper, Sleepy, Gimpy, and Son of Gimpy. He named the one who captures the largest fish Mao. Zong Man describes each individual bird's personality and his or her skills.

Several other accounts throughout Asia describe how the birds reveal their individual personalities. On the Nagara River in Japan, cormorants seem to recognize a form of pack hierarchy. After a session of ukai, the cormorants line up on the gunwale of the boat, by seniority. The senior bird, the *ichi*, stands toward the bow or perches on the sprit. The other birds follow the ichi's example. Like sled dogs, the cormorants become overtly upset if one breaks ranks by not lining up in the proper order.[42]

Cormorants are highly trainable. They learn quickly. They can fish collaboratively both for themselves and for people. They have individual traits and a social structure. They can count.

Not everyone is enamored with ukai. To many animal lovers it appears a cruel process for both the birds and the fish. After I chomped into that ayu at the restaurant, the special ukai-caught fish, I saw scratches by the head: from the cormorant's beak. Tourist brochures claim dubiously that ayu are stunned when caught by the cormorant and dead before hitting the basket.

Part of the objection of animal rights advocates is how the cormorants are first captured. The Japanese cormorants trained for ukai are gathered in late spring from the rocky shores of Japan. Yamashita's birds come from over two hundred miles (320 km) away, from the cliffs of Ibaraki Prefecture overlooking the North Pacific. With wood or live decoys, men attract the cormorants to a section of cliff that they have spread with sticky lime—like flypaper. After they catch the birds and tie their beaks shut, the men stitch through the skin on the top of the birds' head in order to sew up the eyelids for the journey back to Gifu. The men try to catch cormorants that are about one year old.

Once back in Gifu, Yamashita and his fellow usho clip the tip of the birds' beaks and snip a few feathers from one wing. They slowly bathe the birds in warm water, clean off the lime, and remove the eye stitching. The usho hand-feed them fish and gradually adjust the birds to foraging in fresh water and being on a leash. It can take a couple of years until a cormorant is ready for an evening's ukai demonstration.

Chinese fishermen, on the other hand, raise their birds as chicks, hatched from eggs, and have thus completely domesticated them. The men often place the cormorant eggs under the bellies of chickens until they hatch. For genetic diversity they occasionally go into the wild to get new eggs.[43]

Yamashita thinks the Chinese method uses animals strictly as tools, similar to the keeping of cows and pigs. He believes the traditional Japanese method is a more equal one, since the birds are caught wild and not cultivated. Yamashita claims that even if he left his cormorants' wings un-clipped they would not fly away. He says that he removes a few feathers to make it easier to put them into the basket and, he claims, to "remind them that they are members of the usho family."

Those who disapprove of ukai point out that the cormorants are leashed and swimming amid streams of sparks. They think the birds ap-pear nervous, and people worry about the animals getting tangled. Some complain that the fishermen handle the cormorants too roughly, that the birds have been kidnapped from the wild. One author compared the rais-ing of these wild birds to that of a relationship between a hostage and its captor.[44] In response, there is now an independent cormorant fisherman in Gifu, Mr. Riki Nakane, who for over a decade has been seeking to alter ukai by adopting the Chinese method of raising the cormorants from eggs. He fishes with a ring around the birds' necks, without a leash. Nakane's cormorants come back to him by a whistle.[45] (During the off-season, Ya-mashita and his fellow usho let their birds off leash, too. They free them into the Nagara to catch their own food.)

There is a small history of distaste for ukai, particularly among Bud-dhists. An early Japanese folk song, dating back to before the twelfth cen-tury, confirms that some have always had their doubts:

> Woe to the cormorant-fisher
> Who binds the heads of his cormorants
> And slays the tortoise whose span is ten thousand æons!
> In this life he may do well enough,
> But what will become of him at his next birth?[46]

At the turn of the fifteenth century, Japanese author Enami no Saye-mon wrote a play titled *Ukai*. This tells the story of a cormorant fisherman who is a doomed sinner. The usho says in soliloquy:

> Then, in the dreadful darkness comes repentance
> Of the crime that is my trade
> My sinful sustenance; and life thus lived
> Is loathsome then.
>> Yet I would live, and soon
> Bent on my oar I push between the waves
> To ply my hateful trade.[47]

As the play continues, the cormorant fisherman goes to a shrine to rest his birds while he waits for the moon to go down so he can return to fishing. He meets Buddhist priests who are traveling through. He tells them a story of how villagers drowned another cormorant fisherman for cruelly killing ayu with his birds. The villagers had shouted "one life for many."

Mr. Yamashita is a Buddhist. Sitting in his veranda, he explains to us that the Japanese character for cormorant means "little brother bird." But, he grins, he does not know who is the big brother. He explains that with animals "what you see is what you get — there is no front and back."

The more we talk with Mr. Yamashita, the more he really gets going. Thanks surely to Munro's tactful translation, he seems to begin to trust us. He speaks about a sincere love for the birds. He said that he would not be able to do the work if the cormorants did not accept him. He lives with these birds. He gets up in the morning with them. He goes to work with them. Yamashita says the cormorants are on a different tier than other animals. The analogy that best comes to mind is that of a hunter with his retriever or a sheep rancher with his collies.

Yamashita looks out into the garden at a cormorant roosting on a tall limb. He says: "The cormorant is the bridge between God and man."

Munro, Hoss, and I crane our necks up at a Godzilla-size mural of a cormorant painted around three sides of a fourteen-story building. Then we separate to canvass the place. In Gifu City, cormorants appear in various forms on buses and street signs. They are embossed into the iron of sewer caps. (I take a charcoal rubbing.) Cormorants are welded into the stanchions of bridges and etched into the glass of phone booths. We find ukai scenes and cormorants painted on murals, created in mosaics on the walls of public transportation centers, and depicted on the sidewalk in front of an athletic facility. Usho and cormorants are honored in bronze statues in prime, central locations. Bashō's poem is etched into stone by the river. There is a cormorant burial ground. The city has erected signs in both Japanese and English to identify the usho's homes for tourists. Illustrators have drawn cartoon cormorants and caricatured usho on bank posters, business advertisements, and at the post office. Gifu City's mascot is a big furry cormorant dressed in usho clothing. Tourism brochures proudly declare how the silent movie actor Charlie Chaplin was so enamored with ukai that he came back to see it a second time.

Volunteer guides like Mr. Hino and his daughter wear long shirts decorated with ukai prints as they greet tour buses. The guides also hold paper fans with matching ukai scenes painted on them. In the souvenir shops throughout the city, we find a dizzying range of ukai mementos, including hand-painted ukai paper lanterns, ceramic bowls with glazed ukai scenes, and bean pastries wrapped in fancy paper with ukai patterns. The shops sell ukai key chains, ashtrays, bottle openers, bracelets, T-shirts, and toothpick holders. I go on a shopping spree. Munro finds miniature wood models of the boats and cell phone straps with an attached plastic cormorant dressed as an usho. Hoss finds in another shop an ear-pick with a plastic cormorant icon on the handle. He finds a dashboard ornament of a cormorant bobbing in an usho's basket.

While in Gifu City we spend several evenings viewing ukai from boats and on the rocky shore. We go to the art museum, where an exhibit features contemporary artists who painted cormorants and ukai scenes.

The history museum maintains a permanent exhibit on ukai. Here we find on a back wall a gigantic photograph of a small boy dressed as an usho. In the photograph, the straw skirt extends almost to the boy's feet. He holds his little hands toward a glow, presumably a fire. The curator tells us the origin of this image, explaining that this is a boy during a holiday where children dress up as what they want to be when they grow up.

"I'm not absolutely positive," the curator says. "But I'm fairly sure that is Mr. Junji Yamashita."[48]

The Gifu City government subsidizes ukai. The imperial household donates a small amount for the usho's salaries, but the city covers the majority of everything else. Over the last decade, one hundred thousand people a year on average have come to watch the usho and their birds. The boat trips bring in about 300 million yen ($3.6 million) a year, but it costs the city 500 million yen to run the entire operation. The city government and the taxpayers front the difference. Mr. Hino, who works for the city, says they estimate that the "extended economic effect" — hotels, restaurants, shopping, and all other spending that tourists bring — more than makes up for this investment, perhaps earning the city of some 415,000 people well over 4 *billion* yen each year.[49] The city also earns tax money from ukai and employs directly about 170 people in ukai-related jobs.[50]

Ukai has changed subtly over the years. Hino marks 1963 as the year when the ayu population decreased to a level where fishing with cormorants for subsistence was no longer possible. Factory pollution and chemical input increased during a time of rapid economic growth. No longer would the cormorant fishermen sail downriver for trips lasting several weeks, even several months. They used to travel more than thirty miles to the Ibi River. The usho used to sail in small flotillas, living together on their boats with their cormorants, catching enough fish to sell in order to support themselves and their families.[51]

After 1963 usho had to begin feeding their cormorants more often with frozen fish. They began to use synthetic line for the tanawa, instead of the split inner bark of the Japanese cypress. They discarded the use of sails, preferring to use outboard engines to save time when transiting to the fishing grounds. Gifu City organized the introductory talks for tourists and the volunteer guide system, making sure some of them spoke a little English, such as Mr. Hino. Though the usho maintain traditional dress and their assistants wear Japanese coats and head cloths, some in the boat now wear wristwatches, khaki shorts, and even neoprene booties. They use a flashlight as a running light and drink from plastic water bottles.

In Gifu City some believe that ukai helps the ecological health of the Nagara because the government and its people want to care for the river, motivated by ukai's importance to the local economy. Others believe ukai harms the river because the tourists on the shore and in the boats pollute the water.

The Nagara River was once part of a large, clear, free-flowing system until a large dam was completed in 1995, about twenty-five miles (40 km) south of Gifu.[52] Fish gates are installed in the dam, but potential damage to ayu populations was at the center of the significant public protest against this massive construction project. Dr. Shiro Kasuya, a professor at Gifu City University, remains opposed to the dam and has quantified the resulting ecological damage. Kasuya told me that ukai tourism has declined sharply. He attributes this to the dam. During the decades before the construction, Gifu attracted more than twice as many visitors as today.[53]

Kasuya wrote in 2007: "The hotels in Gifu City, which are located along the banks of the Nagara River, have been forced, one by one, into closure. There is no attraction in watching the cormorant catch and swallow

THE DEVIL'S CORMORANT

sweetfish which are no bigger than a few centimeters in length. Sweetfish of 20, sometimes 30 centimeters [1 foot] glowing on an open fire has become a sight of the past. The flavor of the fish, which was the lifeblood of the tourism industry, has also inevitably been lost. The cormorant fishermen see their living fading away."[54]

Yamashita has fished his cormorants throughout this transition. He agrees that the river's health has declined during his lifetime, but from his perspective, this is the wrong way to look at ukai, his profession, and the birds.

"The primary role of ukai is its cultural importance," Yamashita says. "It is not for keeping a river clean or even for catching fish. It is a cultural tradition. It is spiritual nourishment. To learn about sharing, work, and cooperation between animals and humans."

Sitting in his garden, Junji Yamashita tells us that he has been an usho for fifty years. It was his father's career. He was forced into it. Eventually he established his own way to approach the work and the cormorants. Now he believes he has the best job in the world. "I was born to do this work," he says. Yamashita believes the passing down of the usho title from father to son, for so many generations, is part of what has helped ukai survive. Each time it is passed down, it changes a little. He added that cormorants have also changed over the fourteen hundred years of ukai's history. Invoking the metaphor of the tanawa, the cormorant's leash, Yamashita says: "There is a thread from me to the first usho."

Yamashita explains that Confucius said that up until a man is sixty he does what other people say. After that, he can do whatever he wants. Now that he is over sixty, he has started listening to the cormorants. Yamashita tells us that it has taken him most of his life to realize the value of an usho. He was just going through the motions of a cormorant fisherman without understanding its true meaning.

Yamashita has one son. He rides in the same boat with his father, as the optional fourth boatman. "He's not looking forward to the work," Yamashita says. "But he *will* do it."

He laughed when he said that his son, like him, might be sixty before he realizes what a wonderful profession it actually is.

Our visit begins to wind down. He declines our invitation to lunch because, he jokes, "I have false teeth."

We say our goodbyes. We bow and give him our gifts. But as we walk onto the street, Mr. Yamashita hustles out and asks us to return for a moment. He leads us back into the garden. A couple of cormorants are waddling about. One splashes in the pool. The old one slumps with his head still on a stone. Another dark, regal cormorant sits motionless on the highest tree limb. We follow Yamashita to the very back of his garden-aviary. In an alcove he reveals an entire diorama display about ukai. The exhibit is filled with photographs, mounted artifacts, boat models, illustrations of cormorants, and written descriptions. He explains that his father made it after he had a stroke and could no longer go out and fish. His father died the very day after completing the private family display.

April

Male cormorants stand in different areas of the small island to claim spaces for nesting. They dart out their necks and spread their beaks wide to warn others who come close. Males try to attract the females by lifting their tails vertically and rearing their head and neck toward their tail so their crests almost touch their back.

One glossy black cormorant flies into the colony. With a croak like the sound of a creaky old door hinge, he lands on one of two large boulders. He hops down and stands beside a female on a nest. These two boulders, at the high point of the island, provide some shade and shield the nest site from the wind. With his beak the male passes to the female a stick that he has just retrieved from the nearest shore. She tilts her head and places the stick into the nest from the outside, forming the structure around herself. She had begun her work with a depression in the soil and a few nesting materials left over from the previous year. It was one of the only nest remnants that survived the winter. She has added more sticks, seaweed, dry grass, and a couple of other materials found on the island, including parts from two spider crab shells, a skate's egg case, and some green ribbon once attached to a child's balloon. With the male's help, she has been building this nest for two days. She will be done in a few more. She and the male regularly splatter their white guano over the top to help solidify it all.

The male flies back out across the channel and returns a short time later with another thin branch. The female worries this into the nest. They both stretch their necks and open their beaks slowly, as if yawning, revealing the bright blue color of their gapes. They will be the first pair to breed on the island this year.

She stands up in the nest. She expands her orange gular pouch and opens her beak again, arching her back. The male mounts her. He holds her neck with his beak, spreading his wings occasionally to keep his balance. He presses his swollen cloaca to hers.

Henderson Harbor

UNITED STATES

*I'd pull the troops and carriers out of Afghanistan
and napalm the [cormorant] rookeries. For years, they've
done fighter runs over the Great Lakes to practice
bombing. Why not the real thing?*

MARK CLYMER,

sport fisherman, 2002

aptain Ron Ditch has spent the last twenty-five years leading the fight in his small community of Henderson Harbor, New York, against the local population of cormorants. He founded an organization called "Concerned Citizens for Cormorant Control." To build his case, he amassed reams of biological and policy research on the birds. He and a local television producer, who is also a fisherman, created a short film to demonstrate the need for managing the local cormorants. His most significant act—and he will tell you "achievement"—was when he organized a small group of fellow fishermen to go out and kill thousands of cormorants on the night of July 27, 1998.

Ditch and a few other men, including his two sons, had already gone out a couple of times in previous years to kill cormorants on Little Galloo Island in eastern Lake Ontario.

"But we were losing," he tells me. "So I gathered a few boys together and told them what I wanted to do. We knew it was illegal, but we had to send a loud message. I told them it's a fifty-fifty chance we were going to get caught. But we either lose our business or take the chance. I asked if they were in or if they were out. Five said they were in."[1]

Thanks to a good idea from his wife, they went out on a night when there were fireworks over in Oswego. As Ditch tells it, they took four

24

boats and powered the ten or so nautical miles over to the island. It's a flat island with a rocky edge. Only a few dead trees remain, which are close to the water. The cormorants like to nest in the limbs. A couple of men from the deck of one boat shot as many as they could out of these trees. Once they started shooting, though, the adult cormorants flew away. Tens of thousands of gulls flew overhead and shrieked and cawed.

Little Galloo Island is not an easy place to tie up, but it wasn't that windy that night, and it was clear enough so they could see enough of what they were doing. Anchoring to keep their boats off the rocks, the men stepped ashore. With the thousands of shrieking gulls still overhead, the men shot at any adult cormorants that had not flown away. The cormorants gathered in the water just off the island, near enough so one of the fishermen still in a boat could pick more of them off. Ron Ditch is a duck and deer hunter, as are the other men. They are good shots.

By this midsummer night nearly all the cormorant eggs had hatched. Most of the chicks were old enough to walk around. Those that had learned to fly had tried to escape to the water with the adults. There were also thousands of chicks that could walk but could not yet fly. These fledglings huddled into what biologists refer to as nursery groups or "crèches." Ron Ditch and his men walked over to these groups and opened fire. The young birds' high-pitched peeping was impossible to hear over the noise of the gunfire and the gulls.

"We went out and killed about twenty thousand cormorants," Ditch tells me. "And the feds got all upset about it. They fined us $100,000. But it did get their attention and got the attention of all of North America. Had we not done that I don't think the situation ever would've changed. We were told that the federal Fish and Wildlife Service's attitude towards the cormorant situation would never change. In other words, the tree-huggers and the bunny-huggers didn't want you to do anything as far as birds were concerned even though they were destroying a way of life and a fishery and affecting millions and millions of dollars in a fishery that was dying because of the birds. So, we did. We took the law into our own hands."

"Did you say you killed twenty thousand birds?"

"Yes, they prosecuted us for two thousand, but they had no idea what was going on. I wish we got another twenty thousand."

Ditch adds: "I guess I kind of became a local hero."

The government did indeed take action after this incident, both on the

Red-faced Cormorant
(*P. urile*)

Pelagic Cormorant
(*P. pelagicus*)

Brandt's Cormorant
(*P. penicillatus*)

Neotropic Cormorant
(*P. brasilianus*)

Great Cormorant
(*P. carbo*)

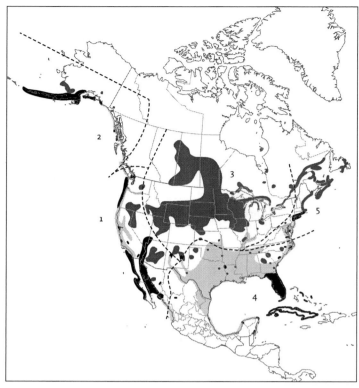

Double-crested Cormorant (*Phalacrocorax auritus*)

Six species of cormorants live and breed in North America. In the map of the double-crested cormorant's range (above), dark gray represents breeding grounds, light gray wintering grounds, and black both breeding and wintering grounds. In maps of the ranges of the other five species (left), dark gray represents the extent of their combined breeding and wintering grounds. In North America, the double-crested cormorant can arguably be divided into five populations: 1) Pacific Coast, 2) Alaska, 3) Interior, 4) Southern, and 5) Atlantic. For management purposes, USFWS identifies four groups, combining the Pacific Coast and Alaska populations.

After Birds of North America Online 2013 [1999 data];
USFWS 2009; Wires and Cuthbert 2006.

state and federal levels. State wildlife managers have since slowly reduced the size of the cormorant colony.

Over a decade later, I meet Ron Ditch at his house once again. He says to me: "I guess everything has come out pretty good, huh?"

Six species of cormorants live in North America.[2] (See map.) The double-crested cormorant (*Phalacrocorax auritus*), the object of Ron Ditch's ire, is the most prevalent and the only one found appreciably inland. Double-crested cormorants nest in the spring in forty-three states, in all of the ten Canadian provinces, as well as Cuba, Mexico, and the Bahamas.[3] During migrations or winter roosting they spend time in every state except Hawaii. They travel as far north as Alaska and Labrador and as far south as the Yucatán Peninsula.[4]

The double-crested cormorant gets its common name from two tufts of feathers on its head. The species name *auritus* — Latin for "eared" — comes from these crests. To the human eye the crests look mostly comic. When viewed from the front with binoculars, the bird looks like a mad professor. The crests are only visible for several weeks a year when the cormorant is attracting mates, coinciding with an increased gloss and sharpness to all the bird's colors. The bare skin under the double-crested cormorant's beak is bright orange. This is called its gular pouch, a much smaller version of the same on its pelican cousin. Like Mr. Yamashita's Japanese cormorants, the double-crested cormorants' eyes are emerald green. When attracting the opposite sex, double-crested cormorants have a vivid cobalt-blue lining inside their mouth.

Ron Ditch and other charter captains and sport fishermen throughout the Great Lakes think there are too many cormorants in their area and across the continent. How many cormorants are actually in the water catching fish and how many there *should be* is at the base of almost every debate about these birds. Local, national, or international management discussions inevitably turn to numbers. So if you're not an ornithologist or a wildlife biologist, let me give you some background before we get into the population figures.

First of all, it's hard to count birds. They fly around when you approach. Cormorants migrate and spend months in different parts of the continent, sometimes in several different regions. Cormorants may adjust from year to year where they nest, based on food availability and human disturbance.

In different parts of the country and the world, if counting is done from afar or above, one species of cormorant might look like another cormorant species that might overlap the same range, or they might look like other birds, such as anhingas or loons.

Official bird counts usually refer to "breeding pairs," which means the number of active-looking nests in a breeding colony. All agree this counting of nests is relatively simple, accurate, and can be normalized across different areas and over the years. Biologists and managers do not all agree, however, on how exactly a given nest translates to how many adult cormorants there are at a colony by the end of the summer when all the chicks of that breeding season are full grown. Sure, there are two breeding adults. But how many of their chicks will survive? Double-crested cormorants, similar to most of the other cormorant species, can lay from one to seven eggs. A clutch of three or four is most common for this species.[5] The survivorship of these eggs into the first year is hard to predict, based on all manner of environmental and genetic factors, especially the age and experience of the individual parents.[6] Also at colonies are nonbreeding adults who are too young or who don't mate for one reason or another. Typical of the other cormorants around the world, double-crested start breeding most commonly toward the end of their third year.[7] They can then continue to live several years in the wild afterward, breeding each year.

The average life expectancy of these cormorants in the wild might be about six years old.[8] Mr. Yamashita's thirty-five-year-old Japanese cormorant couldn't survive out on his own at his age, of course, but a man in 1984 did find a dead wild cormorant at Inks Lake, Texas, which had been banded and registered in Saskatchewan more than seventeen years earlier. There surely are cormorants still older out there.[9] There was a cormorant that was shot in the British Isles that had lived to be more than thirty years old.[10]

With all of this factored in — clutch size, age of breeding, life expectancy, and all sorts of threats to survival and reproduction — conservative estimates that compile nest counts multiply by three to get to the total number of individual cormorants.[11] For example, ten nests or ten breeding pairs at a colony roughly means thirty individual cormorants by the end of a summer.

Thus bird population numbers are estimates. The good scientists and wildlife biologists are careful to reveal their level of confidence, but their numbers are often misinterpreted, mishandled, and easily confused by journalists and members of the public.

I once worked as a volunteer for the Audubon Society on an island in Maine. Despite the diligence of our supervisors and the earnestness of me and my partner, I can assure you that we did a poor and inaccurate job counting egret and ibis chicks. Our numbers still went into official reports. Generally this is OK. What biologists examine is relative change given a consistent methodology over a significant amount of time, allowing for the squish of human error due to volunteers like me or due to circumstances like torrential downpours, for example, on the one day that year allotted to count the nests.

So with this primer, know that the best figures estimate at least one to two million double-crested cormorants in North America today, living in over 850 colonies.[12] This most recent total is from nest counts around 1997, the year before Ron Ditch hit his local island. (For reference, there are about one million herring gulls on the North American East Coast, and there are about five million Canada geese across North America.[13])

Now to put the double-crested cormorant in perspective within the cormorant family, you should know that this is not the most populous cormorant species around the world. Nor is Ron Ditch's Henderson Harbor the only place on earth or the only time in history when fishermen went out one night to kill a bunch of cormorants. North America and its double-crested cormorants do represent, however, more so than anywhere else in the world and during any time in history, the most significant *Human v. Cormorant* conflict in terms of geographical area, monetary investment, the sheer numbers of people involved, the diversity of groups with a stake in the management, and the different types of industries, organizations, and community groups who care where these birds are and exactly how many are paddling about. Throughout the continent thousands of American and Canadian citizens are so pissed off at cormorants that they either want to blast these birds out of the air with a legal blessing or they want their tax money to pay to have someone else do it. People like Ron Ditch and his wife hate cormorants because they believe the birds eat too many of a particular kind of fish. Other citizens are outraged about cormorants nesting and roosting in places where the birds' guano kills the trees and foliage, altering an island or a shoreline's landscape. People also worry that cormorants endanger the prosperity of other native fauna or flora that is more rare or more preferred. A few others are afraid of the diseases and parasites that cormorants can or purportedly carry.

Taxonomists divide the double-crested cormorants into five subspecies, based mostly on physical characteristics.[14] Federal managers, more the concern of Ron Ditch, divide the North American cormorants into four populations based on where they nest and migrate. These four groups are the Interior, Southern, Atlantic, and Pacific Coast–Alaska populations.[15]

In this book, between migrations around the world to places such as Gifu City and Cape Town, we will periodically return to North America. We'll travel to all four of the double-crested population areas. This enables a broad case study of one particular cormorant species and an examination of situations in which (1) sport fishermen want to kill cormorants, (2) commercial fishermen and local conservationists want to kill cormorants, (3) aquaculture fishermen want to kill cormorants, and (4) wildlife biologists charged with protecting endangered fish want to at least move cormorants someplace else — if not kill a few hundred here and there.

Ron Ditch's wife, Ora, says it took several months before authorities identified her husband and the other men who shot the cormorants on Little Galloo Island.

"It was a scary, emotional time," she tells me. "Everyone was whispering to each other. It was all over the news."

Mrs. Ditch points outside her window over to a dock. "Right over there was a van with antennae sticking all over. They were listening to everything we said. Men with too many creases in their shirts were coming up to our place asking for a brochure to go out and fish with Ron. But you could tell they weren't fishermen."

As the Ditches had hoped, the event went national. Newspapers and television from Alaska to Florida carried the story. The *New York Times* ran it on the front page with a photograph of bloodied dead birds. This was the article I first read over a bowl of Cheerios with my housemate Munro. A spokesman for the U.S. Fish and Wildlife Service declared at the time that the act on Little Galloo was the most significant mass killing of a federally protected animal species in at least twenty years.[16]

"They caught us eventually," Ditch says. "It's my understanding that two young good-looking women were undercover at the bar. Feds. I don't know how stupid you have to be, because there haven't been two young good-looking women at our one bar — in January, the off-season, a snowy night — since the start of this town."

Ora Ditch scoffs.

"Well, two of the boys," Ron continues. "I'm not naming names. They had too much to drink, so I heard. And they spilled the beans."

Throughout the Great Lakes and on the inland bodies of fresh water in the bordering provinces and states, sport fishermen catch a large variety of fish species. Anglers cast lines from the shore, they push off in their own small boats, and they pay to go out on larger boats with charter captains who know the area and who will, if you want, bait your lines, teach you how to reel in the fish, and even fillet your fish for you. Though steadily trending downward, since the 1970s over eight million fishing licenses have been sold annually on the American side of the Great Lakes.[17] Sport fishermen visit the local towns and buy fuel, eat at restaurants, stay at hotels, and buy these licenses. They buy bait, fishing gear, and boating equipment. Some love an area so much that they buy lakeside property. Over the last thirty years this has represented thousands of jobs and regularly brought in tens of millions of dollars to various communities. According to one study, this income peaked in 1988 at about $100 million in "angler expenditures" for the U.S. communities bordering all the Great Lakes.[18] Henderson Harbor and the nearby towns around eastern Lake Ontario have consistently been one of the more popular of Lake Ontario fishing destinations.

The bulk of the catch for Ron Ditch and the forty or so other licensed Henderson Harbor fishermen who run charters is salmon, trout, bass, and walleye. Ron Ditch's favorite fish to catch is smallmouth bass (*Micropterus dolomieui*), also known locally as black bass. For him this is the prize of Henderson Harbor. He runs some charters to target only these fish. Smallmouth bass is a good fighter, and it tastes excellent.

"This part of Lake Ontario was in years past the number one spot in all of the Americas for smallmouth bass," Ditch says. "It was great fishing here in better days. Eddie Bauer came up to fish for bass. And I could go on with a whole bunch of other people who over the years have come up just to catch bass."

Henderson Harbor is a good place to cast your lines because it is in a long protected inlet in which you can still motor out even if it's rough on Lake Ontario. A few distant beaches are good places to have picnics and cook up your catch. Ron's father, Ruddy Ditch (pronounced Rudy), was a charter fisherman in Henderson Harbor, too. Ruddy began as a charter captain in 1934, then with his son Ron started "Ruddy's Fishing Camp" in the 1950s. For over half a century now, families have come up to stay in their waterfront cottage or one of their rustic apartments. Visitors grill their fish outside at night beside the dock and watch the sun go down.

Ron Ditch is now in his eighties. He still runs Ruddy's Fishing Camp with Ora. He still takes people out fishing on his own boat. One of their sons helps run the business, too. Ron and Ora have four grandsons, one of whom is already, as Ron says, "a fishing fool — he's been fishing since he could crawl in the boat."

Ron Ditch has held a U.S. Coast Guard captain's license for over sixty years. "I would have to guess that I am probably the most experienced and oldest continuous captain on Lake Ontario in history," he says. "That's not to pat myself on the back. I don't know if that makes me smarter or dumber."

Part of the attraction to Henderson Harbor is that it is a hamlet. There's no town green or any central area. It is just a serene, gorgeous, lakeside community. Within Henderson Harbor's boundaries and in the neighboring area, the recreational fishing industry these days employs perhaps a few hundred people directly, although few if any can live on the seasonal income all year round.[19]

Ron Ditch passes by Little Galloo Island almost every time he takes his guests out to fish. He sees Little Galloo when he casts lines at Calf Island Shoal, one of his favorite fishing spots for bass. Cormorants sometimes nest on Calf Island and regularly roost there. Although tens of thousands of gulls live on Little Galloo, they are barely visible from the waterline because the cormorants nest primarily on the outer rim of the island. At the height of the colony's cormorant population, all you could see from the water were cormorants. Ditch's charter fishing season is generally from late April to September, which coincides almost exactly with the arrival of the cormorants, their nesting and raising of chicks, and then the flock's departure to migrate south.

Cormorants were first known on record on Little Galloo Island in 1974, when wildlife biologists counted twenty-two cormorant nests. A decade later there were over 1,400 active nests.[20] Residents of Henderson Harbor, particularly those out on the water often, watched as this local cormorant population continued to grow rapidly in the 1980s and into the 1990s. Ron Ditch and his fellow fishermen complained to the New York State Department of Environmental Conservation—who had already been monitoring the situation and had begun to count birds annually. The state biologists kept cormorants from nesting on the other potential colony sites on Eastern Lake Ontario, within American waters.[21] In 1992 the state wildlife biologists counted 5,443 pairs of breeding cormorants spending the summer on Little Galloo. In 1996 the Little Galloo colony peaked at 8,410 pairs of cormorants, thus over 25,000 total birds.[22] At the time, this was the largest single colony of this cormorant species in the United States and almost certainly all of North America, which means in the world.[23] In addition, almost 2,500 more cormorant pairs were living on five colonies on the Canadian side of this eastern basin of Lake Ontario—a short flight away.[24]

Over the years, Ron Ditch watched these birds gulp down some mighty large fish right in front of his eyes. He saw the shape of fish inside a cormorant's neck as it swallows its prey. He watched thousands of cormorants swarm on a school of fingerling trout released to supplement lake stocks. Ditch's front porch looks out onto a boat channel in which he has seen cormorants dive and eat large bass the size of his shoe.

Meanwhile, state biologists were compiling a report, published *after* the raid on Little Galloo, which suggested the smallmouth bass population in eastern Lake Ontario had indeed decreased to the lowest level it had been in two decades.[25] This confirmed what Ron Ditch had been complaining about. The scientists published what they found to be "convincing evidence" pinning the decrease in schools of smallmouth bass and yellow perch to the growing number of cormorants in this part of the lake.[26]

"I tried for thirteen years to get the government to listen to us because the cormorant is not a native bird to this part of the world," Ditch tells me. "You see when they did away with DDT it allowed these birds to propagate in such numbers that they got way, way, *way* out of balance. We had at one time about 150,000 cormorants on this side of Lake Ontario. And those birds eat a pound of fish a day."[27]

To help pay for the total of $100,000 in fines for which Ron Ditch and his men were collectively slapped, members of the Henderson Harbor community held pasta dinners and other fund-raisers. The fishermen began calling themselves "The Ten." This was the number of people prosecuted for the cormorant slaughter in 1998 and for previous runs out to the island. They received donations in the mail from all over the country. Ron and Ora Ditch had anti-cormorant posters and buttons made up. They made flags, which a few people flew from their boats. One man stapled the flag beside his garage door with a long black feather. "I support cormorant control," the flag reads, featuring a circle and a red slash over an illustration of a cormorant. I can tell you that the feather stayed stapled to the flag for at least three years, and, though yellowed and brittle, the anti-cormorant flag was still stapled beside that garage the most recent time I was up in Henderson Harbor, which was almost fifteen years after "the shoot," as Ditch calls it.

Various marinas and fishing supply shops in town sold the anti-cormorant gear that Ditch made to raise money for their fines. You could buy white hats with red circle-slashes across cormorant silhouettes. The hats said: "Concerned Citizens for Cormorant Control" or "I support Cormorant Control." Another hat style had "Little Galloo Island Shootout" stitched on the front. On the back: "Fisherman [sic] 850, Cormorants 10." The marinas also sold T-shirts. One design had a circle-slash over a cormorant with the words underneath in caps and bold: "Zero tolerance." Another had a picture of a bass, a perch, and a walleye on one side and a cormorant on the other. It said underneath "You Choose," with a rifle's crosshairs through the first "o" in "Choose." I bought one of everything.

In addition to the difficulty of counting birds in the wild (it is even harder to count fish), it is also a challenge to know what exactly cormorants are eating and how much. Biologists use three primary methods.

The most accurate way is to shoot cormorants and slice open their stomachs to see what is inside. Every fisherman in Henderson Harbor has a story about what has been found in a cormorant's stomach. Ron Ditch knows of an account by a local conservation officer who found a 17.25-inch (43.8 cm) catfish in a cormorant's belly.

A second method to determine what cormorants are eating is to collect their "regurgitants." These are barely digested fist-size balls, mushed up with fish or whatever else they have been eating. A regurgitant, also know as a "bolus," can be difficult to find, however. They usually can only be collected on colonies, around the nests, and at a certain time of year when the adults are feeding their chicks. Sometimes chicks and adults barf up these regurgitants when they are frightened, a strategy used by cormorants around the world to defend against attacks from gulls, eagles, frigatebirds, and skuas, who all want to eat the cormorants' recently caught fish or the cormorant chicks themselves (a sort of "here, take my wallet" defense behavior). One biologist told me that on Little Galloo he once found a half-digested fourteen-inch (13.5 cm) bass coughed up beside a nest.[28]

Because finding regurgitants is so limited and shooting the birds isn't always ethically acceptable or even practical, another method biologists often use to estimate cormorant diet is to collect "pellets." These are nut-size, gray, gummy blobs that cormorants cough up about once a day.[29] Cormorants hack up pellets throughout the year. Cormorant pellets are convenient in that, similar to those of owls, they remain intact for a few days and contain bones, eye lenses, scales, and other parts of fish or other organisms that the birds have eaten and cannot digest.

I'm not a biologist, but I've spent my share of time over the years searching for cormorant pellets on islands of New England. So one of the high points of my professional career was when I was helping to collect pellets on Little Galloo Island. One of the state biologists said to his colleague as he placed a few pellets that I had just found into baggies: "You know, these coastal guys are almost as good as you, Irene. And you're the cormorant puke magnet."

Analysis of these pellets focuses primarily on tiny otoliths, the white oval ear bones of fish. Each fish species has a signature shape to its otolith. In theory, each pellet then reveals the species of fish and the number eaten from about one day's worth of feeding. Unfortunately, there are some factors that can diminish the accuracy of pellets. For example, it can be hard at times to figure out if a given otolith wasn't already inside a larger fish when the cormorant ate it, or even how many meals or days are represented in a single pellet.[30]

Indisputably, based on all these methods conducted in various parts of North America for over a century, double-crested cormorants are opportunistic feeders, just as are most other cormorant species. Cormorants dive underwater after whatever fish is the easiest to catch at a given time. Biologists have identified as prey for double-crested cormorants across North America more than 250 species of fish from over sixty taxonomic families. These birds prefer slow-moving, schooling fish, most commonly less than six inches (15 cm) long. Cormorants also eat eels, and the occasional crustacean, amphibian, insect, or even small mammal. I read a report of a double-crested cormorant eating an entire snake.[31] One study in North Dakota in the 1920s found that owing to environmental conditions, cormorants shifted their diet to eat primarily salamanders.[32] On rare occasions, cormorant species around the world have been observed eating small rodents and birds.[33]

All cormorants around the world eat live food. Cormorants do not eat dead fish, carrion, or human trash.

Biologists try to determine an average of how many fish, or the weight of fish, a cormorant will eat on a given day. This is challenging because birds have different needs based on time of year. They are going to eat the most, for example, when they are feeding chicks.

The two most cited experts on double-crested cormorant biology — Jeremy Hatch, an Englishman who spent much of his career at Boston University, and Chip Weseloh, a biologist for the Canadian Wildlife Service, reported that double-crested cormorants eat on average about 0.7 pounds (0.32 kg) of live food per day.[34] A couple of years later Weseloh revised this average up to 1.0 pounds (0.47 kg) in a study conducted with the cormorants of eastern Lake Ontario.[35] A healthy adult cormorant is bigger than a duck and smaller than a goose. It typically weighs about four to five pounds (1.8–2.3 kg).[36] So according to these studies, a double-crested cormorant's average daily consumption is about 20 to 25 percent of its mass.

So do cormorants have a bigger appetite than other birds?

Over a century ago in *The Birds of Ontario* (1886), Thomas McIlwraith wrote (which would hardly be the last word on the matter): "All the Cormorants have the reputation of being voracious feeders, and they certainly have a nimble way of catching and swallowing prey, yet it is not likely that they consume more than other birds of similar size."[37]

Ron Ditch and his fellow fishermen are not the only men to go out and kill cormorants around the Great Lakes. In the 1930s ornithologist T. S. Roberts wrote that cormorants got into the pound nets and gill nets of commercial fishermen around Lake-of-the-Woods, Minnesota, which is beside the Canadian border. Roberts wrote: "Fishermen grant [cormorants] no quarter and destroy them whenever possible by shooting the adults and breaking up their nesting places on islands in the lake."[38]

More recently, a Canadian colleague told me: "I've heard some crazy stories of locals in Manitoba attacking colonies with weed-whackers, and men shooting at conservation officers who try to stop them."[39]

In June of 2000 more than five hundred cormorants were shot on Little Charity Island in Saginaw Bay, part of the Michigan Islands National Wildlife Refuge. Some person or group gunned down half the breeding population—many of the birds were sitting on eggs and newly hatched young.[40] Authorities never caught the offender(s).

Two years later, recreational fishermen from Cedarville, Michigan, made their own national news complaining about their cormorants. Tourism was down. Fishing resorts were closing. Local fisheries managers documented a near crash in the perch population. As cited in this chapter's epigraph, Cedarville residents proposed cormorant culls, and—hopefully with hyperbole—advocated calling in the U.S. military to "napalm the rookeries."[41]

In 2007 men shot birds and crushed eggs and nests at several cormorant colonies on Canadian islands in Georgian Bay, Lake Huron. Controversy about cormorants and pressure from sport fishermen for government action had been under way there for at least a few years.[42]

When two decades earlier, in 1987, some eighteen hundred cormorant eggs were illegally, clandestinely destroyed on an island in northern Lake Michigan, poet Judith Minty responded with verse. Minty wrote in her poem "Destroying the Cormorant Eggs":

> the cormorants now emitting faint squawks,
> flapping their wings over this darkness,
> the albumin and yolk, the embryos shining on dull rock,
> the small pieces of sky fallen down—Black
> as the night waters of a man's dream where he gropes
> below the surface, groaning with the old hungers.[43]

Double-crested cormorants are native to the North American continent. They are not "alien" or "invasive" or "exotic" by any wildlife management definition. There is, however, not as much evidence of cormorants nesting in the Great Lakes before the twentieth century, as compared to other records in other regions of the continent.[44] Cormorant advocates argue, however, that there are several early records into the nineteenth century of abundant breeding in the Prairie provinces, Ontario, and the surrounding midwestern states. It also turns out that there isn't that much archaeological or pre-twentieth-century anecdotal information for *any* of the now common seabirds of the Great Lakes.[45]

And a few records of cormorants in the bordering areas do exist. These accounts suggest nesting of cormorants across most of the Great Lakes.[46] For example, in 1887 a naturalist near the Minnesota-Iowa border heard the locals declare (thinking the cormorants were loons): "The air is jist black with em' [*sic*] an they're nestin' on the island so yer can't see it for eggs."[47]

Twenty years earlier, a man described cormorants living around a man-made reservoir in Ohio: "About seven miles from Celina, was the 'Water Turkey' Rookery. Here I used to go to shoot them, with the natives who wanted them for their feathers; I have helped kill a boat load. One season I climbed up to their nests and got a cap full of eggs. The nests were made of sticks and built in the forks of the branches. The trees (which were all dead) were mostly oaks, and covered with excrement. I found from two to four eggs or young to a nest. The young were queer little creatures — looked and felt like india rubber."[48]

In addition to a handful of anecdotal accounts like these within a few hundred miles of the Great Lakes, we know of huge populations nesting in the Canadian Maritimes at the time of European contact, as well as a few early records of staggering numbers of cormorants transiting up and down the Mississippi.[49] It is hard to imagine that these highly mobile animals did not, if the food was available, regularly nest and roost in Lake Ontario, perhaps even on Little Galloo Island, in the centuries before human disturbance.

The first ornithologist to do a comprehensive study of double-crested cormorants — or of any single cormorant species in the world, as far as I know — was Harrison Flint Lewis in 1929. While serving as chief migratory bird officer for Ontario and Québec, he worked toward his PhD at Cornell University. He did his field research on cormorants in the Gulf of St. Lawrence but compiled as much information as he could about the bird's life history throughout North America. He corresponded by mail with colleagues all over the continent. He dug through libraries and archives. Lewis's research was sound and foundational. Today doing work on cormorants in North America and not mentioning Harrison Flint Lewis is akin to working on chimpanzees and not mentioning Jane Goodall.

Lewis's writing was tempered wonderfully with the tone of his time. It was in this way that he concluded his natural history: "In ending for the present this paper about the Double-crested Cormorant, I do so with the hope that it may show that that bird, though unfortunate in some respects, is by no means as unpleasant as it has often been painted, but is actually a reputable avian citizen, not without intelligence, amiability, and interest."[50]

Though he sounds tepid there, Lewis was an active cormorant sympathizer. He looked into the possibility of harvesting cormorant guano for fertilizer in the Canadian Maritimes. He thought cormorants' killing of trees was "local and limited." Lewis's research, following that of previous biologists, showed that the cormorants did not eat much if any of the Atlantic salmon or the trout of the Gulf of St. Lawrence, as some fishermen had accused the birds of doing.[51]

In 1929 Lewis calculated that in North America there lived fewer than forty thousand double-crested cormorant individuals of what we now refer to as the Interior and Atlantic populations, known collectively as the subspecies *P. auritus auritus*.[52] Of that continent-wide total of cormorants, fewer than three thousand lived within the borders of the United States, mostly in colonies around inland lakes in Minnesota and South Dakota.[53] Lewis wrote: "The history of the Double-crested Cormorant during the latter part of the nineteenth century and the first quarter of the twentieth is largely a history of persecution and of the gradual abandonment of one breeding-place after another."[54]

Since then, the graph of the numbers of breeding pairs of double-crested cormorants of the Interior population, mirroring the other populations around the continent in most respects, has been nothing but a roller coaster. By the 1940s, as Lewis was being promoted to chief of the Canadian Wildlife Service and Ruddy Ditch was taking his young son Ron fishing on Lake Ontario, the Interior population of cormorants had begun to increase quickly. This was probably due in part to the creation and protection of reservoirs in order to meet the water needs of an expanding human population in the Midwest and the Prairie provinces.[55] Maybe persecution from people in New England and the Maritimes pushed the birds inland. In the 1940s and 1950s there were perhaps 200 to 240 cormorant nests on Lake Ontario.[56] Wildlife biologists in Manitoba, meanwhile, felt so threatened by population growth in their area, particularly around

Lake Winnipegosis — which had reached nearly 10,000 breeding pairs — that managers began officially shooting birds and destroying nests in 1943. In less than a decade they had halved the local population. The official program was discontinued.[57]

From the 1940s through the 1960s cormorant numbers began to decline again throughout North America, but not solely because of culling programs. Farmers had begun to widely use pesticides such as DDT. These and other industrial contaminants long in use, such as PCBs and DDE, had continually seeped into lakes, rivers, and coastal waters throughout the continent.[58] The chemicals got into the systems of birds through the fish they ate, thinning the cormorants' eggshells.[59] In Green Bay around this time, for example, more than one in every two hundred cormorants was born with a defect, such as a curled, deformed beak.[60]

By 1970 there were again fewer than one hundred pairs of cormorants throughout the Great Lakes. There was only one colony on Lake Ontario. It had two nests.[61] In 1972, Wisconsin listed the double-crested cormorant as endangered and erected nesting structures to help them recover. The National Audubon Society listed the cormorant of "special concern." The Prairie provinces also experienced huge drops. Officials in Alberta also designated the cormorant as endangered.[62]

Then cormorants throughout North America, especially those of the Interior population, began to rebound once again. Environmentalists such as Rachel Carson inspired and nudged the introduction of several green laws in the 1970s. Legislators passed the Clean Water Act and the Endangered Species Act. They banned chemicals such as DDT. The Migratory Bird Treaty Act of 1918 was revised in 1972 to include additional birds, such as the cormorants.

Coincident with these pieces of legislation and the official protection for cormorants and their habitat was the rise of the aquaculture industry in the South. Catfish farmers in the Mississippi Delta constructed thousands of open ponds, all stocked tightly with fish. This supplied migrating cormorants and other fish-eating birds with a convenient and easy food supply in the wintertime.

As the cormorant numbers began to climb once again and the birds became more and more visible, men such as Ron Ditch complained louder and louder about the loss of their fish. Fish farmers, especially in the birds' wintering areas, objected more and more bitterly that their hands were tied as to how to defend their crop against this blatant hit. Pockets of residents and conservation groups around the Great Lakes, inland freshwater bodies, and throughout the Mississippi Flyway began to clamor about the smelly colonies, the killing of the trees, and that cormorants were pushing out other, more-valued birds.

Double-crested cormorants migrate. The birds of Little Galloo might spend the winter months anywhere from Mexico to North Carolina. Thus a federal agency, the U.S. Fish and Wildlife Service (USFWS), is in charge of the big picture of cormorant management. The Migratory Bird Treaty Act is not just for birds flying within the United States, but it also represents in part an agreement among Canada, Mexico, and Japan. For cormorants, the treaty is primarily between the United States and Mexico, since Canada's cormorants are protected by provincial legislation, not federal. The USFWS is mandated to enforce the treaties that protect wild birds — yet right beside the agency's conflicting responsibility to promote aquaculture.[63]

So the USFWS had to respond to this public outcry against the growing cormorant population. In the mid-1980s it began issuing a few control permits in cooperation with the U.S. Department of Agriculture's Wildlife Services division. In 1998 the USFWS granted permission for catfish farmers in several states to shoot cormorants on their ponds. It was that summer when Ron Ditch and his fellow fishermen raided Little Galloo Island. The following spring the USFWS allowed wildlife biologists onto Little Galloo Island to start oiling eggs at the start of the breeding season. To oil eggs, a few technicians or biologists enter a rookery and coat all the eggs in the nests with a sprayer connected to a backpack full of vegetable oil. The chick fetuses suffocate without the ability to exchange oxygen and carbon dioxide through the shell, but the parents are unaware. The adults keep brooding for several more weeks. If the wildlife biologists just went in and destroyed the eggs, the cormorants would likely try to lay a second clutch.

Still, the shooting on the aquaculture ponds and a few local egg oiling and harassment programs in several states under various permits were not doing enough, fast enough.[64] Demands for action continued, especially around the Great Lakes. Ron Ditch's big shoot in 1998 spurred several editorials and official position statements.

The USFWS began a grueling, multiyear analysis and public comment period. Several groups pushed for more oiling of eggs, direct culling, and more states' rights in determining how to manage their local populations without needing federal permits. Some cried out for a hunting season for cormorants. Other interests, including the National Audubon Society, the Humane Society of the United States, and later one environmentalist group called Cormorant Defenders International challenged both the science and the management goals.

In 2003 the USFWS settled on a comprehensive double-crested cormorant management plan. They expanded the rights of citizens and managers to deal with cormorants, handing over much of the decision making to selected local agencies within broad parameters. So, in thirteen states today fish farmers may shoot cormorants on their private ponds. They may call on government wildlife biologists to shoot birds on nearby roosts. These provisions are termed the "aquaculture depredation order." In twenty-four states, including New York, local managers may oil eggs, destroy nests, or kill cormorants that threaten what are considered public resources, such as wild fish, wild flora, and other birds' rookery space. This is termed the "public resource depredation order."[65]

From 2003 to 2006, Canadian wildlife biologists heavily managed growing cormorant populations on two islands of Presqu'ile Provincial Park, on the northern shore of Lake Ontario. They did so in order to protect native trees and leave more space for egrets, night herons, and migrating monarch butterflies. It seems logical that thousands of these cormorants were previously on Little Galloo or other sites on the other side of the border. Managers at the park reduced the breeding cormorant population by egg oiling, nest destruction, and the killing of thousands of adults — including shooting over six thousand individual birds in 2004. In a few years they reduced the number of breeding pairs from 12,082 to 2,819.[66] Managers shoveled the carcasses with ATVs equipped with front-end loaders and dumped the birds in composting bins on one of the islands.[67]

Each year of the cull, environmental groups filed complaints. Canadian citizens took to small boats, Greenpeace style, in order to film and record as much of the culling as they could. They rescued injured birds found floating in the water.[68]

Canada does not as yet have a countrywide management plan or a federal branch that deals with agricultural animal management. But their cormorant conflicts are many, similar, and in almost every province.[69] Chip Weseloh, who agrees with selected cormorant management in certain places in order to preserve space for other nesting birds or maintain rare vegetation, explained to me that despite a few culling programs, like that at Presqu'ile on Lake Ontario or on Middle Island in Lake Erie, "in Canada, we do very little of that."

At the same time, Weseloh tells me: "In Ontario land owners can do virtually anything to protect their property values from harmful wildlife without a permit and without having to report their actions. We usually do not know how much land owner control or management or culling there might be."[70]

Russ McCullough, senior aquatic biologist, has been with New York's Department of Conservation for over thirty years. He loves his job. It places him firmly in the middle of the U.S. Fish and Wildlife Service, environmentalist groups, and the needs of his local community, including the recreational fishing interests and the tourism industry.

When I first met McCullough, a couple of years after Ron Ditch's cormorant shoot, he said: "To me cormorants are an issue like zebra mussels are an issue or like phosphorous levels are an issue. If there is a problem, then you've got to deal with it. But I don't really have the emotional connection to the birds like some people do."[71]

McCullough was one of the people who first discovered the dead birds on Little Galloo. He and his colleagues were going out for a routine trip to collect pellets, which was by chance only two days after Ditch and the others went ashore with their guns. As usual, McCullough and the others anchored off the island. They began to row ashore in an inflatable. They were familiar with the powerful smell of the guano, but this time the place smelled much worse. Something was rancid. Russ and a couple of other state biologists began to walk around the perimeter of the island as they normally did. It is common on a bird rookery to find a few dead chicks. They die naturally. They starve, outcompeted by stronger siblings. Gulls eat them. "But this was different," McCullough says. "There were just too many. We saw some shotgun shells, saw the groups of cormorant chicks all dead together. Some birds were wounded, still alive. Two of my colleagues had to euthanize them with a hammer."

McCullough thinks the killing of the cormorants on Little Galloo

Island was a grossly illegal activity. It was not civil disobedience, he says, because Ron Ditch and the other fishermen denied it at first. At the time of the shooting, his agency was in the middle of assembling a huge report on the matter. They were thoroughly aware of the problem and were in the process of identifying strategies. McCullough wants to think that it was their final report that finally got management approval from the USFWS to begin to oil eggs. But he knows that the shooting raised awareness, particularly with the politicians. McCullough quickly points out that the incident also slowed down Ron Ditch's cause, because certain agencies and private groups now did not want to allow any management at all, since that would only reward this type of activity. With the national press and environmental groups now watching closely what was happening on the island, cormorant management was stifled.

He does not demonize Ron Ditch, however. For McCullough, it is important to recognize that the ideas of scientists are just as value driven as anyone else's. Even the notion of biodiversity is a value, not a fact.

"We biologists need to recognize that we have values, too," McCullough says. "When we make decisions, and we develop ideas, it has got the values attached to them. So let's recognize that and not say 'Well, ours is the truth, ours is fact driven and theirs is value driven.' It doesn't work that way. This cormorant issue, it was such a hot issue. And it still is pretty hot. But it's brought a lot of stuff out in the open. That's been positive."

Value judgments are certainly at the root of whether or not to kill cormorants in order to preserve certain flora or fauna on islands. Some conservationists are in favor of cormorant management on Little Galloo in order to assist the nesting of the few remaining black-crowned night herons and other colonial waterbirds, such as Caspian terns.

What is happening underneath the surface — how much cormorants are actually to blame for diminishing fish stocks — is a far more complex matter, especially in the Great Lakes. McCullough explains to me some of the variety of factors that have affected the fishery in eastern Lake Ontario over his thirty years.

Before he begins teaching me about the ecology of the region, Mc-Cullough says, "Firstly, there is nothing natural about Lake Ontario."

What McCullough means by this is that the Great Lakes have been vastly manipulated by industrialization, fishing pressure, introduced species, and stocking programs. The first records of accidentally introduced

sea lampreys and alewives in the lakes are in the late nineteenth century.[72] Rainbow smelt appeared in the 1930s. Then, coincident with the later twentieth-century rise in cormorant populations, Canadian and American managers in the late 1960s and early 1970s intentionally introduced Pacific species of salmon, such as coho and chinook, and different species of trout. They hoped to counteract the rise of the alewives, which, combined with the sea lampreys, it was believed, had helped decimate the native lake trout and perhaps the native Atlantic salmon. Alewife populations did especially well in the 1970s through the 1990s. It also turns out that the alewife, a small schooling fish, is an ideal food for cormorants. Then in the 1990s zebra and quagga mussels spread throughout the Great Lakes. These invasions, along with other primary productivity shifts, made the water in eastern Lake Ontario clearer, surely shifting the movements of fish, such as smallmouth bass.[73] Finally, by the time Ron Ditch and his gang hit Little Galloo Island in the late 1990s, a fish called the round goby had arrived accidentally in Lake Ontario from Eastern Europe and would quickly become a dominant species.

All that is just to say that in the Great Lakes, there are a lot of fish under there eating each other, and people have been continually manipulating the ecology of this region — intentionally and accidentally — with extraordinary intensity over the last century. Yes, Russ McCullough thinks cormorants affected the local fishery — they eat a lot of fish — but he does not think the birds are one of the most significant of factors causing the depletion of bass or any other given species.

These days Ron and Ora Ditch spend their winters at a condominium in Florida. Off the beach or out on the water, Ron regularly sees double-crested cormorants down there, both those of the Southern population that live in Florida all year round, as well as those from the Interior or Atlantic populations who have also migrated down for the winter. Quite possibly Ditch has seen individuals that spent that year nesting on Little Galloo Island.[74] Directly in front of the Ditches' condo, to their distaste, they regularly see an anhinga — the cormorant's closest relative and a black waterbird that looks quite similar.

Ron Ditch applauds the state's egg oiling and the culling of cormorants. Since the 1999 egg-oiling permit and the state's killing birds on Little Galloo as a result of the 2003 federal plan, wildlife biologists have met their goal of bringing the population down to the predatory impact associated with 1,500 breeding pairs on the island. At least for now, the NYSDEC staff continues to manage the colony in order to maintain this level of predation on fish — to make sure the cormorant population does not just bounce right back up.[75]

"Our intention was never to destroy the birds completely or get rid of them completely or put them on the endangered species list," Ditch says. "Our intent was to get them down to a number that we could live with and that would allow the fishery to rejuvenate. That is where we are right now. The fishery is coming back to the levels of where it was before the cormorants — I call that 'BC.' Before Cormorants. So that's proof in the pudding."[76]

McCullough and his colleagues, including Irene Mazzocchi, who now oversees the cormorant management on Little Galloo, and Jim Farquhar, who previously held the post (and now has a cormorant printed on his regional wildlife manager business card), all seem generally satisfied with the management's progression, too.

McCullough was a coauthor on a study recently that showed that 92 percent of the cormorant diet from Little Galloo Island is now the introduced round goby. Smallmouth bass composes 0.1 percent.[77] Yet the state wildlife biologists are planning to stick with their present level of egg oiling and culling, for at least a few more years.[78]

Each April when Ron and his wife return from Florida to their home in Henderson Harbor and Ruddy's Fishing Camp, they put up bird feeders and several birdhouses. Their home is decorated with paintings of ducks. Beside their dining room table is a large painting of a few scarlet ibis. Ron has always been a hunter. In the early years, he made money during the off-season carving duck decoys out of wood. When plastic decoys took over, he crafted and sold more-decorative bird carvings. These are all over his living room. Some are really quite exquisite, especially his small birds in flight, which have delicately carved feathers and paper-thin webbed feet.

Recently, while I was visiting with Ron Ditch again, he told me that he got it into his head that he wanted to carve a cormorant. Perhaps he would use it in his talks. So he spent several months, much of his downtime in Florida, whittling away at a life-size double-crested cormorant. When home at Henderson Harbor, he worked off a mounted bird that a taxidermist in town had prepared. The wings of Ditch's sculpture are outstretched — every single feather is carved: every rachis and every barb. He painted the bird a sharp black. Its mouth is open. It stands with its wide feet on a piece of driftwood. He made the double crests out of rabbit fur.

Ron Ditch doesn't know what to do with his cormorant carving now. Or even where to put it.

"Maybe I'll put it up on eBay," he says.

"Look at that lousy bird," says Ora. "Bird from hell."

May

Nearly two hundred cormorant nests cover almost the entire island. By this first warm windless day of the year, Gates Island already has a whitewashed, paint-splattered look. Only two chicks have hatched so far. They broke out of their shells over the last two days and are cuddled under their mother, curled inside the nest of this first breeding pair beside the two largest boulders.

The other cormorants have been building nests along the eastern edge of the island, crafting them outward in both directions from the two boulders and this first clutch. The younger and later of the breeding adults constructed their nests more toward the center of the island. This is a more vulnerable location, because they are near the sandy area where the gulls will soon begin to lay their own eggs. The gull nests are loose circles of algae and twigs. Here a tomato plant has sprouted from the seeds brought over by a gull that had eaten a tourist's cast-off lunch.

Boat traffic in the waters around Gates has increased to several small vessels a day, but since the island is off the main channel few if any of the boaters take any notice of the concealed, dense activity on the little rock clump.

Yet today a man paddles a kayak over to have a closer look. As he approaches, the gulls sound a sharp *map-map, map-map*. The juvenile cormorants, with pale brown breast plumage, stand on the southern rocks that are separated from the island by the high tide. Convinced the man is approaching, they fly away.

The kayaker paddles closer, within a few feet of the beach on the gulls' side, as if planning to land. The gulls fly up in the air, complaining louder. In a black cloud, nearly all the adult cormorants fly off their nests and roosts. Their wings together sound like the clapping of wet hands. Several other cormorants sitting on eggs are more reluctant to fly off, but they do, and then the last to leave is the mother by the boulders who had been sitting on her two newborn chicks and three developing eggs.

The kayaker decides not to beach his boat. Perhaps he sees there's nothing on the island but a few rocks — or maybe he was surprised, alarmed even, to see the hundreds of black birds suddenly rising off the island. As the man paddles away the gulls stay airborne. The majority of the cormorant colony is rafted together in the water, just a short distance from the opposite side of the island.

The two newborn chicks lie in the cup of the nest with the three eggs. Each newborn is smaller than a child's fist and weighs no more than an envelope. Lizard-like, liver-colored, and chubby, the newborns are without tails or any trace of feathers or down. They are camouflaged to blend into the shadow of a deep nest, but the sun is directly over them at this point. They can barely lift their head. Their translucent eyelids are clamped shut over dark, protruding eyes. The newborns occasionally issue a high-pitched peep. They cannot last long in the sun. Just hatched, they will be dead from exposure in less than fifteen minutes.

And now beside the newborns, perhaps because of the sudden change in temperature and light, there is a hole toward the narrow end of one of the three chalky, dirt-specked eggs. From the inside, a little white horn at the end of a beak, the egg tooth of a new chick, makes a barely audible scratching sound, expanding the hole.

The kayaker continues to paddle back toward the Mystic River. A heavy male gull flies over to one of the big boulders. He lands, and looks down with cocked head at the cormorant chicks in the nest and the egg beginning to hatch.

The mother cormorant swoops in and lands beside her nest. She sees the crack in the third egg. She looks up at the gull, then quietly digs her webbed feet under the chicks and eggs, pushes her tail out, stretches her wings, and sits down over her brood.

Aran Islands

IRELAND

> As I lie here hour after hour, I seem to enter
> into the wild pastimes of the cliff, and to become a
> companion of the cormorants and crows.
>
> JOHN MILLINGTON SYNGE,
> *The Aran Islands*, 1907

> Nothing can really prepare you for the reality of the shag
> experience. It is an all-power meeting with an extraordinary,
> ancient, corrupt, imperial, angry, dirty, green-eyed,
> yellow-gaped, oil-skinned, iridescent, rancid, rock-hole glory
> that is *Phalacrocorax aristotelis*. They are scandal and poetry,
> chaos and individual rage, archaic, ancient beyond any sense
> of ancientness that other birds might convey.
>
> ADAM NICHOLSON,
> *Sea Room: An Island Life in the Hebrides*, 2001

*L*iam O'Flaherty was born on the largest of the Aran Islands, Inishmore, in a small house a stone's throw from the cliffs. Near this sheer western edge he spent his childhood and adolescent years looking down at sea stacks and ledges and terrifying drop-offs, as if on a masthead in the middle of the cold, wide, lonely North Atlantic. He constantly heard the thud and thrash and seethe of the ocean against the rock. In fall and winter the storm waves relentlessly pummeled Inishmore, shifting and rolling truck-size boulders.

Inishmore is largely flat. The island is practically bare of any sizable trees. Soil is rare and precious. When O'Flaherty grew up in the first decades of the twentieth century, soil was composted out of seaweed and

50

sand. Inishmore was crisscrossed and patterned by fences of stone walls that laced around occasional stone, whitewashed cottages.

After first studying for the priesthood and then serving in World War I, O'Flaherty went on to be a prolific author. He was well known on both sides of the Atlantic and is perhaps best remembered for his novels *The Assassin* and *Famine*. His novel *The Informer* was made into a Hollywood movie. Yet O'Flaherty's most respected literary achievements are his short stories, which focus on animals and Irish peasant life.

In 1925 O'Flaherty published "The Wounded Cormorant."[1] This short story is one of his most famous and most widely anthologized. It is set just off Clogher Mor ("great stony land" in Irish), one of the most towering of the sheer cliffs of Inishmore at about two hundred feet (60 m).[2] Following the island's edge, Clogher Mor is about an hour's walk from O'Flaherty's boyhood home.

"The Wounded Cormorant" begins by foreshadowing dangerous events. As if before a storm, the weather is calm and serene, but heaving swells lift strands of seaweed "like streams of blood through the white foam."[3] O'Flaherty writes:

> On the great rock there was a flock of black cormorants resting, bobbing their long necks to draw the food from their swollen gullets.
>
> Above on the cliff-top a yellow goat was looking down into the sea. She suddenly took fright. She snorted and turned towards the crag at a smart run. Turning, her hoof loosened a flat stone from the cliff's edge. The stone fell, whirling, on to the rock where the cormorants rested. It fell among them with a crash and arose in fragments. The birds swooped into the air. As they rose a fragment of the stone struck one of them in the right leg. The leg was broken. The wounded bird uttered a shrill scream and dropped the leg. As the bird flew outwards from the rock the leg dangled crookedly.

The frightened bird flies into the air. The rest of the cormorants and all the other seabirds scatter.

> The flock of cormorants did not fly far. As soon as they passed the edge of the rock they dived headlong into the sea. Their long black bodies, with outstretched necks, passed rapidly beneath the surface of the waves, a long way, before they rose again, shaking the brine

from their heads. Then they sat in the sea, their black backs shimmering in the sunlight, their pale brown throats thrust forward, their tiny heads poised on their curved long necks. They sat watching, like upright snakes, trying to discover whether there were any enemies near. Seeing nothing, they began to cackle and flutter their feathers.

O'Flaherty returns to the single cormorant. It—he doesn't specify gender—is in great pain, the saltwater stinging the wound. The bird tries to keep up with the rest, but when all the cormorants return to the rock where they had been roosting before, the injured one can no longer stand. It falls. It closes its eyes. It wobbles around. It cannot hold up its wings.

The rest of the flock seems to "suspiciously" appraise the injured bird's chance of living.

"They began to make curious screaming noises. One bird trotted over to the wounded one and pecked at it. The wounded bird uttered a low scream and fell forward on its chest. It spread its wings, turned up its beak, and opened it out wide, like a young bird in a nest demanding food."

The healthy cormorants take flight again. The hurt one tries to follow. The flock circles out to sea and then back to Clogher Mor. They fly up to a ledge on the cliff and stand among black pools and scattered downy feathers. The bird with the broken leg tries to land with them, but has trouble. Eventually the wounded animal makes it onto the ledge. It is exhausted and pitiful.

The healthy cormorants attack their fellow bird "fiercely, tearing at its body with their beaks." They rip out its feathers, poke at one eye, and tear at its broken leg. When it lies trembling without any resistance, the mob hurls the animal off the cliff into the sea.

O'Flaherty finishes the story: "Then it fluttered its wings twice and lay still. An advancing wave dashed it against the side of the black rock and then it disappeared, sucked down among the seaweed strands."

"The Wounded Cormorant" is like many of Liam O'Flaherty's animal tales. It is brief and has a simple plot. He includes no human characters, uses a small amount of anthropomorphism, and narrates in a direct third person that is evocative of a fable.

As far as I have been able to find, "The Wounded Cormorant" is the first piece of poetry or prose in English that not only considers cormo-

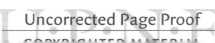

rants carefully, but also has these animals as the moral center of the tale. A thorough reading of "The Wounded Cormorant" also funnels directly toward one of the fundamental questions I am trying to explore in this book: from where derives the anger toward cormorants, which is especially widespread in North America and Europe.

We rank animals. We are all guilty. It is probably unavoidable to some extent. We do it both overtly and subconsciously, determining which animals mean more to us. But where do we get these perceptions? For cormorants, how is it different from place to place, and from generation to generation? What creates the starkly opposite yet equally committed feelings toward cormorants held by Ron Ditch and Junji Yamashita? Perhaps Ron Ditch is not just angry at these birds over bass or even because of the larger economics of the issue. Maybe something deeper has contributed to his hatred of cormorants.

In an introduction to a volume about contemporary issues with cormorants, biologist Jeremy Hatch wrote: "It is appropriate, however, to recognize that neither the economics and politics of these conflicts, nor the biological issues of diversity and changing numbers addressed here, provide a complete background. Another, less tangible context is illustrated by some lines from *Paradise Lost*."[4]

I will get to Milton's epic poem soon, but consider first how in our earliest children's books, in folktales, in the Bible and our other shared cultural texts, we form and shape our judgments of different animals. Consider where and how you have developed your own perceptions of animals, the anthropomorphic "traits" we culturally ascribe to, say, the wise owl, the innocent deer, or the foreboding vulture.

We characterize animals based on a fluid interchange between direct observation and cultural influences. The stories our writers, poets, and filmmakers tell provide portals into the consciousness of their given cultural influences and their time period—both by what is in the creator's work and by what is not. It matters how the cormorant or any individual species is portrayed in stories over time, because this alters, reinforces, and reveals our broader, evolving relationship with the natural environment.

Since our earliest stories, cormorants have been depicted in two primary ways: evil and greedy. O'Flaherty, like all good writers, was a reader and a listener. O'Flaherty surely knew that the cormorant has been a metaphor for darkness and rapacity for centuries. Yet with "The Wounded

Cormorant," Liam O'Flaherty subtly shifts the previous representations of this animal, even after he begins in a way that is predictable. O'Flaherty starts with a dark, gluttonous cormorant, but then does something new as the story progresses. Thus "The Wounded Cormorant," set on the Aran Islands, is a benchmark in how the depiction of this bird has changed. The story serves as a reference point in our cultural relationship with the cormorants. It also provides a nice opportunity to survey the use of cormorants in literature and film in English.

Writers in English, long before Liam O'Flaherty, used cormorants in their stories to evoke evil and greed. Let's first take evil.

The cormorant as devilish or portentous of some dangerous event likely has its roots in the bird's black plumage, its nearly silent voice, and its snakelike neck. The bird is infamous for the way it stands still on a rock or a wharf piling while spreading its wings to dry, as if draping a dark Dracula cape.

The word "cormorant" in English likely morphed through French from the Latin, *corvus marinus*, meaning "sea raven," which appeared as early as the eighth century.[5] Crows and ravens have long been associated with death and ill portent, in part because of their color, their mischievous intelligence, and their feeding on carrion — including dead people.[6] Subsequently the seaborne analogues of these black birds have been anthropomorphized with the same traits.

In Scottish folklore, a gruesome hag named Cailleach eats men's bodies. She usually appears as a raven or crow, but has also been told as a cormorant. The Cailleach apparently has its roots in still older Irish Celtic legends, such as the war goddess Badb Catha, or the "Raven of Battle."[7]

The authors of the 1611 King James version of the Bible wrote of cormorants four times, in two places where it was not specified in early editions.[8] The bird is listed among animals not to be eaten because they are not clean. In the book of Zephaniah, for example, cormorants are one of the few animals that are named to inhabit a land of "desolation" that the Lord has destroyed to punish its people, connecting the cormorant with death.[9]

By the years of the composition of the King James Bible, the evil cormorant had already worked its way into the vernacular. One of the earliest records is by Lady Anne Cooke Bacon. In 1593 she wrote about her son in a letter to her other son, Anthony: "For I will not have his cormorant se-

ducers and instruments of Satan to him committing foul sin by his coun-
tenance, to the displeasing of God and his godly true fear."[10]

Scholar Kevin De Ornellas at the University of Ulster is also a "literary
cormologist." He and I have kept in touch over the years. It's a small club.
(There's just one other fellow; he's at King's College, London.[11]) Dr. De
Ornellas has done a great deal of smart, scholarly work with the depiction
of cormorants in the Renaissance. He reveals that in verse, prose, and on
the stage, over and again, British and Irish writers used cormorants metaphorically to represent oppressive, selfish people — often wealthy men
who take advantage of the poor. De Ornellas studied in particular one
seventeenth-century satire, titled *The Water-Cormorant His Complaint*
(1622), in which the author, John Taylor, argues that a series of sinners,
such as Catholics and drunks, are so base and without morals that each
is lower than even the lowest of animals, the cormorant. De Ornellas explains that by the end of this satire, "long associated with hellish conduct,
the cormorant now is transmogrified into the devil himself."[12]

Most famously, as Jeremy Hatch referenced, John Milton in 1667 likens
the devil to a cormorant in *Paradise Lost*. Milton describes Satan climbing
up to look down on Eden. He is described as like a hungry "prowling wolf,"
then like a robber "bent to unhoard the cash." Then:

> So clomb this first grand thief into God's fold:
> So since into his church lewd hirelings climb.
> Thence up he flew, and on the tree of life,
> The middle tree and highest there that grew,
> Sat like a cormorant; yet not true life
> Thereby regained, but sat devising death
> To them who lived.[13]

It is no coincidence that while he was composing *Paradise Lost*, Milton lived
across the road from the king's trained cormorants in St. James's Park.[14]

Though Milton burned this image into the public consciousness —
the cormorant as the dark angel sitting on a tree, its wings surely spread
forebodingly — it had already been used in folklore and literature to specifically evoke or portend some disastrous event, usually a storm. In *The
Odyssey*, after Zeus hits a ship with lightning, the men swim around the
ship like "sea crows."[15] Some of the biggest names in the business have
since used this bird for their dark, tempest-tossed verse or prose, includ-

ing Sir Walter Scott, P. B. Shelley, Herman Melville, Robert Browning, Jules Verne, and W. B. Yeats.[16] Charlotte Brontë references *Paradise Lost* often in her novel *Jane Eyre*; she describes an evil cormorant within a dark, barbaric shipwreck painting.[17]

In 1841 American poet Henry Wadsworth Longfellow used the metaphor of a cormorant in his ballad "The Skeleton in Armor." An old sailor describes a ship stealing a maiden in the middle of a storm:

> As with wings aslant,
> Sails the fierce cormorant,
> Seeking some rocky haunt,
>   With his prey laden, —
> So toward the open main,
> Beating to sea again,
> Through the wild hurricane,
>   Bore I the maiden.[18]

It is hard to say, of course, whether these works create or merely reflect cultural superstitions, but apparently in 1860 a single cormorant landed one Sunday morning on the towering steeple of St. Botolph's Church in Lincolnshire. This so troubled the local residents that after the "doom bird," as historians called it, stayed there through the night, the church caretaker shot it dead off the steeple the next morning. It was too late, though. News soon arrived that the ship *Lady Elgin* on the other side of the Atlantic had wrecked at the same time that the cormorant landed on the church. Three hundred people died, including one of the church members and his son.[19]

These connotations connecting cormorants to evil, death, and shipwrecks continued into the twentieth century. One writer who employed this symbolism was John Millington Synge, who lived on the Aran Islands in the years when Liam O'Flaherty was a toddler. When Synge was writing at the turn of the twentieth century, a few men still lowered each other down the sheer cliffs to gather seabird eggs, kill chicks and adults for food, and to gather birds, especially puffins, for the feather trade. There are records of "cragmen," or "clifters," going back to the 1600s. These men sometimes went down when it was dark with a candle. Sometimes a man was left on a ledge for the night and lifted up again at sunrise.[20] In his 1907 nonfiction book *The Aran Islands*, Synge wrote of a flock of black seabirds luring men to their death.[21] In his earlier play *Riders to the Sea*, a grieving sister on the Aran Islands cries out about her drowned brother: "Isn't it a bitter thing to think of him floating that way to the far north, and no one to keen [mourn for] him but the black hags that do be flying on the sea?"[22] These black hags are cormorants.

Writers today still show the cormorant as evil or morose or bad luck, even in comic ways. Ted Hughes, Britain's poet laureate for almost fifteen years, penned a little saltwater bestiary, seemingly intended for children. It was published in 1999, the year after he died. His "Cormorant" begins this way:

> Drowned fishermen come back
> As famished cormorants
> With bare and freezing webby toes
> Instead of boots and pants.[23]

Writing about the same time as Synge, and two decades before Liam O'Flaherty's "The Wounded Cormorant," at least one critic pushed back against the trope of the evil cormorant as an omen of storms. This was in response to "The Cormorant" by the Irish poet Emily Lawless, printed in a 1902 issue of the *All Ireland Review*. Here the cormorants steadily, with each refrain, preside over a grim shipwreck:

> Round his bark the billows roar,
> Dancing along to a lonely grave;
> Death behind, and Death before.
> Yo, ho! The breakers rave!
> > *Only the Cormorant, dark and sly,*
> > *Watches the waves with a sea-green eye.*[24]

An anonymous commentator ignores the evil connotation ascribed to the bird by Lawless, and writes: "I wonder poets have not more often celebrated that very interesting bird the Cormorant, whom our coast people in the West call very properly the Cormoral," or sea crow. The commentator cites Longfellow's "The Skeleton in Armor" and then continues on to tell a story of how a few cormorants came to the River Nore near Kilkenny. "A brutal sportsman came along the opposite bank of the river and killed them all in one shot," the commentator writes. These birds were "harmless visitors" and "murdered," and he or she describes how the only way to resolve this sort of brutality is not through laws or newspaper articles but through a system of education, "which will include, among its indispensables, Natural History and love of Nature generally."[25]

Now let's talk greed.

The cultural and literary image of the gluttonous cormorant surely derives from observations of these birds feeding close to the shore, in easy view. Cormorants can bring up surprisingly large prey to the surface and swallow the fish whole. Crows and ravens have also been connected to greed. Consider the very word "ravenous."

As early as the 1380s, Geoffrey Chaucer, in "The Parliament of Fowls," wrote of the "cormeraunt ful of glotonye."[26] In 1544 William Turner wrote: "[The cormorant] ought to be called the Crow that devours. And this it does naturally, since it is endowed by nature with only one intestine straight and without a coil."[27] (For the record, I can tell you from my own dissections of cormorants that their intestines are tightly coiled and overlapping like ours.)

The Swedish historian Olaus Magnus in his *Description of the Northern Peoples* (1555) knew the cormorant as the "eel-crow." He wrote: "These birds are jet-black apart from their breasts and bellies, which Nature has coloured gray, and she has given them unmatched gluttony. They pursue fish, but are sluggish in flight and remain for a long time under water when they dive. Their beaks are toothed like harvesters' sickles, so that they may keep a firm hold on slippery fish, especially eels, which they hunt and gulp down so greedily that they discharge them alive through their bowels as though along a drain-pipe."[28]

Cormorants do not defecate live fish, of course. And they do not have serrated edges to their beaks, as do other fish-eating birds, such as mergansers or anhingas. Magnus did write accurately about cormorants in other ways, however, such as how other seabirds frighten cormorants to get their fish and how cormorants spread out their wings to dry.[29] Magnus described how fowlers caught cormorants and other waterbirds — presumably for food — by using nets "dark or dyed the same colour as the water." The men stretched these nets with pulleys just over the surface of a lake in narrow flyways between rocks.[30]

(It's worth a quick aside to note that *Northern Peoples* is the earliest text I have found that comments on the aesthetic effects of cormorants killing trees over time with their guano, along with the judicious recognition that these birds are not the only ones who do so. Magnus wrote: "They have a most disagreeable habit of excreting on the bark and branches of the trees they inhabit, so that these become contaminated by the droppings and quickly wither. Herons behave similarly."[31])

In everyday speech in English this image of the gluttonous and undesirable cormorant has been perpetuated, as adjective and noun, even more so than the image of the bird as evil. At least as far back as Magnus's sixteenth century, the word "cormorant" used to mean a greedy and rapacious individual.[32] A person could be known as a "money-cormorant." In Milton's seventeenth century, cormorants symbolized "greedy exploitation of the weak," providing an outlet for some people's distaste for the clergy.[33] In 1747 an author used the adjective "cormorous" to mean insatiable.[34] In the nineteenth century, the "cormorancy" meant a greedy and oppressive class of society.[35]

Reinforcing or perhaps informing the vernacular, several British authors describe the bird as gluttonous or used the word cormorant to signify greed, including Sir Philip Sidney, Joshua Sylvester, Michael Drayton, William Congreve, and Alexander Pope.[36] The most famous of them, William Shakespeare, used "cormorant" in four plays, all with the associations of greed and appetite. For example, King Ferdinand says in the opening lines of *Love's Labour's Lost*: "When, spite of cormorant devouring time."[37] In *King Richard II* it is the "insatiate cormorant." Shakespeare has a character in *Coriolanus* declaim: "Should by the cormorant belly be restrain'd."[38] In addition, Shakespeare likely named Shylock, his Jewish moneylender in *Merchant of Venice*, after the biblical Hebrew word for cormorant, pronounced "shalach."[39]

In 1799, shortly after composing "The Rime of the Ancient Mariner," the English poet Samuel Taylor Coleridge wrote "The Devil's Thoughts," in which he combined both associations for cormorants: the evil and the greedy. Coleridge knew *Paradise Lost* well. In this verse from "The Devil's Thoughts," notice his pun with "sate":

> [The Devil] peep'd into a rich bookseller's shop,
>> Quoth he! we are both of one college!
> For I sate myself, like a cormorant, once
> Hard by the tree of knowledge.[40]

The ravenous cormorant has continued to appear in literature in English. In *Dracula* (1897), the Irish novelist Bram Stoker has his virgin Lucy Westenra write letters from the coastal English town in which Count Dracula had bitten her. Her vampire symptoms have yet to emerge, and she, unaware, is now sleeping well again and feeling exceptionally healthy. Lucy writes to her friend: "I have an appetite like a cormorant."[41]

Two species of cormorants live in Ireland and around the rest of the British Isles. One of them, the great cormorant (*Phalacrocorax carbo*), is among the largest of all cormorants and is by far the most widespread, as I mentioned in Japan.

The second cormorant in the British Isles is the European shag (*P. aristotelis*), a smaller cormorant with a thinner beak and a narrower body that lives more along the saltwater edges of the British Isles. Irish ornithologist Oscar Merne, an expert on the birds of the Aran Islands, told me that O'Flaherty's descriptions in "The Wounded Cormorant" better fit the shag.[42]

This European shag tends to be a more greenish black than most other cormorants. When breeding it displays a vivid yellow gular pouch and a wonderful crest of feathers swooping forward off the forehead. European shags range from Iceland and northernmost Scandinavia, within the Arctic Circle, to as far south as the Atlantic coast of Spain. They seem content nesting on the very steepest cliffs and narrow ledges, in caverns, or under overhangs. European shags have a varied diet, depending on where they are living. They tend to eat a lot of sand eels. One study found that European shags eat about a half pound (250 g) of fish a day, which is approximately 13.5 percent of their body weight.[43]

In ornithological circles, the word "shag" is usually attached to the smaller, cliff-dwelling marine cormorants. But in most historic and current settings around the world, the common names of "cormorant" and "shag" are practically synonymous.[44]

The word "shag" probably is derived from the scraggly crest when breeding—but this is present in some form in nearly every cormorant species. A usage of "shag" to mean matted hair, like mussed-up, tangled wool, goes back to at least early seventeenth-century English. The word's usage to describe a cormorant goes back this far, too, but "shag," the bird, rarely appears in literary works until years after O'Flaherty's story, notably in Hugh MacDiarmid's gorgeous little poem "Shags' Nests" in 1934.[45]

Gaelic words for these birds vary. The most common contemporary name, used on nature signs beside the road, is *broigheall*, which means pale, or white, belly; this likely relates to the cormorants or shags in juvenile plumage. As J. M. Synge referenced, the shag has also been known as *Cailleach-bheag-an-dhubh*, meaning literally "small black hag." A variation of this is still in use. In the mid-twentieth century, Irish lexicographer Tomás De Bhaldraithe found the Irish colloquial word *amplóir* to also mean cormorant. This translates simply to "glutton."[46]

More recently, thanks to the Austin Powers movies, the British slang sexual connotation for the word "shag" has been popularized and expanded internationally, as in the comedy film *Austin Powers: The Spy Who Shagged Me* (1999). Following this meaning then, O'Flaherty's story is "the first good shag."

In "The Wounded Cormorant," Liam O'Flaherty seems at first to follow in step with the tradition of the evil and greedy cormorant. In the beginning of the story he tells of two species of birds, "rock-birds" (probably guillemots) and gulls. Then he describes cormorants: the only birds that are still eating. The cormorants have "swollen gullets." After a shard injures the individual and the flock flees into the water, O'Flaherty describes the black birds like serpents: the cormorants swim "like upright snakes." The healthy birds violently, mercilessly kill an injured bird of their own kind.

So at first glance, O'Flaherty seems to depict cormorants in the same way as did his literary predecessors. But his creation of a cormorant protagonist is a first. He creates an isolated, injured, and doomed creature. It is sympathetic. It is an individual killed by chance and the will of its fellows. The eighteenth-century philosopher Jeremy Bentham is often cited for his suggestion that the gauge for the rights of a living thing should not be the person or animal's intelligence or capacity for language, but whether or not a being can suffer.[47] O'Flaherty's wounded cormorant suffers.[48] Life is cruel. We feel sorry for this bird.

In some ways, O'Flaherty excuses the flock for killing its injured member. It is a requirement for their collective survival. When the stone first lands on the rock, the cormorants immediately fly into the water, scanning for enemies. The cormorants are constantly wary of predators, and they are worried about another stone falling. O'Flaherty writes of the injured protagonist later in the story: "At all costs it must reach them [the other cormorants] or perish. Cast out from the flock, death was certain. Sea-gulls would devour it." The wounded cormorant and the others of its species are victims. They are all prey. Their predator is a white bird high on the cliffs. O'Flaherty suggests their group cruelty is due to necessity. If they don't kill a sick or injured bird, it could attract predators, spread disease, or at least drain valuable resources. Without a foot, by the nature of the cormorant's style of hunting, it could not swim appreciably underwater to forage. It would starve. The world is violent and dark, not the cormorants.

Literary scholar Helene O'Connor describes Liam O'Flaherty as a "literary ecologist."[49] He had a keen, patient eye for the natural world. Biographers and his personal letters describe how O'Flaherty often sat for hours to observe the traits of animals, as carefully as a marine biologist. One critic wrote of another O'Flaherty story that is "like something out of a *National Geographic* article or a Jacques Cousteau film."[50]

O'Flaherty is careful with the details in "The Wounded Cormorant." When agitated, cormorants do "shiver" their heads. He describes their swooping and close-to-the surface flight. And the gulls of the Aran Islands, like anywhere, would certainly threaten an injured cormorant.

To be fair, however, a couple of O'Flaherty's zoological details are questionable. He gives cormorants improbable, perhaps impossible sounds in this story. Both the injured and healthy cormorants "scream," "shriek," and "cackle." The truth is that adult cormorants are nearly mute, except in the nesting colony where they can give piglike grunts, little barks, and croaking sounds. Female European shags can click and hiss. The male shags click and have a louder, creaky sort of "*a-aark*," usually grunted when in some danger.[51]

Local ornithologists Oscar Merne and Stephanie Coghlan each also expressed to me that it would be quite a rare behavior for a flock of cormorants to attack and kill an injured member in this way. This mobbing of same-species individuals has been recorded, however, in crows, gulls, and other birds.[52] O'Flaherty either created this behavior in cormorants for the sake of effect, or he was recording a rare incident that few if any have observed.

Regardless, perfect ornithology is not critical to the story. O'Flaherty makes no claim to accuracy in this fiction. "The Wounded Cormorant" is significant because the author bends traditional cormorant symbolism.

He attempts to describe the animal on its own terms, no worse or better. O'Flaherty emphasizes that all life is feeding and being fed upon. Ecologists teach that each species is equally important to a biological system, common or rare, aesthetically beautiful to humans or not. So in this way, "The Wounded Cormorant" has a green, environmental message.

"The Wounded Cormorant" also shows a recognition of chance events in nature.[53] One random kick by a goat causes the death of a cormorant. O'Flaherty emphasizes this by framing the story with an image of the sea heaving and pulling seaweed attached to the rock where the injury and final death occur. In the final sentence of the story, the sea washes the dead cormorant off the rock, reminding us of the immortal ocean and the smallness of individual life.

O'Flaherty writes of the inconsequential individual in nature more extensively in his semiautobiographical novel *The Black Soul*, published in 1924, the year before "The Wounded Cormorant." In *The Black Soul* his main character, a disillusioned veteran of World War I, says: "I am a part of nature." The author continues:

> Before, he had considered himself superior to nature. Now it struck him that he was merely a component part of the universe, just an atom, with less power than the smallest fleck of foam that was snatched by the wind from the nostrils of an advancing wave. . . . He saw others, lean-faced men, with anger in their eyes and hunger in their stomachs, shouting at the fat-bellied men, agitating for revolution and liberty, shouting about ideals and principles, honour, self-sacrifice, brotherly love! They were still more ridiculous. Did the sea have principles? . . . Nothing was assured but the air, the earth and the sea. He fancied that he could see the cormorants sitting stupidly on the Jagged Rock, bobbing their heads lazily. "We have lived here five hundred years," they croaked sardonically. "And we have heard it all, all before now! But tell us what does it end in? In ashes and oblivion?"[54]

As in *The Black Soul*, O'Flaherty does not recognize religion, destiny, or any higher force in "The Wounded Cormorant." Much like the American authors Ernest Hemingway and Jack London, O'Flaherty in his nature stories has events just happen, often by chance. Animals and people merely work to survive in a cruel world.

For my humble part, in the semi-fictional vignettes between the chapters in this book, I follow O'Flaherty's lead in some ways. After years of observation and study, I try to craft a glimpse into a representative annual cycle of life and death for one clutch of cormorants born on a tiny island.

American poet Amy Clampitt, another "literary ecologist," wrote one of the loveliest poems about cormorants. This was the one I remembered from a literature class when eating a bowl of Cheerios and reading about the slaughter of cormorants on Little Galloo Island. Clampitt presumably observed her cormorants on the East or Hudson River off Manhattan where she spent most of her adult life and where she was living when she wrote "The Cormorant in Its Element," published in 1983.

Clampitt chooses not to depict a devilish, gluttonous cormorant. She instead focuses on the dynamic ability of this particular animal to be able to fly and swim. She describes the cormorant's landing on the surface and its quick, tight plummet under the water. Very occasionally a few cormorant species, such as the European shag, will dive from the air straight into the water like pelicans or terns, but much more commonly cormorants land on the water first and then do a little half-flip to dive straight down.[55]

As in O'Flaherty's story, Clampitt's cormorant is indifferent to human concerns, be they cultural, moral, or mercantile. Her reference to "Homo" Houdini asks us to look reflectively at *Homo sapiens*. Here is the poem in full:

THE CORMORANT IN ITS ELEMENT

That bony potbellied arrow, wing-pumping along
implacably, with a ramrod's rigid adherence,
airborne, to the horizontal, discloses talents
one would never have guessed at. Plummeting

waterward, big black feet splayed for a landing
gear, slim head turning and turning, vermilion-
strapped, this way and that, with a lightning glance
over the shoulder, the cormorant astounding-

ly, in one sleek involuted arabesque, a vertical
turn on a dime, goes into that inimitable
vanishing-and-emerging-from-under-the-briny-

THE DEVIL'S CORMORANT

deep act which, unlike the works of Homo Houdini,
is performed for reasons having nothing at all
to do with ego, guilt, ambition, or even money.[56]

A few twentieth-century children's books have also steered clear of the
evil, greedy image of the cormorant. My favorite is the character of Graculus in the British *Noggin the Nog* series published in the 1960s and '70s.
These books also appeared as popular animated television shows. Writer
Oliver Postgate and illustrator Peter Firmin created the "sagas," set in a
Norse-style kingdom. Graculus, "the great green bird," is a European shag.
His name is from the earlier genus name for the cormorants, *Graculus*,
which was an old Latin word for a jackdaw or a cormorant.[57] I wish this
genus name was still in use, because it is even more fun to say aloud than
Phalacrocorax.

In the *Noggin* series, Graculus is the king's loyal and talented guide,
"friend and counselor," leading the king and his party across the Northern Seas to meet his queen. Graculus is a Lassie, a Baloo, a Chewbacca.
Throughout the *Noggin* adventures, he is usually by the king's side. He
often saves the day. In one saga, the *Ice Dragon*, Graculus remembers
where he was hatched. He visits all his ancestors on a steep ledge in the
Glass Mountains, then recruits them all to help the king. In the animated
version, the "*a-aark*" sound that his uncles, parents, and cousins all make is
really quite accurate and very funny. (Graculus's relatives have not learned
how to speak English—as he has.)

Graculus in the *Noggin* stories is loyal, proud, polite, and a problem
solver. At one point in the animated *The Ice Dragon*, he introduces himself
to a tiny warrior: "I am Graculus, royal bird of the Land of Nog, guide and
protector of Noggin, Prince of the Nogs. And anyway, I don't eat people,
even little people like you."[58]

There seems to be just one adult novel written in English—in any language, for that matter—that features our bird as a significant character.
This is *The Cormorant* written by British writer Stephen Gregory in 1986.
Whatever good Liam O'Flaherty and the character of Graculus did for
this animal's public image, Gregory erased. His novel is literally a horror story.

The novel earned good reviews, received the Somerset Maugham

Award, went into multiple printings and translations, and even inspired a BBC television film. The critic for the *New York Times Book Review* wrote: "*The Cormorant* has a relentless focus that would make Edgar Allan Poe proud. . . . This is a first-class terror story that does for cormorants what *Cujo* did for Saint Bernards."[59]

Yet what Gregory's novel actually does is reinforce that enduring, deep-seated, cultural antipathy toward the cormorant in Britain, Ireland, and North America, despite the transition begun by Liam O'Flaherty and others. One cover design for the novel features a close-up photograph of a bronze cormorant statue with its wings spread. The graphic designer has filled the eye with blood that drips over the face. Beneath the bird's beak it says in black text: "Heir to a legacy of evil. Victim of its malignant terror."[60]

In *The Cormorant* an English couple and their young son inherit a cottage in Wales from an eccentric uncle. Fixed into the will is one catch: they must take care of the uncle's pet cormorant. He had rescued it from the wild. Some say the cormorant killed him.

The narrator names the cormorant Archie. The bird is, in the man's words, "a Heathcliff, a Rasputin, a Dracula."[61] As the novel plays out, the cormorant emerges as more like Satan himself, the image of the dark-winged fallen angel. Though Gregory never mentions Milton or *Paradise Lost*, the influence is unmistakable. For example, when the narrator builds the cormorant a pen in the backyard, it immediately attracts a "snowstorm of gulls," squawking and flying around the cage; Gregory writes: "The cormorant stood with its chest pressed against the wire, its neck extended and the murder-beak jutting through. It had outstretched its wings and hooked them somehow onto the wire, gripping there like some prehistoric bird with clawed fingers. Archie stood erect, croaking and hissing, a black, malignant priest in a multitude of angels."[62]

The man's wife, Ann, watches the scene. She clutches their son to her breast. The blond boy, however, is entranced: "suffused with the malice of the sea crow."[63]

As the days turn into weeks, the family learns to live with the bird. The narrator becomes taken with the cormorant, even though it is "a lout, a glutton, an ignorant tyrant." He brings Archie to the beach to fish with a leash around its neck. It works. Yet slowly the narrator begins to lose his mind. He struggles with differentiating if the bird's behavior is just the cormorant's animal nature or if this individual cormorant is a malicious, supernatural beast.

Gregory weaves in much Christian imagery and tilts the novel toward the Gothic. Every once in a while he inserts a sex scene, perhaps to keep things moving and to shift the tone. Even this, however, is connected to the cormorant's witchcraft. For example, here is the narrator with his wife in the living room before the fire; he has just cut his hand on the bird's beak and is trying to coerce her into accepting the animal: "She could not have seen what I was doing. Her body was marked with the blood from my hand. Every tender touch against her throat and face, over her breasts and silken stomach, among the heat of her thighs, each caress branded her with blood. I smeared her marble body. My whispered endearments numbed her into a stupor. Soon she was asleep in the falling colors of the fire, stained with the wounds inflicted by the cormorant. All this so Archie would be forgiven."[64]

By the end of The Cormorant the first-born son is fully possessed by Archie, and the man acts gluttonous and evil like the satanic cormorant. It does not end well. In the epilogue it is spring once more, and white birds return to the backyard.

In 1993 Ralph Fiennes, the award-winning English actor who would make a name for himself in films such as The English Patient, played one of his first roles as the lead character in the BBC television version of the novel, also named The Cormorant. The film is more of a drama than a fantastical horror story. In the film the boy is saved in the end. And the screenwriters renamed the wife Mary.[65]

In addition to reading Liam O'Flaherty's "The Wounded Cormorant" as a shift of traditional depictions of the cormorant "traits" of evil and greedy, and as a story with an ecological moral, it is reasonable to also interpret this tale with a more political, social, or even racially charged lens. These are all important perspectives to consider today when examining our cultural antipathy — or at best collective ambivalence — toward the cormorants.

O'Flaherty shows the cormorant as equal to other animals and exemplary of a disillusionment with a cruel, random system. By the time he wrote this story, O'Flaherty had been an active member of the Communist Party, had written for socialist newspapers and journals, had fought for Irish Republicans, and had been a member of the Industrial Workers of the World. O'Flaherty grew up in cold, painful poverty on Inishmore. He suffered. Like so many of his stories and novels, "The Wounded Cormorant," like a fable, asks for comparisons between animals and humans. O'Flaherty wrote with a respect for the peasant figure and, as one scholar put it, the peasant's "historical and ecological significance."[66]

Perhaps in "The Wounded Cormorant" these birds symbolize the working class. What better animal to represent the downtrodden Irish laborer? The cormorant is a common bird accused of being greedy because it is hungry. The cormorant had rarely been sympathetically depicted in literature. This bird is nearly always quiet. Perhaps O'Flaherty knew of the shag's silence, with its occasional croak. Maybe he wanted the reader to hear its "voice," which society never does.

Liam O'Flaherty responded to theories of social Darwinism, in which Darwin's concepts of natural selection and survival of the fittest are loosely, incorrectly, invoked to explain the behavior of man. O'Flaherty's birds attack their weaker fellow, injured only by being in the wrong place at the wrong time within a harsh, indifferent system.

The injured cormorant is a lonely, unfortunate creature like any other animal or person. Literary critic Angeline Kelly writes: "The cormorant is more of a symbol than a separate entity. The story is an illustration of 'Nature red in tooth and claw,' with no mercy for the weak, and the long strands of seaweed in the opening paragraph point forward to the cormorant's torn and bloody body at the end, lynched and rejected by its fellows."[67]

Kelly used the word "lynched" precisely. Critics of current public policies that cull cormorants suggest that the intolerance to cormorants is often driven by an intrinsic racism, extending feelings about the color of people's skin to the color of the bird's plumage. If the cormorant had white feathers, many argue, there would be less distaste for this animal. Quick: do you feel differently when you see a photograph of a black bear cub vs. that of a polar bear cub?

O'Flaherty was not, by today's terms, what we would call an environmentalist. He did not write about conservation or lecture about protecting animals. He certainly did not aim to alter public opinion about cormorants. He did, however, surely have a soft spot for the bird. Besides choosing this animal as a subject for one of his best stories, there is some evidence that "The Wounded Cormorant" was his favorite tale. O'Flaherty also wrote after a visit home to Inishmore two years after publishing the story: "The people are sadly inferior to the island itself. But the sea birds are almost worthy of it. The great cormorants thrilled me."[68]

In another sense, however, Liam O'Flaherty's messages about cormorants are more complicated, as shown in his *The Black Soul*. He shows cormorants as a mirror for humility against the larger forces of nature, but not as intelligent animals. Even he cannot entirely shake the Miltonian, deeply set literary and cultural image of the dark, ominous, voracious cormorant. In *The Black Soul*, the narrator hears a cormorant "dismally" flying past the house on Inishmore after he remembers being hit by a shell in the war. Later, the narrator has a premonition of his own death, reminiscent of Prometheus.

O'Flaherty writes: "The wind would sing a cunning hissing song, trying to calm his fears so that the sea would crawl up unawares and devour him. Then all those black cormorants that he had seen on the Jagged Reef would strain out their twisted long necks and tear pieces from his carcass. They would swallow the pieces without chewing them and tear again. Then he discovered himself counting the number of cormorants that were tearing at his body and he tried to shout. But he was too agitated to shout. He crept down under the blankets and commenced to cry."[69]

June

Thick fog. Low tide. Cormorants fly on and off the island so low that their wings almost touch the glassy surface of the water. The high-pitched peeping of hundreds of chicks is steady and more noticeable because the gulls are quiet, as if dampened by the fog. A long shrill comes from an oystercatcher that has finished its meal on the limpets exposed at the southern rocks, where the juvenile cormorants roost. The smell of the exposed seaweed mixes with that of guano and warm fish.

On Gates Island dozens of the nests are now empty, especially those around the two boulders. Black woolly cormorant chicks waddle around or stand on some of the rocks. The four oldest chicks from the first clutch are by the two boulders. (The fifth and last of this clutch hatched but starved within a few days.) The four chicks are now almost the height of the adults but are more slender. They are still covered with fleece-like down, except that their wings and tail have begun to grow true feathers. Their feet are caked with white guano. They can swim if pressed but cannot fly. Younger, downier cormorant chicks stumble around in little gangs. From the fog each has glistening webs of white water droplets lacing its fleece.

At the moment, over one hundred adults are standing beside nests with small chicks or sitting on their nests with newborns and the last laid eggs of the season. Under one male cormorant are the freshest eggs, a clutch of three that are the product of a pair trying again after the first were crushed and eaten by a gull.

Through the fog both the male and the female of that first pair of cormorants by the boulders have flown off. The male forages by a harbor breakwater some distance away. He flies back directly to the colony, navigating somehow in the fog. When he lands, his chicks immediately leave their group and run, half fly, over to him, shrilling and jabbing at his beak and gular pouch. He resists.

Eventually the father allows the most persistent chick to probe its head entirely into his mouth and down into his neck. The chick's beak is visible poking within the father's throat. He pushes the chick's head down and regurgitates a few warm whole cusks into the chick's mouth. The chick pulls out and swallows visibly.

The other three chicks begin to pursue their father once again. But he flies back out into the fog.

Soon a motorboat anchors beside the northern rocks of the island. One gull swoops toward the head of the driver, who turns away. The man has brought what looks to be his two children with him. Wearing high boots, he eases into the water and then picks the children up and carries them close enough to the island where their own boots will be high enough to walk the rest of the way through the water. Startled by the aggressiveness of the gulls, one of the kids slips on algae. She stays quiet as instructed and doesn't get too wet.

As soon as the people set foot on the rocks of the island, all the adult cormorants fly off. The cormorant chicks that can swim tumble into the water. The younger chicks frantically scurry closer together and hide behind the boulders or rock crevices. They peep more loudly and more urgently.

The man and the two children walk around the outside of the island. At first both the girls hold their nose and laugh, but they begin to smile less as they look around. Dead chicks, moldy eggshells, and even the parts of a dead adult gull are scattered around the colony. Dead chicks have been pressed into the formation of nests. Small phlegmy pellets and half-digested pockets of dead fish, finger-length eels, and red shrimps are on the ground, dried and covered with guano before a gull or a crow from the shore got a chance to eat them. One child points to a leathery carcass of a large cormorant chick between rocks. Its spine is mostly exposed, its muscles still red under the peeled black skin. The man redirects their attention and shows them a nest in which cormorant chicks squeal and stab up at them with open beaks. The girls lean over at another cormorant nest in which there is a newborn. The man says it's ugly. He points over to a downy speckled gull chick, which he says is much cuter.

Hundreds of cormorant adults and chicks swim quietly just off the island, but closer than they would have a month ago, since they are worried about their offspring. When the man leads the children toward the gull chick, the gull parent rushes down at them out of the air with its beak and wings. The two children scream.

The man leads the girls off the island, crouching, back into the water. He carries the two at the same time over and into the boat. While he climbs back in himself, while he starts up the motor and flips on the radar, while he pulls up the anchor, and while the two children giggle about their adventure as the boat roars away back into the fog, a half dozen of the gulls land back on the island. They stand over the uncovered cormorant nests. Each gull pauses, looks around. One reaches down with its stout yellow beak and lifts up a bald cormorant newborn. He swallows it alive and whole. The other gulls tilt their heads back to help them get cormorant nestlings down. The mother of a gull chick hops over to a nest and nabs a cormorant newborn, too. She brings it over to teach her offspring how to tear it apart.

South Georgia

ANTARCTICA

The shags and soundings were our best pilots.

CAPTAIN JAMES COOK, 1775

Niall Rankin became a millionaire by inheritance. The story goes that he fell in love with one of the queen's consorts at Buckingham Palace, but the wedding was forbidden because he barely had a quid to his name. He had never met his own father, Sir Reginald Rankin, so he decided to approach him. Niall's hopeful bride, Lady Jean, managed to strike a dazzling first impression, melting old Rankin, who decided to leave his son all his money when he died—which he did just a few months later.[1]

By the end of World War II, Niall and Lady Jean lived on a massive Scottish estate he had purchased on the Isle of Mull. She traveled down to London to serve the Queen for weeks at a time. Rankin had already achieved some notoriety as an amateur naturalist, as a professional nature photographer and filmmaker, and he'd been a member of an expedition to the Arctic. Yet he had long dreamed of traveling to Antarctica to see the birds and to actually have some real time to look around without having to answer to a captain or the dictates of other higher priorities of an expedition. Rankin wanted to spend some time looking carefully and patiently at the natural world at its wildest, at its farthest from civilization. Having watched Europe emerge from two wrenching world wars—participating in the latter as a lieutenant colonel in the Emergency Reserve—Rankin was also perhaps a bit cavalier with his personal safety.

Rankin found himself a neglected Royal National Lifeboat that was forty-two feet (13 m) long. It had weathered decades of faithful service in the north of Scotland and had a hull planked twice over with the sturdiest mahogany. He had the engines replaced. He updated the electronics and interior. Then he managed to sweet-talk one of the Scottish whaling companies into loading his craft onto their ship, southbound.

Rankin persuaded two merchant mariners from the Shetlands to be his volunteer crew, and they all boarded the whaling ship *Southern Venturer*. During the passage down they scraped and painted and prepared his boat for the rigors of the region as he anticipated them. He christened his boat the *Albatross*. Rankin's adventure would be the first known private yacht expedition to Antarctica.

The *Southern Venturer* arrived in Leith Harbour, South Georgia, in November 1946. The Norwegian and Scottish whalers with their thick sweaters and wool coats crowded around the dock to see the arrival of the ship and to watch the little *Albatross* lowered into the harbor. In a few days, Rankin wangled a local pilot. He found a Norwegian whaling captain whose ship was idle and persuaded the man to come along for the trip.

Thus expedition leader Niall Rankin, Captain Konrad Olsen, and able seamen Campbell Gray and Robert Inkster motored out of the harbor one windy day to spend five months with the primary goal of observing the birdlife of this remote island.

South Georgia is a narrow, mountainous place. At about one hundred nautical miles long, it is near the length of Long Island, New York. South Georgia never supported a native human population. It is too bitterly isolated at well over one thousand nautical miles east of Cape Horn and just a few miles closer than that to the Antarctic Peninsula. South Georgia is given an Antarctic address because it is south of the Antarctic Convergence, a fluid oceanographic and zoological border where frigid, less saline water from the continental ice meets the warmer, saltier waters of the South Atlantic, Pacific, and Indian Oceans. South Georgia's vertebrae of sky-scraping snow-piled peaks rises up in the middle of a band of the Southern Ocean known as the Furious Fifties: often fifty knots or more of icy wind blasting in the latitudes between 50° and 60° south. Waves smash onto the island's western face from seas and swells that circle the earth unchallenged by a single major land mass. The taciturn author of a mariner's guidebook of Rankin's era writes: "The climate of South Georgia is uniformly dismal."[2]

Thus most of the thick-palmed whalers in the harbor chewed on their pipes and considered this Rankin chap and those he persauded to go with

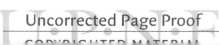

him to be a small band of quixotic fools.[3] For Rankin and his crew, an encounter with South Georgian blue-eyed shags (*Phalacrocorax georgianus*) would be a highlight of their voyage. These cormorants are famous in the region as a delicacy for the table, as a fascination for biologists, and as a navigational tool for early mariners.

My fellow travelers considered me a quixotic fool. The blue-eyed shags were not only the highlight of my voyage to Antarctica: they were my mission. When we anchored off the beach at Paulet Island in the South Shetlands — at the edge of the Antarctic Peninsula — I could barely contain myself. With my spotting scope and tripod steadied on the foredeck, I could see that those were not just penguins nesting along an entire hillside, but that a significant swath of the birds were blue-eyed shags! Sure, I said to my fellow travelers, yes, neat, penguins. But look at those Antarctic cormorants! Just look in this scope! Hundreds — no, likely thousands of blue-eyed shags!

I was already just so jazzed to be in Antarctica. On the previous stops in the South Shetlands I barely knew what to do. All at once I wanted to take photographs, sketch, record the sound of the place, write in my journal, focus on one animal or plant, and at the same time walk around and see everything. I was on an ecotourist ship, a converted Russian vessel welded strong for the ice. But we just never seemed to get enough time ashore. They wanted us back for mealtimes. Trips back in the inflatable boat seemed to be based around the first person to get cold or bored.

By the point in the trip when we anchored off Paulet Island, I had barely slept. How could I sleep when we were steaming down the Beagle Channel, observing the cormorants of Tierra del Fuego in the wake of Captain FitzRoy and Charles Darwin? How could I sleep as we barreled across the Drake Passage, beyond Cape Horn, with albatross soaring behind our ship? How could I sleep as we first approached these islands, and I tried to imagine early sealers and explorers sailing down here in wood ships without engines, radar, sonar, without charts, and maneuvering around with the wind blowing, the snowcapped mountains obscured by snow streaks, and the currents swirling across pitiless, hidden rocks? You don't go down to your cabin when you can see icebergs! It only got dark for a few hours each night. And this trip only lasted ten days!

What I'm getting at is that I had been drinking a lot of coffee.

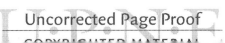

So with my camera, binoculars, spotting scope, audio recorder, and sketchbook I got in the boat with a group of passengers to go to the beach on Paulet Island and view what the naturalist said would be our best look at blue-eyed shags for the voyage. We were told that we had only two hours ashore.

Maybe it was the feeling of the cold water around my boots, but as soon as I stepped onto the beach, I had to go to the bathroom. Very badly. I was with sixty others on this stretch of sand. There was nowhere, and I mean *nowhere*, to crouch behind and go. We had been listening to mini-lectures about leaving no trace. I thought about doing it in my pants. This would have been just too extreme, because this was to be forceful.

I swallowed my pride out of desperation. I asked the ship's naturalist to radio for a boat to bring me back to use the head onboard. Mortified, frustrated, I was to waste half of this precious time at one of the most significant blue-eyed shag colonies in Antarctica. It was certainly the best and largest colony I will ever get to see, unless I get my own mystery millionaire's inheritance. I had gone to the very hollows of my savings for that trip to Antarctica. And here I was motoring away in a boat with a Russian crew member sneering at me because this was to be his time off.

Blue-eyed shags have nearly the same color pattern as the local penguins. From a distance without a scope it's tough to tell who's who. Blue-eyed shags have a wide black stripe of feathers on the back of their neck that comes over the head, around their eye, and then blends into their black beak. When breeding they have a black spiky crest that swoops straight up. The "blue-eyed" of their name is actually a royal-blue fleshy ring *around* their brown eyes. Unlike any other group of cormorants, these blue-eyed shags have two clumps of yellow "caruncles" — fleshy masses that look like a dollop of scrambled eggs — between the eyes, at the base of their beaks.

Perhaps a dozen different species of cormorants with blue orbital rings, caruncles, and white breasts and bellies live on the Antarctic Peninsula and on the desolate Southern Ocean islands. Several of these blue-eyed cormorant species live on their own island or on little isolated clumps of islands. Entire populations of certain species number fewer than one thousand birds. Maybe over millennia the shags were blown from west to east.[4] Several of these blue-eyed cormorant species have since lived in varying degrees of isolation.

The dozen or so species of Antarctic blue-eyed cormorants all look quite similar. Biologists differentiate them by the width of their "alar bar"—a white stripe across the wing—and by how much white they have on the cheek. There are subtle differences in their behavior, too. Thus the scientific names and their zoological distinctions are far from settled. A few ornithologists over the years have suspected that there are probably fewer Southern Ocean cormorant species and the rest should qualify only as subspecies.[5]

The common names are equally variable. Individual Antarctic cormorant species at different locations have been called blue-eyed shags, carunculated shags, rock shags, imperial cormorants, king cormorants, Antarctic shags, emperor shags, or any variation of these. This is a problem when trying to compile historic records, even those carefully transcribed by earnest ornithologists. Even Bryan Nelson, or "Lord" Nelson, as I call him—who wrote the leading biological reference book on cormorants— became exasperated by this confusion. He wrote: "The common names are like myriad footprints criss-crossing in the snow and about as easy to disentangle."[6]

When Niall Rankin ventured from Scotland to South Georgia in 1946, he knew the island's cormorant by the common name most agreed upon today, the South Georgian blue-eyed shag, and the most-accepted scientific nomenclature at the time, *Phalacrocorax georgianus*, which remains the case in most circles. If he made shag jokes with his crew—the sex definition existed in the 1940s—he had the good taste not to publish the allusions. Rankin did not have too many chances, though. He had difficulty finding many of these shags with, as he wrote, the "remarkable plumage." He did not find a single colony that was accessible enough for him to observe closely. Rankin heard that there were "huge numbers" only on Shag Rocks, a set of islets a couple hundred miles to the northwest. He wrote of the South Georgia shag: "So far as known at present this particular bird is peculiar to South Georgia and the Shag Rocks. Another sub-species of the genus is found on the other island groups farther south as far as latitude 65°."[7] The cormorants of South Georgia feed primarily on bottom-dwelling nototheniid fish, known as "ice fish."[8]

"Lord" Nelson places the South Georgian blue-eyed shag, *P. georgianus*, as nesting on all the islands of what is called the Scotia Arc, which includes South Georgia, South Sandwich, and the South Orkneys. Today there may be 9,000 pairs in total. Perhaps 4,000 pairs live on South Georgia, seemingly many more than Rankin roughly observed in the 1940s.[9] Rankin believed that there had been more cormorants in the area before he arrived, even in preceding decades. He wrote: "How different from days gone by, when it is said that they used to float in the air alongside whale-catchers so that the lookout in the masthead barrel had only to lean out and knock them to the deck for the next meal!"[10]

Large numbers of blue-eyed shags have been recorded in other parts of the Southern Ocean. Two explorers in 1905, for example, observed "innumerable cormorants" in a few locations near the Antarctic Peninsula, including, one day, "a flock of thousands of cormorants [that] flew close over our heads."[11] A second account in the South Orkneys around the same time described "a flock containing several thousands" of blue-eyed shags.[12]

Regardless, the first few weeks aboard the *Albatross* were going well. Captain Olsen was teaching the small crew how to work the vessel in the blustery inlets and fjords of South Georgia. Rankin's photography had been progressing. He had begun to catalog his observations of what would amount to descriptive records of nearly thirty species of seabirds, including six species of penguin, four albatross, the Antarctic skua, and the Antarctic petrel.

Yet on December 23, Rankin began to worry. Preparing for their Christmas, the whalers stationed at South Georgia planned to hold their annual collection of albatross eggs, which would disturb the colonies. Rankin had yet to get any good photographs of the albatross mating rituals, which was one of the major goals of his expedition.

He wrote:

> We also had another weight on our minds. Olsen had told us that by far the best bird for eating was the Blue-eyed shag — even more tasty and tender than a young albatross. Our Christmas dinner, we had determined, must make up in excellence for our lonely surroundings and we had given it considerable thought. For the last few days everyone had kept a sharp look-out for a shag, but never one had we seen; in fact I had only caught a glimpse of one or two in the distance ever since we came to the island, so I was far from optimistic. I had visions of a whale steak or a tinned meat dished up in a

THE DEVIL'S CORMORANT

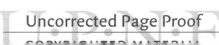

strange sauce. We therefore began our day with a thorough search of all the rocks along both arms of the Undine Bay, but not a shag was to be seen. We decided to return in the evening in the hopes that one might turn up.[13]

Since the earliest European and American exploration of the high southern latitudes, men have been interested in seabirds for food, including cormorants. Sir Joseph Banks, the naturalist of James Cook's first voyage, wrote in his journal about the Maori eating all sorts of seabirds in New Zealand. In 1769 Banks and the rest of his crew shot and ate cormorants up a river on the North Island.[14] He wrote: "A tree in the neighbourhood on which were many shaggs nests and old shaggs setting by them confirmd our resolution; an attack was consequently made on the Shaggs and about 20 soon killd and as soon broild and eat, every one declaring that they were excellent food as indeed I think they were. Hunger is certainly most excellent sauce, but since our fowls and ducks have been gone we find ourselves able to eat any kind of Birds (for indeed we throw away none) without even that kind of seasoning."[15]

Sometimes seamen shot cormorants in the southern seas just for sport. In December of 1811 the ship *Beaver*, sailing toward Cape Horn and bound for the Columbia River, approached the Falklands. A passenger from New York lamented in his journal that his own Christmas dinner would be on the east side of the Cape, because the winds had not been favorable. He was further disappointed because the captain told him that their holiday meal would be no different from the normal boring fare. With the sea rough and "the inkstand [having] been over once or twice," the man wrote of the Falklands:

The land . . . rugged & barren & the only inh[abitants it] possesses are Seals & Birds. . . . We shot though from the ship fifteen or twenty shags, a bird that greatly resembles some of our species of Wild Duck & which the Captn. says are equally good for eating. It was no satisfaction to kill them, for we are going 8 or 9 knots through the water, & to *heave too* with a fair wind to get a bird, could not be thought of, so we resolv to keep our powder & shot. . . . [Later] the shags came off to us in great numbers, & Mr. Clarke & one or two of the clerks shot some of them.[16]

I'm not suggesting that killing a cormorant holds any mythical bad luck, compared to, say, the slaying an albatross, but consider this was in the same part of the world as the setting of Samuel Taylor Coleridge's "The Rime of the Ancient Mariner." What inspired that poem was a sooty albatross killed in this region during George Shelvocke's eighteenth-century voyage. About a week after shooting these cormorants for sport and two days before Christmas, two of the *Beaver*'s sailors drowned when the bowsprit snapped during heavy weather. Perhaps they should have strung a cormorant around Mr. Clarke's neck.

In Tierra del Fuego, early explorers and settlers found that the Yaghan Indians had been eating seabirds regularly—cormorants in particular. Lucas Bridges was born and raised beside the Beagle Channel in the late 1800s. He recounted how in 1876 a party of natives, in an attempt to help starving shipwrecked mariners on that barren outer coast, offered them a shag to eat.[17] The Yaghans, Bridges reported, captured cormorants in different ways. They had developed an ingenious wooden hook, like a tiny toggle harpoon, that would expand and stick into the bird's neck. The native men also caught cormorants—primarily rock shags (*P. magellanicus*) and blue-eyed shags (*P. atriceps*)—by paddling out to the colonies to catch the birds. "The natives would cover over the fires in their canoes but would have prepared torches of bark beforehand," Bridges wrote. "Suddenly these were ignited and at the same instant the covers thrown off the fires. When the astonished birds were thus startled from sound sleep they would flop down in a semi-conscious state into the sea, where the occupants of the canoes would kill all they could."[18] Sometimes the Yaghans paddled out and hid on an island while the cormorants were away for the day. After nightfall they would sneak up on one bird at a time, holding its wings, "thus preventing a flutter or a cry, and kill it by biting its neck or head."[19] On a successful night they would kill several hundred cormorants.

Natives of Tierra del Fuego not only ate cormorant flesh, but the animals figured in their mythology.[20] They also used parts of the birds. They used cormorant stomachs to hold oil or sausage. They used the feathers for decorations and for fishing gear.[21] In a museum collection in Ushuaia, Argentina, I've seen a cormorant bone that had been sharpened for use as an awl to help make native boats and baskets.[22]

During their voyage with the Scottish National Antarctic Research Expedition at the turn of the twentieth century, explorer W. S. Bruce and botanist R. N. Rudmose Brown wrote of how easy it was to approach cormorants: "These shags nested on islets off the coast [of the South Orkneys] which were easily reached over sea-ice. The carefully constructed nests contained much moss and lichen among which the two eggs nestled. The birds were very fearless and confiding and allowed us to approach and even to stroke them as they sat on their nests."[23]

Robert Cushman Murphy was the first major ornithologist of the seabirds of the Southern Hemisphere. As a young naturalist aboard a whaleship he studied the birds of South Georgia.

Here Murphy observed the cormorants' placidity in their rookery:

> I sat beside one shag that was brooding a naked, black, newly-hatched young and one green egg. (I had to lift her—or was it him?—off the nest to find out what was underneath.) She settled back and watched me with blue-rimmed eyes. Her only note was a barely audible croak—such as Keats calls "a little noiseless noise." She kept her bills parted, the mandible and throat trembling as when one's teeth chatter, but I doubt that she was afraid. At any rate, she had no cause to be. I can shoot them at shotgun range, but they are safe when I'm a guest in their homes![24]

The "throat trembling" is a common cormorant behavior, most likely related to thermoregulation.[25]

Thanks in part to this seabird's fearlessness, in the early twentieth century the blue-eyed shag became a special food for the Antarctic mariner. As Niall Rankin and Robert Cushman Murphy described, the whale catchers seemed to have found it easy to nab cormorants, and the men particularly enjoyed their taste. In 1929 L. H. Matthews wrote: "They fly up towards the vessel and, keeping the same speed as it, fly alongside the crow's-nest at the foremast head. This is often the cause of their destruction. . . . They are much sought after as a table delicacy in South Georgia."[26] In 1932, A. G. Bennet in his *Whaling in the Antarctic* wrote that he was eager for any change of the normal diet but was not "able to face this dish" himself, conceding that "shag is another delicacy (!) among whalers."[27]

Cormorants of the region still fly around the upper levels of ships. I witnessed this myself from the ship's topmost deck as we approached the South Shetlands.

And one more proof of this cormorant's value as a table food in Antarctica. This is important not just for understanding Rankin's quest, but for later when I show you how disgusted so many people have been about eating cormorants in the Northern Hemisphere. An anonymous author of a 1950s-era cookbook at a British base just within the Antarctic Circle wrote this: "My advice is if you see any [blue-eyed shags] around, take a . . . rifle and knock a few off. It is a very meaty bird, and one is enough for about six people."[28]

The blue-eyed shags of the Antarctic are a small part of the region's ecology. Figuring by weight, it is the penguins that represent about 98 percent of *all* seabird life in the region.[29] It is worth examining these shags closely, however, because this blue-eyed tribe is unique to the rest of the world's cormorants in several ways and has led to some compelling pathways of inquiry into bird biology. Studying cormorants in this extreme environment is a reminder of just how little we know about how these birds function from day to day.

First, most of the blue-eyed shags of Antarctic waters make their nests taller than do other cormorants around the world — generally as high as the average person's shin or knee. The Stewart shag, nesting on the coast of New Zealand's South Island, makes its nest as tall as five feet (1.5 m).[30] Antarctic cormorants craft earthen pedestals, like little volcanoes, by constructing their nests with mud, tussock grass, twigs, dead birds, dead fish, seaweed, or, as in the case of the *georgianus* shags that nest on the South Sandwich Islands, entirely with the tail feathers of penguins.[31] The cormorants of this region normally build their nests on steep slopes and on grassy cliff sides — no trees in Antarctica — and occasionally on ice or snow. Some nest on cliffs and even breed inside deep caverns.[32] On South Georgia, Niall Rankin observed from afar a small colony of half a dozen shag pairs inside "what was virtually a cave high above the sea."[33]

So why would cormorant nests look so strikingly different from, say, the penguin nests, which are really only rings of pebbles? Wouldn't penguins also want a nice, towering sturdy nest to which they could return and build on year after year? For penguins this might not be worth the calories. It takes a lot of effort to collect materials and build a nest, especially if you cannot fly. Penguins are also better equipped to protect their eggs and chicks. They put their energy into laying a newborn that can almost hit the cold ground waddling. Cormorants need a tall, dry nest, among other

THE DEVIL'S CORMORANT

strategies, to help their offspring survive against the elements. Unlike the downy penguin chicks, all cormorant newborns are what biologists call "altricial," meaning naked of feathers, entirely helpless, and unable to regulate their own body temperature. (The chicks of most songbirds are altricial, too.) Blue-eyed shags also tend to raise more chicks than penguins: on average two to three in the Antarctic region compared to the penguins' one or two. A larger nest might simply be necessary for cormorants to keep the eggs and chicks safe.[34]

In the early 1980s Neil Bernstein and Stephen Maxson, two young biologists from the University of Minnesota, traveled down to Antarctica for two southern summers. Here they established a good deal of the groundwork research on blue-eyed shag biology. Blue-eyed shags on the Antarctic Peninsula switch on and off the nest less often than their cormorant cousins in other parts of the world. For example, double-crested cormorants breeding on the Îles de la Madeleine, Québec, tended during daylight hours to shift their watch on the nest every two to three hours.[35] Bernstein and Maxson observed, by contrast, a surprisingly rigid schedule where the male and female shags of the colony sat for twelve-hour shifts on the nest, switching at approximately midnight and noon.[36] Cormorant expert Timothée Cook told me that he thought the long shifts are most likely driven by how far the individual birds have to travel in order to find food.[37]

A second unique aspect to blue-eyed shags is the growth of the fleshy caruncles at the base of their beaks. At first glance, caruncles seem to have minimal biological utility. They are present in both males and females. Since the caruncles physically expand and the vividness of their color increases before they breed, they are surely connected to mate attraction in some way. But that is about all biologists know with confidence. Maybe the vivid color is an indicator to a potential mate, a beacon of good genes and physical health. Perhaps this has or once had some biological function that we have yet to understand.[38]

A third distinction that the blue-eyed shag species share is that, unlike almost all the other cormorants, they almost never stand out in the air and spread their wings to dry.[39] When we go to South Africa, I'll discuss more about why cormorants spread their wings after fishing, but Bernstein and Maxson determined with a microscope that the blue-eyed shags are like the rest of the cormorants in that their outer feather structure is not terribly water repellent.[40] They theorized that the blue-eyed shags use their inner downy plumage to trap a thicker layer of air than the other cormorants do. This keeps the water from getting all the way to their core when they swim in such cold seas. Whatever advantage there would be to spreading their wings and drying their feathers out once they have emerged from the water perhaps just is not worth it, because the Antarctic environment is too cold and windy.[41]

Almost thirty years after his initial studies, Professor Bernstein explained it to me this way: "The contour feathers and flight feathers could be soaked through and through, but the bird could still have little contact with cold water near its core." He compared it to how a dog gives a shake after coming out of a lake and is suddenly almost dry again. "In Antarctica I watched blue-eyed shags bathing for five to ten minutes," he said. "They popped out of the water, gave a shake, and looked dry as a bone. They were only washing their outside feathers. This is also true of penguins. When they're swimming underwater at aquaria, you can see the trail of bubbles coming out of their down feathers as they swim by."[42]

A fourth difference between the blue-eyed shags of Antarctic waters and the rest of their global family is that despite the frigid water, these cormorants dive significantly deeper. By taping depth gauges to the birds' backs, scientists recorded blue-eyed cormorants in the waters around Crozet Island swimming down as far as 475 feet (145 m). The birds could remain beneath the icy surface for over six minutes.[43] At South Georgia a study monitored blue-eyed shags fishing as deep as 331 feet (101 m).[44] By contrast, Liam O'Flaherty's European shags do not seem to go farther down than 165 feet (50 m).[45]

The reason these blue-eyed shag species dive deeper than other cormorants, and in such cold water, is wrapped up in a variety of biological traits that remain perplexing to biologists. Part of it must be that simply they *have to*. That is where the food is. They do seem to me, visually, a bit hardier a cormorant than those up north, perhaps because their wings are a bit more squared. I clearly remember one shag that flew close over my head: it looked like a cormorant linebacker. None of the cormorants, however, has a blubber layer of any kind, and they all have hollow bones. Both these characteristics — in opposition — affect their buoyancy underwater and thus the amount of effort needed to fly and dive beneath the surface.

Timothée Cook, who is based at the University of Cape Town, explained to me that studies have confirmed that blue-eyed shags do indeed have an insulating layer of air in their feathers, but not necessarily any thicker than the other birds around the world. "One of the secrets of blue-eyed shags," Cook said, "might be the ability to temporarily drop their individual body temperature when in the water. This could save energy, because the bird is not working to keep all its organs permanently at 40°C [104°F]."[46]

The trade-off for the extra energy that blue-eyed shags expend in order to swim so deeply is that these birds need a much longer recovery period back up on the surface before diving again. Compelling for evolutionary biologists, all the deep-diving cormorants are in the Southern Hemisphere.[47]

A fifth biological trait in blue-eyed shags, and the last one I'll discuss here, is that of their white and black plumage. This is present to some extent in several other Southern Hemisphere cormorant species. The guanay cormorant of Peru, for example, has a white chest, but no white on its face. Temperature regulation might be part of the adaption, and, certainly, camouflage might be part of the plumage's utility, too. Penguins and killer whales have similar black backs and white underparts. Even the hue and value of the black-and-white plumage of the South Georgian blue-eyed shag seems exact to that of the local Adélie and chinstrap penguins. The white alar bar on the cormorant wing even matches the white line under the flipper wing of the penguin. Both penguins and shags also have pink feet. The major color difference between the two — other than the blue eye-ring and the caruncle — is that the underside of the wings of the cormorant is black, while the penguin's is white. Perhaps this similarity of coloration evolved convergently for hunting fish underwater. A widely held theory explains that predators and prey have difficulty seeing the white belly of a bird (or a fish) when looking up from the water since it is lost in the white of the sky. And predators and prey have difficulty seeing the black or deep blue of a bird (or a fish) when looking down in the water, or from the sky, as it is lost in the darkness of the sea. Perhaps blue-eyed shags have black underwings, therefore, because it is not important for this part to be camouflaged, since they normally swim with their wings held tightly against the body. A penguin "flies" underwater, showing its underwings. It's also important to consider that the chemicals that make wing and tail feathers black provide stiffening agents — making the cormorants better fliers.[48]

In sum, it seems logical that Niall Rankin's blue-eyed shags of South Georgia and all those of the Southern Ocean build tall nests to protect their altricial chicks from the cold climate. We do not know with any real confidence, however, why these birds have caruncles, why they do not stand and spread their wings as do most other cormorants, why they dive deeper than other species in their family, or even why they have the distinctive black-and-white patterning.

The Cincinnati Zoo and Botanical Garden is the only zoo in the world that keeps a blue-eyed shag. The bird was brought up from the coast of southern Chile as an egg. He lived with a female that also survived and hatched from the same trip, but she has since died at age ten.

So these days Cincinnati's blue-eyed shag often stands alongside a pair of double-crested cormorants. David Oehler, the director of animal collections, has studied these birds at the zoo and in the wild. He told me that the blue-eyed shag does not appear to become as waterlogged as the other cormorants. There is a tank where the public can watch them swimming among the penguins. Docents like to point out how the penguins use their wings and the cormorants do not. The blue-eyed shag gets a twice-a-day feeding of small herring, sand eels, and Spanish sardines — with a supplement of vitamins. Twice a week he gets to go after live fish, typically shiners, for both his own enrichment and the enjoyment of the visitors.[49]

After a visit to the zoo, you might like to stroll over to Eden Park by the Ohio River. Within the park is a bronze sculpture of an usho and his cormorant identical to one that is prominent on the bridge over the Nagara River. Cincinnati and Gifu are sister cities. Who knew?

On my own trip to the Antarctic Peninsula I was not surprised, of course, when the cooks aboard our ecotourist ship did not serve blue-eyed shag for Christmas dinner. Anchored off Port Lockroy on the Bismarck Strait, we had a barbecue on the bow featuring chicken, beef, and salmon. Several passengers sang Christmas carols. One passenger brought a French horn for the occasion and gave a short concert that echoed off the icebergs. The cooks at the grill were young Scandinavian guys. A couple of days earlier, they let me sneak into the galley and scramble a penguin egg that one of the crew members had found floating and gave to me. It wasn't rotten as I expected, but the egg fried up small and dry. The scramble was very bright orange and fishy tasting. I struggled to finish it, even with toast and butter.

That afternoon we had gone ashore to visit the Port Lockroy station. The warden, Dave Burkitt, was a British fellow who wore a festive red flannel shirt. For six years he had been coming to Port Lockroy and running and restoring the station. Burkitt knew of two dozen nesting pairs of blue-eyed shags around the bend. He believed the shag populations were declining overall on the Antarctic Peninsula, along with the Adélie penguins. Climatic shifts and warmer temperatures have increased the amount of snow in the area.[50] Since cormorants and penguins do not like to nest on snow, perhaps their available rookery space has been shrinking.

"In the Peltier Channel there is an island called Priest," Burkitt told me. "Now in the past when this base was operational, in the '40s and '50s, the guys used to row there and kill the young shag chicks for fresh food. Of course, it was allowed in those days. And there were a number of nests there. Dozens. But I went round there two years ago, and you can see where the old sites are, see the old nests on the cliffs. There were only five nests with three chicks between them."[51]

Burkitt hadn't eaten a blue-eyed shag himself but said he heard they were very good from an older guy who was around in the 1940s—the same time that Rankin was visiting South Georgia. "Supposedly very tender," Burkitt said. The shags' eggs were also taken for food.

It was Burkitt's opinion that today there is minimal direct local human impact on Antarctic seabirds in the area. The taking of birds for any reason is now strictly outlawed internationally. With the shore-based whaling industry gone, any human effect on the animals of Antarctica currently comes from larger, global forces happening elsewhere—such as from the Southern Ocean fisheries or climate change.

Port Lockroy is a British base that has a little Antarctica museum. They send mail from the station, which helps them make some money from the occasional visiting ship. I bought blue-eyed shag stamps and a blue-eyed shag special edition envelope that illustrates the tender neck-stroking mating behavior of these birds. The woman tending the small store at the time told me that many people wanted to see the shag colony, and quite a few had bought postcards with the blue-eyed shag photograph.

"Unfortunately," she added, "there has not been a great interest in the blue-eyed shag stamps."[52]

The blue-eyed cormorants of Antarctic waters do not seem to migrate far, if it all.[53] Those living on the distant solitary islands do not stray any real distance from their respective frigid coasts. For this reason the cormorant was a favorite animal, a reliable navigational aid, for those early mariners who dared to sail in this treacherous, uncharted part of the world.

On Captain Cook's return home from his second voyage in January of 1775, he charted South Georgia. Cook spent several days sailing around the coast, confirming it to be an island. He named various rocks and bays (including Bird Island) until he got tired of the hazards of the weather and judged the entire island of little value. He anchored at the head of the same bay that would be one of Niall Rankin's stops over 150 years later, just around the bend from the Bay of Isles where Robert Cushman Murphy named a Shag Islet. Captain Cook planted a flag on the uninhabited beach and named the island after his king. He claimed the entire "savage and horrible" place for Great Britain. Cook wrote of albatross, shags, terns, and gulls. To replenish their fresh provisions, his men captured seals and huge emperor penguins.

Back aboard the HMS *Resolution*, Cook and his men continued exploring around the eastern coast of South Georgia. He observed a "Sugar-loaf Peak," had a narrow escape from the rocks during a sudden storm, and then sailed to the southeast, where he observed another set of rocks on which lived solely birds in "vast numbers, especially shags."[54]

Cook named these latter islets "Clerke's Rocks" after a crew member. As he navigated around to examine their extent he almost got his ship into trouble once again because of the weather. The cormorants and his knowledge that they only fly near to land enabled him to steer clear. He would write elsewhere during his voyages, after a cormorant circled his ship, that these birds rarely "fly far out of sight of land."[55]

Trying to sail away from South Georgia and clear of Clerke's Rocks, Cook wrote:

Thus we spent our time involved in a continual thick mist; and for aught we knew, surrounded by dangerous rocks. The shags and soundings were our best pilots; for after we had stood a few miles to the north, we got out of soundings, and saw no more shags. The succeeding day and night were spent in making short boards [sailing back and forth against the wind]; and at eight o'clock on the 24th, judging ourselves not far from the rocks by some straggling shags which came about us, we sounded in sixty fathoms water, the bottom stones and broken shells. Soon after we saw the rocks.[56]

Cook had already named an island "Shag Island" a few weeks earlier, on December 23, 1774, just before Christmas and a rounding of Cape Horn from the west.[57]

Confirming indirectly the observations of Captain Cook, a study set on South Georgia in the 1990s found that the shags would rarely venture even a nautical mile offshore.[58] Few if any of the historic accounts mention the sensory signals of any colony, both in terms of smell and sound, but I am certain that Cook and other Antarctic mariners before the age of radar, GPS, and accurate charts would have used these aids, too. The smell of guano even in Antarctica is pungent, and the shrill of chicks in the nest, even if there are only a few, can be audible for a great distance over the water if there is little wind.

As the early sealers began to arrive in Antarctica after Cook's explorations, it is clear they used these landmarks named for birds and used the birds themselves, including cormorants, as navigational aids. Today's nautical charts identify several locations in Antarctic waters with "shag" or "cormorant" in their name, demonstrating the prevalence of the bird and perhaps its utility to mariners as a clue to signal a coastline. The islands well to the northwest of South Georgia, which Cook seems to have missed, would later be named "Shag Rocks" — the same of which Niall Rankin referred to anecdotally.[59] Because of the smell of cormorant guano, men named a small, ice-free island in the South Orkneys "Shagnasty Island" in 1947. The British Graham Land Expedition of the 1930s named an island near the Antarctic Peninsula "Shag Rock." Bernstein and Maxson did their pioneering blue-eyed shag studies on "Cormorant Island," just southwest of the United States' Nathaniel Palmer Research Station.[60]

Aside from Cook, perhaps the most famous navigator to use cormorants to find his way was the expedition leader Ernest Shackleton, or, as I like to call him, "Blue-eyed Shagleton." He and his ship's captain, Frank Worsley, lost their *Endurance* to the Weddell Sea ice pack in 1916. After the ship's company struggled to Elephant Island at the tip of the Antarctic Peninsula, Shackleton and Worsley took a few of the men and set out in a small boat for the whaling station on South Georgia. After one of the most daring open-water voyages in history, traversing in a decked whaleboat some of the most treacherous waters in the world, they finally saw a bit of floating plant life. Then they spotted the saving vision of a blue-eyed cormorant.

Worsley wrote later: "Fifteen miles offshore we saw the first shag. The sight of these birds is a guarantee that you are within fifteen miles of the land, as they hardly ever venture farther out."[61]

The cormorant confirmed to them that their navigation had been correct and they might actually survive their journey, instead of having passed South Georgia to be left floating down-current and downwind toward a certain death.

In his book *South*, Shackleton compared the cormorants to a beacon of deliverance: "An hour later we saw two shags sitting on a big mass of kelp, and knew then that we must be within ten or fifteen miles of the shore. These birds are as sure an indication of the proximity of land as a lighthouse is, for they never venture far out to sea. We gazed ahead with eagerness, and at 12.30 p.m., through a rift of clouds, McCarthy caught a glimpse of the black cliffs of South Georgia."

Blue-eyed Shagleton added: "It was a glad moment. Thirst-ridden, chilled, and weak as we were, happiness irradiated us."[62]

As their Christmas in South Georgia neared, Niall Rankin and his crew remained unsuccessful in their plans to have blue-eyed shag for their holiday dinner. They continued their search but could not seem to find a cormorant close enough to shoot. As the day approached they began to look for other options. Rankin suggested elephant seal might suffice. Captain Olsen said he had heard that those weren't even edible. Anchored in a fjord at the north end of South Georgia, Rankin decided to go ahead and kill an elephant seal anyway. They skinned it and then cut it into strips and pieces that they could cook. (As an odd anecdote, Rankin explained that he cut off the sea elephant's penis in order to bring it home to Scotland, but a skua stole the organ when he put it on the ground for a moment.)

Resolved at giving the sea elephant a try for their holiday meal, they brought the meat back to the *Albatross*. The load delivered, they prepared

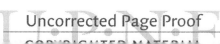

to set out again in their small boat to observe what they thought might be a colony of cape hens.

Rankin recalled:

> The dinghy was already in the water and two of us about to embark when past the boat, not 20 yards away, flew three white-breasted, blue-eyed shags. So taken aback were we that not a word was spoken. We just followed those three figures gliding in line ahead till they alighted on a rock barely 200 yards away. It seemed almost too good to be true. Then I grabbed my gun and we pulled towards the rock, making a slight detour so as to approach as much as possible from the seaward; they are apt to take wing sooner when danger threatens from the landward.[63]

Rankin finished the hunt: "The three just sat and watched us coming nearer, till I could hardly bear to shoot. But it had to be done, and eventually our blue-eyed Christmas dinner was assured."[64]

On their Christmas Eve it snowed, covering the ground all the way to the surf line. Niall Rankin, Captain Olsen, Campbell Gray, and Bob Inkster spent the day on the boat and shared the cooking between them. The menu as published:

M. Y. ALBATROSS
Christmas Day, 1946

MENU

Breakfast	Wheat flakes
	Fried Sea Elephant's liver
	Boiled Wandering Albatross eggs
Lunch	Available on request
Dinner	Stewed Blue-eyed Shag
	Red Whortleberries
	Chutney
	English-made Plum Pudding
	Golden syrup

Whiskey, Gin, Rum, Port, Brandy, glacier water.[65]

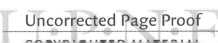

The adult cormorants begin returning to Gates Island in large numbers before sunset. They fly by the half dozen or so in short, rough lines. Nearly all of them come from the west, from upriver. As they land, most are chased by fledglings and chicks, all pecking at the adults to be fed.

The four fledglings from the first clutch by the boulders pursue their parent into the water after food. They are able to fly about half the length of the tiny island—several airborne beats of their new wings. Taking off or landing they slap the water's surface. When they seek food from their parents now they are so aggressive that it appears they are attacking.

As the sun lowers, the gulls stand around on the island, perhaps in hope of some regurgitated fish from the cormorant chicks. The colony grows busier as more and more cormorants return.

The sky deepens to purple. The sun eases beneath the horizon. The navigation lights of the river flash green and red. The black cormorants blanket almost the entirety of the island. They stop only at the gulls' patch of nesting space in the low scrub brush above the little beach.

The fledglings seem satisfied. Activity lulls. The steady single squawks of the gulls slow but do not cease. The cormorants spread themselves evenly over their territory, wrapping their webbed feet over driftwood and rocks, or splaying them flat and wide on the ground. The wind has lulled. The cormorants face in different directions. The fledglings of the first clutch stand together with their parents close to the two boulders. Each curls its neck back around to tuck underneath a wing.

A few cormorants preen. Others stretch out their wings. At the border between their territories, an adult cormorant and an adult gull snap at each other halfheartedly. Another cormorant feeds a fledgling once more, then closes its eyes. A gray downy gull chick, barely discernible, pitters within the cormorant colony and pecks around in the dirt.

By the time the stars are all sharply visible, the cormorant colony on the island is individually and collectively asleep.

East Sand Island

UNITED STATES

We got the big rhubarb here on the West Coast
about salmon runs. They say the salmon runs are being
depleted because of the dams and on and on and on. Control
'em, as far as I am concerned. The skuzzy birds [cormorants]
are eating up all of the salmon. And the tree-huggers want
to blame the lack of salmon on everything else, except
for things like that. Are you a bird hugger?

VISITOR TO THE TILLAMOOK CHEESE FACTORY,
Oregon, 2002

t is just after 11 p.m. on East Sand Island, Oregon, site of the largest single double-crested cormorant (*Phalacrocorax auritus*) colony in
the United States.[1] A light rain has been falling steadily, day and night,
for nearly all of March and well into this April. Several cormorants are
roosting in what the research team calls the "dissuasion area." The scientists and managers that run this project want to prevent cormorants from
sticking around to nest on this side of a fence, one that they built explicitly
this year to delineate a new preferred border to the rookery. They want
to see how the cormorants will respond if the ground for the colony is
reduced on one side by a fraction, and they want to know where the birds
will go once scared away. The researchers want to scare them on this side
of the fence before the cormorants get fully settled for the season and start
laying eggs.

Since the researchers are already planning to bother the birds on this
part of the island, it makes sense to capture several individuals in the process. It is a chance to place identification bands on the cormorants' legs,
as well as to affix to the tail feathers a small radio transmitter that can be

93

used locally to figure out if the birds that are "dissuaded" still find their way back to nesting on East Sand Island — in the approved area on the other side of the fence.

Megan Gensler is a new member of this seasonal research team. She is a little nervous, tired, and the standing around as they prepare is not helping things. An intelligent, poised young woman who is one year out of college, she grew up in the suburbs of Philadelphia. She has some field experience working with pelicans and terns on the coast of North Carolina, but she hasn't done anything like this before. Megan has handled chicks, but not a full-size adult bird or any animal in a manner anything like what they are about to do. She tells herself they are just birds, but the veterans have told her stories of cormorant bites that cracked one researcher's wrist, cut the skin of another through a glove, and an event a couple of years ago that sent one woman to the hospital for a puncture wound in her right bicep.

Megan waits within the tent. Their camp consists of two tents, a latrine, and a plywood blind that that looks out onto the main colony over the fence. This blind is covered in gull poop, and during the day it is an oven inside and chock-full of flies. Someone scrawled in Sharpie marker over the top of the entrance: "Outpost of Filth." The research camp is connected to the fence on the dissuasion side. They call the fence "the Great Wall," since it goes the width of this narrow section of the island.

Tonight Megan can hear the varied calls of gulls. The cormorants are quiet and are all sleeping. The windows of the tent are zippered closed so that the glow of the hanging propane lantern does not spook any of the birds outside.

Megan is starting to overheat as she waits to begin the cormorant capture. She has already put on all the recommended gear: a headlamp, safety goggles, a bandanna that she'll pull up to protect herself from the birds' beaks and the smell of their droppings, as well as knee pads and thick gloves with the fingers cut off so she will have some dexterity. Megan has several layers on to keep her warm but also to protect her arms. She checks to make sure her hair is still secure up in her hat. Megan is one of seven people working tonight. Two others have the specific job title "cormorant capture technician." She listens as the leader of the group, a postdoc whose specialty is cormorant diet, explains the plan. He tells them to be careful because the birds sometimes go for people's eyes. Megan feels like she's in a locker room. She reminds herself: "Keep the goggles on. Don't get bit. Try to get at least two or three birds."[2]

At this moment it is too distant a connection for Megan Gensler to consider that the primary reason she is getting paid to scatter and capture cormorants in the middle of the night, presumably for the long-term good of the bird population, is owing to how this tiny East Sand Island has emerged from the river, then been stabilized, built upon, and managed — which traces back to the planning, interests, approval, and funding of individuals working for more than a dozen local, state, and federal government, tribal, research, and private organizations — who in turn are all institutionally and often personally concerned from one perspective or another about the massive, bewildering, contentious, interlaced near-century-old single question that has dominated the Pacific Northwest: How can we restore the historic runs of salmon?

The Columbia River estuary is a notoriously rainy, blustery part of the world. Despite the best efforts of the U.S. Army Corps of Engineers, the river mouth just a few miles beyond East Sand Island is something of a nightmare in its force and unpredictability: sands shift, swells and waves come from opposing and senseless directions, and the tidal currents can ramp around at eight knots or more.

When Robert Gray first found for the Americans the rumored entrance to the Columbia in 1792, East Sand Island did not exist as it does today.[3] Sand Island seems to have been an underwater shoal that over the decades was washed around and then emerged and shifted after various storm events. It was not until the 1850s that Sand Island began to appear on charts, although in a different location.[4] Over the decades a thin hand formed from the elbow and began to separate, which in the early twentieth century started to be identified as East Sand Island.[5]

Cormorants as well as gulls, terns, murres, pelicans, and a variety of shorebirds, sea ducks, eagles, and falcons have surely been fishing these waters and nesting on islands, on cliffs, on beaches, and among the trees and shorelines along the Columbia River for millennia.[6] It was not until 1836, however, that John K. Townsend gave the first description of the region by a trained naturalist who focused on birds. Townsend, in his early twenties, had traveled overland to the area from Philadelphia — the same part of the country coincidentally that Megan Gensler is from.

Townsend wrote: "The [double-crested cormorant] is common on the river, & there are at least three more species of the same genus not yet indicated, which reside near the Cape [Disappointment] & probably nest upon the surf washed sides. They never ascend the rivers & are in consequence very difficult to be procured."[7]

Today there are indeed four species of cormorants flapping around the Pacific Coast–Alaska region—although only three of those live in the Columbia River estuary: double-crested, Brandt's cormorants (*P. penicillatus*), and pelagic cormorants (*P. pelagicus*). Townsend sent skins of two species of cormorant that he shot at Cape Disappointment to John James Audubon, who returned the favor by calling one of them a "Townsend's Cormorant." But this name for the Brandt's did not stick, even though Audubon wrote of Townsend as "that zealous student of nature."[8] The fourth cormorant species of the Pacific Coast–Alaska region is the red-faced cormorant (*P. urile*). It is possible that Townsend saw one of these near the Columbia, but according to current records they rarely make it south of Alaska.[9] From afar all four of these species of cormorant look fairly similar, but with a little patience it's not impossible to identify each, because of the slightly different coloration, size, and the posture while flying. Though their nesting and foraging often overlap, their ecological niches are different.[10] Three or more species of cormorants overlap throughout most of the Pacific Coast and up along Alaska and out to the Aleutians. (See map on page 000.)

The double-crested cormorant on East Sand Island belongs to the subspecies *P. a. albociliatus* and prefers calmer waters. As Townsend observed, it is the only cormorant of the Pacific Coast–Alaska region that lives inland from the coast. (Taxonomists recognize a second West Coast double-crested cormorant subspecies, *P. a. cincinatus*, which lives only in Alaska; there seems to be little overlap in their range. Federal managers clump the two together and call it the Pacific Coast–Alaska population.[11]) Double-crested cormorants *P. a. albociliatus* nest along the coast from Baja to Vancouver and beside freshwater lakes or rivers, including beside the farthest reaches of the Columbia River watershed. There are little gangs of double-crested cormorants nesting and diving right now in lakes and rivers in the middle of British Columbia and colonies in Idaho, Montana, Nevada, Utah, Colorado, and Arizona.[12] The cormorants nesting at Yellowstone Lake in Wyoming live at over 7,700 feet (2,245 m) above sea level.[13]

Double-crested cormorants nesting away from the ocean within the western states is nothing new. In 1808 explorer and naturalist David Thompson observed cormorants beside Kootenay Lake in British Columbia, near the source of the Columbia River north of Idaho. Thompson wrote: "There are many Cormorants, we killed one, they are very fishy tast[ing] and their eggs almost as bad as those of a Loon; [its] eyes a fine bright green ... the head and neck of a glossy black, with a bunch of feathers on each side of the back of the head."[14]

Western double-crested cormorants when in breeding plumage have crests that grow to a variety of sizes, shapes, and shades: huge white crests that stick straight out, black crests that curl, or even sometimes a salt-and-pepper mix. It seems most double-crested cormorants west of the Continental Divide stay fairly close to their rookeries all year, although those of East Sand Island generally disperse in the winter.[15]

Double-crested cormorants, although they get in the most trouble with people, are actually the least populous cormorants along the extensive Pacific Coast–Alaska region. Red-faced cormorants are probably the most numerous cormorants in the West, with something like one hundred thousand breeding pairs. There are roughly sixty-five thousand pairs of pelagic cormorants in a similar range but extending farther south.[16] Some seventy-five thousand Brandt's cormorant pairs range from British Columbia to Mexico. And lastly there are about 35,000 double-crested pairs in this Pacific Coast–Alaska population (32,000 *P. a. albociliatus*; 3,000 *P. a. cincinatus*), including those on the coast and breeding inland.[17]

To further put the Pacific Coast–Alaska double-crested cormorant population in perspective, in 1913 a biologist surveying San Martín Island, off Baja, California, estimated a single colony of double-crested cormorants at almost 350,000 breeding pairs. This was likely overestimated, but even so it remains by far the largest single double-crested colony ever recorded.[18] By the end of the century there were but six hundred pairs of cormorants living on San Martín.[19]

So East Sand Island today, with fewer than 15,000 pairs, is the largest double-crested cormorant colony on the West Coast, and one of the two largest in all North America.[20] My point is that it is safe to conclude that cormorants were once far more common on the West Coast than they are in the twenty-first century.[21]

Around the Columbia River estuary, the Brandt's and pelagic cormo-

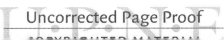

rants are not appreciably on the public radar, partly because their numbers are much smaller and because they tend to fish more along the ocean coast. There are about 1,500 pairs of Brandt's that breed within the double-crested colony on East Sand Island. Dozens of pelagics nest on the cliffs outside the river and underneath the bridge to Astoria.[22] All the cormorants probably eat a salmon smolt or a dozen when available, but owing to habitat and local population size it is the double-crested cormorant that is known these days as the "salmon killer."[23] Megan Gensler says that once when she was in line at the marine supply store in Astoria, a local guy told her: "They should burn that East Sand Island with all the birds on it."[24]

The American poet Robinson Jeffers wrote regularly about the wildlife of the Pacific. His poem "Birds and Fishes" was published posthumously in 1963. The setting is in Monterey Bay, California:

BIRDS AND FISHES

Every October millions of little fish come along the shore,
Coasting this granite edge of the continent
On their lawful occasions: but what a festival for the sea-fowl.
What a witches' sabbath of wings
Hides the dark water. The heavy pelicans shout "Haw!" like Job's
    warhorse
And dive from the high air, the cormorants
Slip their long black bodies under the water and hunt like wolves
Through the green half-light. Screaming the gulls watch,
Wild with envy and malice, cursing and snatching. What hysterical
    greed!
What a filling of pouches! the mob —
Hysteria is nearly human — these decent birds! — as if they were
    finding
Gold in the street. It is better than gold,
It can be eaten: and which one in all this fury of wildfowl pities
    the fish?
No one certainly. Justice and mercy
Are human dreams, they do not concern the birds nor the fish nor
    eternal God.
However — look again before you go.

The wings and the wild hungers, the wave-worn skerries, the bright
   quick minnows
Living in terror to die in torment —
Man's fate and theirs — and the island rocks and immense ocean
   beyond, and Lobos
Darkening above the bay: they are beautiful?
That is their quality: not mercy, not mind, not goodness, but the
   beauty of God.[25]

The capture method that Megan and her colleagues are about to employ
in running into the dissuasion area is rarely done on East Sand Island
because it is so disruptive, though it has been carefully reviewed by the Institutional Animal Care and Use Committee at Oregon State University.
At the start of her job Megan and the other new researchers were required
to pass a course to be certified to handle animals.

Later in the summer Megan might get a chance to capture *within* the
main colony, instead of in this dissuasion area at the far edge. To catch the
birds within, the researchers use an entirely different method. After the
cormorants have begun nesting, they do not want to disturb birds in the
way that they are going to do tonight in the dissuasion area.

Over the course of previous winters, the research team on East Sand
Island has erected and expanded stretches of aboveground tunnels constructed out of wood and black tarp. From these tunnels the researchers
facilitate cormorant capture without widespread disturbance, they observe
birds closely, and they get access to an observation tower in the heart of
the colony into which two or three people might squeeze and spend long
hours and even overnights.

The triangular tunnels enter the colony from the beach, then snake into
the middle of the nesting area. I can tell you from experience that they are
just large enough for a small man to maneuver on his hands and knees
over the sand, feathers, guano splatters, bones, and other detritus. The
guano smell is aggressive. Before we went in, we each pulled on a construction face mask with an air filter. Inside the tunnels we were inches away
from the birds without the animals seeming to take much notice. We were
so close that I could hear a bird nibbling at its feathers. At one point when
I was by myself and crawling in one of the branches of the tunnels, I had
to sneeze. I couldn't help it. I tried desperately to hold it in, and then I

buried it in my elbow. A cormorant pecked in at me through the construction cloth.

As with the cormorant capture in the dissuasion area, the method from the tunnels within the colony is done at night. To allow this to happen, early in the season the researchers place rubber tires alongside the tunnels just outside openings in the tunnel walls. Double-crested cormorants like to nest in these tires, so the parents unsuspectingly lay their eggs within an arm's length of a trapdoor.

The cormorant capture within the colony usually involves two researchers crawling in the tunnels out to a row of tire nests. They do not risk using a headlamp, as the light might wake the birds, even through the cloth. It can be so dark in the tunnels that the researchers determine where they are by counting the wood frames by feel. Once at their spot, one person gently pulls back the fabric. He or she finds the shape of a sleeping cormorant in a tire nest. After a couple of deep breaths and a nod, the person reaches out and grabs the base of the bird's head, just underneath the beak. Sometimes the bird does not even wake up until it is in the tunnel and its mate has stepped into the nest to cover the eggs, as if he or she were in on the plan. Ideally the person grabbing the cormorant keeps the beak away while his or her partner gets the wings under control. If the project requires the banding of chicks, they do this immediately inside the tunnel, in the dark. If it is an adult cormorant that is to get a band or tracking device, they put the bird in a canvas bag and scramble with the animal to a little windowless outpost off one of the tunnels where they can use a light and have slightly more room to do the job. Before they built that outpost, there used to be a "surgery hut" back outside the colony and practically half the island away. The older researchers tell Megan tales of when veterinary surgeons came on the island to insert satellite tags. The researchers had to slog in the middle of the night through the water, sloshing up to their waists in the bay while carrying terrified cormorants in bags over their heads to the surgery hut.

Tonight, however, in the dissuasion area—on the edge of the main colony and far from the tunnels and several weeks before any nesting has begun—the team that Megan is working with for her first cormorant capture will employ a sort of bum rush method, sprinting in among the sleeping cormorants with handheld spotlights. Megan has been given a huge dip net. Other researchers will grab the birds by hand.

Despite the Columbia River's dangerous entrance—which in large part kept Astoria from becoming a Seattle or a San Francisco—the river has remained a crucial artery for commerce. In order for the large ships to move up and down the river to Portland and beyond with any degree of safety and predictability, the channel needs to be dredged. The ships steam close to East Sand Island on their way in from the bar. Megan can look off the porch of the research team's rental house in the hills of Astoria down onto the Columbia River. Nearly every morning she sees at least one or two massive ships from countries such as China and Korea—empty of ballast and standing by to make the trip upriver.

The Army Corps of Engineers has had several major pushes over the decades to maintain the navigability of the Columbia to accommodate bigger ships. In the 1950s and '60s, the Army Corps was digging the channel and needed a place to dump the dredge spoils. So they used these materials to stabilize and extend East Sand Island.[26] About fifteen miles farther upriver, the Corps also created a brand-new island out of dredge deposits. They named this Rice Island.

Stick with me here because, as digressive as this all sounds, it is a compelling example of how convoluted wildlife management problems can get, giving even Russ McCullough's "nothing natural" Lake Ontario a run for its complexity. And I promise it spirals back to the cormorants and salmon and Megan Gensler getting paid to stand there in the middle of the night with her safety glasses on and a bandanna over her face while she is holding a net wider than a bicycle tire and hoping not to get a beak puncture in her bicep. It also helps illuminate why a man from Portland at a cheese factory in Tillamook, Oregon, would like the government to thoroughly eliminate all these "skuzzy birds."

Hydroelectric power from the hundreds of dams along the Columbia and its tributaries supplies a significant portion of the electricity for the homes in the Pacific Northwest. The dam system also pumps irrigation for millions of acres of farmland, provides the security of flood control to properties, and allows the safe navigation of commerce all the way into Idaho. I don't need to tell you, though, that dams significantly alter a river's ecosystem, especially to anadromous fish such as salmon. For the First Peoples of the region, salmon held *the* central role—culturally, ecologically, and economically. Scholars believe that when Lewis and Clark arrived, salmon had already provided physical and spiritual sustenance to native peoples for some nine thousand years.[27] The early fur traders and settlers soon moved into the business of salmon when they recognized the enormity of the resource and had a method to preserve and ship it. Astoria supported at its peak in the 1880s as many as forty salmon canneries supplied by over a thousand commercial boats, using a variety of methods to fish on the Columbia. In 1883 Astoria packed over thirty million pounds of salmon into cans.[28] By then men had constructed almost fifty fish traps, fish wheels, and canneries just within and around Sand Island and the surrounding Baker's Bay.[29]

By the 1920s the runs were already declining precipitously because of continued fishing pressure, logging practices that destroyed riverside habitat, and industrial pollution. The early network of dams had also already cut off, by one estimate, fully half of the entire watershed to salmon migration.[30] This was even before the construction of the New Deal hydropower behemoths.

Today salmon are still significant to the Columbia River region in many ways, but represent only a medium-size recreational fishing interest and a small commercial fishery.[31] The salmon you buy in a restaurant outside of the immediate area rarely comes from the Columbia region.[32]

Yet salmon still represent a deep cultural connection, both as a symbol for First Peoples and a reminder of "the good old days" for current residents of the Columbia. This is similar on a much larger scale to the way some Gifucians feel about their anadromous ayu and ukai, or how residents of Henderson Harbor feel about smallmouth bass. Salmon makes the Pacific Northwest different from other parts of the world. Salmon has become the iconic animal and the indicator species for environmental health in an area that prides itself on conservation and green living.[33]

Salmon — and I'm talking *salmonids* here, which include sockeye salmon, chinook salmon, chum salmon, steelhead trout, and other closely related species — travel upriver to spawn. Fertilized eggs grow through a variety of stages into smolts, which are large enough at a few inches to make the trip downriver and eventually out into salt water. Columbia River salmon runs have a wide variety of migrations, some voyaging out for several years into the North Pacific, ranging as far north as the Bering Sea. Regardless, they then return to the river of their birth, swimming up to the place of their hatching to continue their species. This upriver run of the adults, many of which die after spawning, not only once provided for a new generation of salmon but also once helped fertilize and feed a diverse web of plants and animals upriver.

Spurred by the mandates of the Mitchell Act in 1938, hydropower companies showed some effort to help maintain salmon runs by installing fish ladders at the dams and funding salmon hatcheries. But none of these strategies were all that effective in overcoming the extent of the disruption. In the 1970s, along with other major environmental awakenings around the country, resurged a public recognition of the significant role and vast diminishment of the salmon populations. Meanwhile federal courts affirmed the nineteenth-century treaties that protected the salmon rights and allocation of the caught fish to First Peoples. Then in the 1990s a few distinct populations ("evolutionary significant units") of Columbia River salmon were placed on the endangered species list. Others were federally designated as threatened.[34]

In response to all this, dams were required to install still more fish ladder systems. Fish hatcheries were more carefully managed. Scientists delved more thoroughly into the effects of hatchery fish on wild populations. Today some 70 percent of all salmon caught commercially in the Columbia basin originated in a hatchery of some sort.[35] The Army Corps of Engineers and the Bonneville Power Administration fund each year the barging of millions of salmon smolts to get them safely through the dams when heading downstream. They also run a program that pipes the young salmon into trucks to deliver them downriver by highway, thus skipping the dams. The Corps has constructed guidance screens to keep the smolts from being sucked into the turbine blades on the way downstream. These screens funnel the salmon into a bypass channel from which they can continue downriver. At certain high migration times of the year, hydropower plants such as Bonneville Dam, the one closest to the mouth of the Columbia, stop some of their turbines completely to give fish an easier time of getting through.

I've watched cormorants and osprey hang out beside the bypass channel release point at Bonneville Dam. They wait to catch fish. Also waiting for salmon smolts are northern pikeminnows, a native fish that has prospered thanks to the placement of dams, because pikeminnows thrive in still water. Trying to protect the salmon, the Army Corps offers a $4 to $8 bounty for each pikeminnow caught that is longer than nine inches (23 cm). You'll earn $500 if you hook one that has been tagged by the

Bonneville Power Administration and the Oregon Department of Fish and Wildlife.[36] (Yes, cormorants eat pikeminnows.) To further protect migrating salmon smolts, workers for Bonneville Dam occasionally use water jets aimed at the hungry birds and the sea lions that swim all the way upriver to enjoy the easy, concentrated food supply.

Now here is where it really returns to the seabirds. When explaining the threats to the salmon in the Pacific Northwest, wildlife managers typically use the mnemonic of the "four H's": habitat, hatcheries, harvest, and hydropower. Yet in the late 1990s they needed to add another "H" — for *Hydroprogne*, the genus name for the Caspian tern (*Hydroprogne caspia*). After the endangered species listing of the salmon, managers began to examine what portion of the smolts that made it through and beyond the dams — by their own means or by barge or truck — were not making it all the way to the Pacific. Early studies suggested that birds, mostly terns, were eating a surprisingly large percentage of smolts before the young salmon swam their last miles of the Columbia toward the Pacific. Soon the dams, the federal government, the tribes, the water companies, the agricultural interests, and the environmental groups all began to be invested, figuratively and literally, in understanding what impact the birds were having.

The research project on East Sand Island in which Megan Gensler is involved is run by a collaborative organization called Bird Research Northwest.[37] Cormorant dissuasion is a small part of their expansive study of the entire region, examining over a dozen seabird colonies in the three contiguous Pacific states. It is funded with the primary mission of getting at this question of bird predation on salmon and learning ways to resolve the problem.

The entire project has grown to become one of the best-funded long-term bird studies in the world, certainly of those that involve cormorants. It is not at the level of funding that goes toward studying Antarctic penguins or building blinds in the canopy of the rain forest, but where else on earth do cormorant researchers have two elevated observation structures to observe these birds, accessed by a network of tunnels so as not to disturb the colony? Megan Gensler is not out eating Dungeness crab every night or buying a new hybrid hatchback, but during her first summer, Bird Research Northwest paid for her to go on aerial surveys and funded her travel to observe colonies as distant as Puget Sound and the Salton Sea. Bird Research Northwest runs a detailed and accessible website, has funded and supported almost a dozen PhD and master's projects, and has produced scores of scientific papers and government reports. Over the last two decades Bird Research Northwest has compiled an exceptionally impressive and useful pot of biological knowledge about coastal seabirds, as well as advancing our understanding of which management actions and tools regarding these animals "work" — and which do not.

The initial bird "problem," the previous "salmon killer" before the cormorants took over this title, was this Caspian tern, a white bird with a black shock of head feathers and a scarlet dagger-shaped bill. This is the largest of the terns. Terns don't dive underwater appreciably, but instead plummet from the air, like gannets or pelicans, and nab fish from just under the surface.[38] While attracting mates and raising chicks, terns fly conspicuously with a fish in their beaks. Flapping around with waggling salmon smolts didn't help their public image.

The first records of Caspian terns nesting in the Columbia River estuary was in 1984 on East Sand Island, immediately after some dredge materials were deposited.[39] After a couple of years the nesting habitat was apparently better upriver at newly formed Rice Island. Here the tern colony really took off, growing into the largest Caspian tern colony in the world at over 8,700 breeding pairs.[40] Between 1997 and 1998 Bird Research Northwest found that salmon made up almost three-quarters of the diet of the tern colony on Rice Island. The terns ate almost 13 percent of all outbound smolts, which totaled an estimated 12.4 million juvenile salmonids.[41]

After a few years of study and public hearings, as well as navigating suits from several environmental groups including the National Audubon Society and Defenders of Wildlife, the Corps contracted Bird Research Northwest to encourage the tern colony to move back to East Sand Island.[42] The thinking was that East Sand was still a safe, mammal-free island but closer to the ocean: the terns would diversify their diet. On East Sand Island the managers cleared vegetation so it looked as it did after the deposition of dredge spoils, thus making an appealing sandy nesting area. Terns are pickier about their bedding than cormorants. The researchers placed tern-shaped plastic decoys in the sand. They blared a recording of tern colony sounds. On Rice Island, they dissuaded.

The project worked. Within three years, the entire tern colony moved back downriver to East Sand Island, and the result was a substantially reduced percentage of the tern diet composed of salmon.

Over the years, however, as the tern colony thrived on one end of East Sand Island, so did the cormorants at the other end. From one perspective, the cormorants gummed up a perfectly good plan that seemed to be moving along nicely to help the salmon, the terns, and most of the humans along the Columbia coexist.[43]

Double-crested cormorants first started nesting on East Sand Island in the late 1980s.[44] Both cormorants and terns eat a wide variety of fish from the river, including anchovies, surfperch, sardines, herring, sculpin, and flounder.[45]

The double-crested cormorants of the Columbia River basin forage for a large variety of fish in a range of salinities, from ocean salt water to river fresh water. Within a range of less than ten miles (16 km), the East Sand Island cormorants dive around the stone jetties leading to the ocean, across the tidal flats of sand and mud in the river, in the waters of marshes, in the local creeks and tributaries that feed into the Columbia mouth, and even right in the middle of the shipping channel.[46] The cormorants seem to especially like to forage around pilings and sometimes collect downstream of a line of these dock structures, perhaps taking advantage of fish that are either a bit disoriented by the pilings or are more relaxed due to the lessened current.[47]

I posed the question of whether cormorants have a bigger appetite than other seabirds to Daniel Roby, professor at Oregon State University and one of the lead investigators at Bird Research Northwest. He explained to me that one cormorant needs on average three to four times more small fish each day than that required by one tern. "Though this amount of fish as a proportion of a cormorant's body weight is not larger when compared to tern food requirements," Roby says, "cormorants do need considerably more food than other seabirds their same size." The larger a bird gets, the less food it needs in relation to its mass, but cormorants have a high metabolic rate, likely in part because it takes a lot of energy to dive underneath the water to hunt and to stay warm. Cormorants also raise more chicks than most other birds their size, which also requires a lot of energy and food.[48]

Certainly cormorant and tern diet varies within the year based on the birds' energetic needs and the larger movements of fish. The birds' choice of prey varies from year to year, too, notably in the coastal Pacific in response to short- and long-term climatic changes, such as during El Niño years.

Bird Research Northwest continued its studies. To further reduce salmon depredation, the organization was contracted to help relocate the terns to still other places outside the Columbia. Meanwhile, the cormorants overtook the terns in numbers on East Sand Island. Soon cormorants were eating more than the terns in total numbers of smolts. In 2011 the double-crested cormorant colony of about 13,000 breeding pairs ate an estimated 20.5 million salmon smolts. This was roughly one in every seven smolts that had survived to swim outbound to sea. Salmon made up almost 20 percent of the cormorants' diet.[49] The cormorants ate samples from endangered and threatened wild runs, but a majority of the salmon gobbled up were hatchery-raised fish.[50]

From the perspective of a wildlife manager looking to "restore" the historic fish populations, every single one of these outbound smolts represents a tiny silvery piece of the regional cultural identity: each young salmon has been raised with an immense investment of emotion, money, time, and labor — not to be stolen by the cormorants.[51]

Just south of the Columbia River region, various other Oregon groups have been angry about cormorants eating salmon. Studying this phenomenon has been an ornithologist named Range Bayer. One morning he kindly took me on a birding trip around Yaquina Bay, during which he told me how he once had a job for several years at a private fish farm. At one point the boss had him driving a truck around to scare cormorants off a fish pond.

"An impossible job," Bayer said. "There was just no way."[52]

In 2000 Bayer put together an impressive forty-page report titled "Cormorant Harassment to Protect Juvenile Salmonids in Tillamook County, Oregon." In order to protect salmon smolts released from hatcheries in three estuaries, the federal government along with the Oregon Department of Fish and Wildlife gave license to private individuals to harass cormorants — a more intense form of dissuasion. The men used speedboats. When it was too shallow for boats, they used hovercraft. They fired at the cormorants with cracker and screamer shells. They played cormorant distress calls over loudspeakers. They installed human-like dummies, which you could call "scare-cormorants." Bayer found that over a decade none of this had any positive or negative correlative effect on the survival of smolts.[53] He said the program was largely driven in the first place by the interests of a few recreational fishermen.

For my humble part I conducted an informal and unscientific survey of twenty-five workers and visitors at the Tillamook Cheese Factory, Tillamook, Oregon, soon after Bayer's study came out. I wanted to find out if anybody had heard of the controversy of cormorants eating salmon along their coast or even up in the Columbia. To be honest, almost everyone with whom I spoke, both locals and tourists, did not even know what a cormorant is. All were familiar with salmon issues. Several people identified the federally protected sea lions as the major threat. Yet one man spoke to me about an island off Portland that he understood contained "millions" of cormorants, which he called "skuzzy birds." He saw the cormorants as the primary cause of salmon mortality on the river, more so even than the dams. He would not give his name.

He asked me: "Are you a bird hugger?"[54]

Another attempt to restore salmon populations and human jobs on the Columbia River, instituted in the 1970s, was the installation of net pen facilities below the dams, near the Columbia's mouth. Workers at the net pens get a load of small fish brought down in trucks from upriver hatcheries, and then raise them to smolts. After several months in the water of the estuary, the juvenile salmon are released. If all goes well, a couple of years later the fish return to the mouth of the Columbia and the net pens area, instead of trying to travel all the way upriver to spawn above the dams. Here at the mouth by Astoria the recreational and commercial fishing fleet wait with gill nets and fishing rods.

The net pens themselves are a series of floating frames with netting below and above to contain the little salmon and to prevent the birds and larger fish from eating them. When the net pen fish are released, the facilities managers have developed methods to avoid having them all being chomped immediately by the birds. They have learned to release the smolts at night and on an ebb tide.[55]

The Youngs Bay net pen facility, for example, releases millions of smolts each year. Occasionally a sea lion or river otter tears into their pens. The workers there told me a story of when a guy forgot to replace the top netting one morning, and five cormorants got into the pen and ate so many salmon the birds could not fly away. East Sand Island is about the same distance to the rearing pens as is Rice Island, so the moving of the terns did not make much difference for their interests.

I asked the men at Youngs Bay if they or the locals viewed the terns and the cormorants any differently. Keep Jeffers's poem "Birds and Fishes" in your mind, because one worker explained to me: "The terns have a certain notoriety to them. They are one of the farthest migrating birds in the world, they are a pretty bird. But they sure sound obnoxious. Boy, they just 'Blaah!'" He pointed out into the bay and gestured with one arm:

> They just fly over and they just dive, just bam, just one after the other. The cormorants, when they are coming in, it's amazing. You can see these big lines of birds, a black mass, in the Bay. They'll come down and see a group of fish, and they'll all land and they'll all dive together and come up with fish in their mouth. The cormorants will line up. On the mudflats up here they'll dive down to get 'em, they will come up, they will crowd the fish, or they scare the fish right up onto the beach, and the fish are flopping around on the beach, and you'll see silver masses of our salmon. And the cormorants get up there and they are just wolfing them down. It's horrible.[56]

Like so many other interests in the area, the workers at the Youngs Bay net pens monitor closely the numbers published each year as to the percentage and volume of salmon that the cormorants and terns eat. These figures are mostly determined by biologists at Bird Research Northwest. For both terns and cormorants on East Sand Island, biologists collect at the end of the nesting season all of the tens of thousands of passive integrated transponders (PIT tags) that are left on the ground in a given area. These tags are tiny cylinders placed in a sample of hatchery and wild fish, identifying exactly which hatchery or net pen or river the fish came from and when. The birds poop or spit these tags out with other materials they naturally purge. In addition, for terns, Megan and her colleagues on East Sand sit in a blind, in shifts, and observe and record the fish species that the terns feed their young. The researchers also occasionally shoot screamer shells to get the adult terns to drop the fish they are carrying back in their beaks. The screamer gun doesn't really work to get cormorants to barf up their fish, so to determine cormorant diet, in addition to the PIT tags, the researchers hire a few professionals to help them shoot about 140 cormorants a year in order to provide stomach samples. They pick the birds off with shotguns from a boat as the birds are returning to the colony. (Cormorants float, so the marksmen scoop up the dead birds afterward with a dip net.)[57] Combining all these methods, the Bird Research Northwest biologists estimate predation on salmon and other fish species for the breeding season, using population and consumption models derived in part by Don Lyons, the same biologist who is leading Megan's first cormorant capture. The predation estimates can never be foolproof or 100 percent accurate, but trying to get these numbers correct is a full-time job for several people, working upon decades of trial and error.

Now the net pen guys watch these numbers each year to see how many of their salmon are ending up on East Sand Island. And they see with their own eyes the cormorants eat the salmon. These men are not unreasonable as to birds. They recognize that this problem is largely man-made. But when asked if he had an answer, one worker told me: "I have a radical theory. It involves napalm."[58] He was kidding—mostly.

As with the situation on Ron Ditch's Little Galloo Island, the solution is unlikely to be as simple as moving or culling the cormorant colony. One local newspaper editorial proposed sneaking onto East Sand Island and introducing a couple of pigs—as vigilantes did on an island with an unwanted cormorant colony in Lake Michigan.[59] The Pacific Coast–Alaska cormorant population on the whole has been increasing over the last couple of decades by about 3 percent per year—which is substantial when you think about it.[60] The double-crested colony on East Sand Island currently represents almost 40 percent of an entire subspecies.[61] The growth of the cormorant population on East Sand did not just begin with their reproducing like mad. It expanded too fast at first for that alone. This colony is composed of a large portion of birds that emigrated years ago, presumably for the food, from other parts of the coast and from inland, from perhaps as far away as British Columbia.[62]

Like the double-crested cormorants on the other side of the Continental Divide, the Pacific Coast–Alaska population of double-cresteds seems to have been on a roller coaster since European contact. They rebounded during the latter third of the twentieth century for reasons similar to those of the other cormorant populations in North America: environmentally friendly legislation, banning of chemicals such as DDT, improved health of water qualities and fish stocks, increased open air aquaculture, and man-made reservoirs and protected islands.[63] Wiping out the colony on East Sand Island would do serious damage to double-crested cormorants in this part of the world. It is worth noting that in the current federal cormorant management plan (2003/2008), none of the twenty-four states that are now allowed to issue permits to kill cormorants or oil eggs is to the west of the Rockies.

Bird Research Northwest has been working on nonlethal solutions for East Sand Island. The researchers have been experimenting with moving cormorants to man-made islands in other less salmon-centric locations outside the Columbia River. They are trying the same methods they deployed with terns, including playing cormorant sounds from speakers and placing plastic cormorant-shaped decoys at the new sites.

East Sand Island is owned and managed officially by the Army Corps of Engineers. Local fishermen tell stories of camping out there as children, but now if they land there unescorted they would break several laws and regulations. Amid the Bird Research Northwest tunnels, towers, temporary walls, and campgrounds is a curious mix of human artifacts, including half-sunk railway track leading into the water and wood pilings in various states of barnacled, worm-eaten decomposition. With the land artificially stabilized, the terns introduced through decoys and recorded calls, and the cormorant population rising and nesting opportunistically for a variety of anthropogenic reasons, East Sand Island is a peculiar place.

And it gets still more complicated.

When Megan Gensler was on the island, over 14,000 brown pelicans roosted on East Sand, arriving after they had finished nesting as far south as Mexico. East Sand Island is now the largest non-nesting roost of brown pelicans on the West Coast.[64] So far it seems the pelicans are not yet appreciably eating a lot of salmon, but this bird has only just recently been lifted off the endangered species list — which means the managers conducting their cormorant and tern activities must be careful not to bother the pelicans. Not coincidentally, the American Bird Conservancy and the Audubon Society have recognized East Sand Island as regionally and internationally important.[65]

Megan also witnessed an unprecedented amount of predation on the rookeries. It is common for a few terns to be taken by owls and falcons and perhaps even a river otter. Resident gulls often opportunistically eat the eggs and chicks of cormorants and terns. But this season, bald eagles started preying on East Sand Island birds in a way not seen in recent memory. One or two eagles flew in at dusk or after dark to take adult terns. This so freaked out the tern colony that not a single chick successfully fledged over the summer. Megan and a stunned research staff watched a complete failure of this tern colony. Every time an eagle swooped in after sunset, the entire flock of terns took flight and did not return for hours or even until the next day. This allowed the gulls to hop in and eat up all the eggs and chicks.

The eagles also went into the cormorant colony at the other end of East Sand Island. Juvenile eagles alighted smack in the middle of the cormorant rookery, sending off those that could fly. The eagles then marched along and ate cormorant chicks and eggs, something that even the veteran researchers had never witnessed or heard about. Neither the double-crested cormorants nor the Brandt's cormorants on East Sand saw any sort of collapse at the level of the terns. Cormorants returned to their nests as soon as the eagles left. But it certainly wasn't a banner reproductive year. One theory is that there was not enough fish for the eagles, so they were turning to other prey sources. Then again, as eagle populations recover across North America, their attacks on seabird colonies, including on cormorants, have been recorded elsewhere on the Oregon coast, beside the waters of Victoria, British Columbia, in the Gulf of Maine, and on Lake Champlain.[66]

During Megan Gensler's season on East Sand Island the cormorant colony also got hit with Newcastle disease, a virus better known to infect poultry. Newcastle disease seems to afflict a percentage of the chicks or juvenile cormorants every couple of years on the island.[67] It has been recorded in double-crested cormorants in various other parts of North America since the 1970s.[68] Cormorants with Newcastle suffer a sort of palsy. They cannot lift a wing. They walk at an angle. They hold their neck crookedly. Though the disease is not the same strain that can tear through farm poultry — the East Sand researchers confirm this each time — the island foreman still gets panicked calls from a variety of state and federal disease control officials to make sure certain steps are taken. This was one of the reasons we wore masks when crawling in the cormorant tunnels. During Megan's summer, dozens of birds were stricken with Newcastle. She found it difficult to just watch them die.

From inside the blind, aka "The Outpost of Filth," built to look over the dissuasion fence, I watched one of the cormorants with Newcastle dying. Along a line of rocks, it was lying on its back beside another that had died earlier, one that was now gray and splattered with droppings from the other birds. The fading cormorant could only stroke its right foot, as if paddling. It did this constantly, with the one foot paddling over and over for several hours. The animal grew more gray and white-splattered as the birds around it went about their business. Yet it kept paddling with the one foot, usually at the same rate. The dying cormorant occasionally tried to lift its head. It did this for so long that I had to leave, to go to bed, and even when I climbed up to check on it the next morning at first light, the bird was still paddling, but slower.

The cormorant capture team creeps out of the tent, past the dissuasion fence, and around by the latrine. The leader turns on the spotlight. He gives a shout. The people run out at different angles. Hundreds of gulls riot. Megan Gensler forgets to turn on her headlamp. She trips over a log. Most of the cormorants flap, croak, fly off. Megan recovers herself, turns on her headlamp, and picks her net back up. She is surprised to see how passive and frozen several other cormorants still are. The soft ground is uneven. She hurries forward, climbing over another log. She spots a cormorant sitting there staring into her light with one eye. She slips the net over the top of the bird. She pulls out one of the cotton bags from her belt, reaches inside the net, and clasps the cormorant around its neck just under the beak. She places the struggling bird in the bag. Ties it up. The cormorant gives a sort of honk, tries to walk but falls down, and then lies motionless and quiet.

Within two minutes from their turn around the outhouse, Megan suddenly realizes everything is done. The gulls are still raucously flying overhead. She looks back and sees several other gulls standing on top of the tents, cawing urgently.

The researchers shout to each other. "Anyone free? Give me a hand here!"

The spotlights are switched off. Megan leaves her bagged cormorant as her eyes begin to adjust. She pulls down her bandanna and rushes over to help, crouching low. A few cormorants croak. The sound of rustling: feathers against cotton.

Someone laughs and says: "Hey, stop that!"

"I'm free, who needs help?" another shouts.

"Need another bag over here!"

The lead technician had got his net around three cormorants at the same time. He is now lying there, also holding two more cormorants that he had somehow reached over and grabbed with his free hand. In all, they have fifteen birds.

Back inside the tent, Megan and the others carefully lift the cormorants out of the bags to place them into ventilated cardboard pet boxes. They turn off the propane heaters inside the tent. The leader directs two birds to be released because their tail feathers are not in good enough shape to take a radio tag. They release another bird that is too young to breed. Megan hears one cormorant regurgitate some fish into its box.

Now the researchers go to work in the way that you do in the middle of the night: slightly loopy because you're tired, but also diligent because everything feels more intense when it is dark out and you want to finish the job and get back to bed. Megan and the two other new people are charged with holding the birds while the more experienced researchers place bands on the ankles and affix the radio tags. This involves a drop or two of Super Glue and a few small zip ties around the base of two tail feathers. (The tags will fall off with the first molt of these feathers.) After each cormorant is banded and tagged, Megan walks the bird out of the tunnel and onto the beach. She releases it. She waits a few minutes to make sure that it flies away safely. After a pause, a dazed stumble, each of the tagged cormorants that night does indeed fly off.

By about 1 a.m. the capture team is cleaning up. Everyone has bird droppings on their clothes. Splatters are on the floor of the tent. Nothing more than expected, however, and all are actually feeling pretty good. The process went smoothly. No birds or people were hurt. No birds got loose inside.

As do most of the rest, Megan walks into the other tent to try to get in a couple hours of sleep. She strips to her long johns, hangs up her gear, and lies down in her sleeping bag. They are planning another cormorant capture before dawn, if the birds return.

Megan Gensler can't sleep. Her adrenaline is still pumping. Though she isn't sure yet if anyone saw her trip over the log, her pride is hurt. The propane heater hisses. A couple of others are snoring. The colony just on the other side of the tent's fabric is quiet again, because the gulls have settled down. She wonders if the cormorants have returned. Megan replays in her head the rush into the dissuasion area—remembering the sideways, green-eyed stare of the cormorant in the beam of her headlamp. She imagines texting a friend. She will describe the whole thing as feeling like some kind of military operation to go abduct aliens.

August

In the boat channel between two shipyards, a cormorant methodically fishes against the tide. The bird dives for thirty to forty seconds, surfaces for about ten, and then dives again. The cormorant does this for a long stretch of water, submerging for over a dozen repetitions for almost exactly the same amount of time. Every few dives the bird is successful in catching a few small silversides, which she swallows while underwater.

Just to the south hunts one of the juveniles from that first clutch by the boulders. Paddling with his feet, he floats on the surface beside the hull of an old steel workboat. He is the same size and looks the same as the adults, except the feathers on his chest and along his throat are pale brown.

He surfaces beside the hull with a fish from the rocky bottom. It is a black-fish the size of a small football, almost twice the length of his head. The fish flaps and whirls until he drops it in the water. The cormorant swims in the channel for a while as if indifferent.

He dives again with the tidal current and comes up downriver holding the same spiky blackfish in his beak.

Three people are having a barbecue on their boat at a mooring. One says, "Look at that. There's just no way that bird can swallow that thing."

Another takes out a phone and snaps a photograph.

The cormorant does not seem to hear them. He drops his head in the water to reorient the fish. He still can't do it.

Once more the young cormorant lowers his entire head into the water. He manages to keep hold of the fish with the sharp tip of his beak. He reemerges, head titled back and stretched upward. He works to bend the fish's body and swallow it whole and headfirst so as to keep the spines from catching inside his throat. The skin of his neck expands around the shape of the fish as he swallows. After he gets the fish down, the cormorant shakes his head. His whole body trembles.

Then the cormorant looks around and sees the people on the boat. He takes off after several beats on the surface, flying low downriver and toward the island.

6

Tring

ENGLAND

*Cormorants of more than forty species range along
the hundred thousand leagues of earth's shore-line,
well distributed . . . they constitute the mightiest race
of fishers ever known, save those born of the teeming
waters themselves. The piscatorial peculations of men
are as a dot beside their unceasing pillage.*

WILLIAM LEON DAWSON, 1923

*T*he bird research collection held by England's prestigious Natural History Museum is not in London, but at its smaller suburban site, the Natural History Museum at Tring. Access to the birds is not available to the general public. The collection is open to researchers who prearrange their visits and bring proper identification with a photograph — because there have been a few thefts in recent years. Known formally as the Bird Group of the Department of Zoology, it is inside a drab four-story cement building, an edifice that is a bit out of place in a quiet town in Hertfordshire that traces its charter to the fourteenth century. Tring is mostly brick and stone, with Tudor and Victorian architecture, a fifteenth-century church, and pubs with names like "The King's Arms."

The three main floors inside the Bird Group are clinical. They are devoted almost exclusively, aside from a few offices and work desks, to housing aisle upon aisle of tall white cabinets with wide sets of drawers inside. Each of these drawers holds preserved stuffed skins of dead birds. Most of these pelts, some over two hundred years old, have been meticulously gutted, the tendons pulled out, the eyes removed, and the brains scraped out from the inside. Each has been treated with some sort of preservative, most commonly borax or a specially prepared soap that has been laced

116

with arsenic.[1] The eyes of the bird are stuffed with cotton wool, the body is filled with the same, and then the skin is sewn back up so that if done properly the cured bird replicates the girth of the original animal. If the bird had been intended for some sort of public display, as hundreds are for the public space of the Natural History Museum at Tring next door, the carcasses would have been sewn around some sort of armature and crafted in a vertical or active posture to evoke the bird in the wild. These specimens in the trays, however, are mostly placed on their backs, their wing against their bodies.

The feathers are surprisingly, unexpectedly, soft and delicate — even lifelike to the touch. If the bird is a little perching bird, such as a warbler, its head has been tilted back. If it is a large long-necked bird, such as a swan, the head has been tucked down toward its stomach. In most cases the collector chose these postures to take up less space when stored for travel aboard ship or in a box. An open drawer, whether lined with hummingbirds or herons, smells like a box of mothballed old sweaters.

All the birds in these drawers have labels tied around their ankles that identify whatever provenance details are known, such as where the bird was collected, when, and by whom, as well as another label or two that might update the taxonomy or give other information not provided by the person who captured it or donated it to the museum. Some of these labels are the originals written by the collectors themselves, punched out on a card with a typewriter or scrawled on brown paper with a fountain pen or even a quill.

The Bird Group holds approximately three-quarters of a million total skins. There are Galápagos finches collected by Darwin on his voyage with FitzRoy. Dozens of John James Audubon's specimens that he collected in America are here. Nineteenth-century birds from India gathered by Margaret Cockburn are in the collection. By way of the London Zoo, two skins of penguins from the expedition of Niall Rankin are held in one of the drawers.[2] The research collections at Tring hold stuffed falcons, stuffed birds of paradise, and stuffed chickens. There are preserved pelts of titmice and jackdaws and tropical toucans. There are dozens of extinct birds and at least one sample, but normally several — and sometimes even *thousands* — of nearly every bird that is currently living on earth. To be more exact, the Bird Group holds in its trays and on its shelves specimens of 95 percent of all bird species on the planet. The building also holds about three hundred thousand sets of bird eggs, about fourteen thousand bird skeletons of various completion, some two thousand nests, and about seventeen thousand dead birds kept in jars, preserved in methylated spirits or some other pickling liquid.[3] The Bird Group holds reconstructions of extinct birds of which there are no remaining skins, like the dodo. Connected to the Bird Group building is a vast ornithological library and archive of scientific volumes, papers, books, original field notebooks, journals, and letters. Curated and preserved here is a large collection of original bird artwork — including massive, nearly floor-to-ceiling oil paintings of birds — as well as an assortment of odd bird objects, such as a mounted albatross head under a glass dome and ostrich eggs painted with African landscape scenes.

Dr. Robert Prŷs-Jones is in charge of this collection, a position he has held for more than twenty years. His accent is polished: he says "buuhrd" and "re-suuhch." Prŷs-Jones earned his doctorate in avian biology at Oxford with a comparative study of the reed bunting and the yellowhammer. He then taught university students and did field research in Australia, in South Africa, and in the Seychelles before he settled back in England to work at the British Trust for Ornithology. Five years later he took a position as the collections manager of the Bird Group, where he has been ever since. Prŷs-Jones publishes prolifically in peer-reviewed journals on all manner of bird topics, has an encyclopedic memory of the biological literature of bird research, and he was even asked to write the foreword to an important book about artists depicting birds.

Prŷs-Jones does not suffer fools (e.g., me). He strides through the aisles and hallways of the Bird Group at a rate in which he is technically walking by definition, but few people can keep up with him without actually getting real air under their feet.

"I always wanted to study birds," he tells me over his shoulder. "There was never any real doubt. I still have an essay that I wrote when I was about nine saying for my career I wish to study birds and travel. Not that I knew what that involved at the time."[4]

Prŷs-Jones and the Bird Group hold the rarest of all specimens of cormorant. It is not kept with the other cormorants in the white cabinets in the aisles. It is in a specially locked area. He promises to bring it out.

"In fact," he says, "there are two of them."

And off he goes.

When I was browsing through the cormorant cabinets in the Bird Group in Tring I found at least one specimen of every living cormorant, except for one. I couldn't find a skin of a pygmy cormorant (*Phalacrocorax pygmaeus*), a little-studied freshwater cormorant with a short beak and a long tail that was first officially described in 1785 as the "Dwarf Shag."[5] True to its name, this is the smallest of the cormorant species. Each wing, at an average of about eight inches (20.5 cm), is shorter than a takeout chopstick.[6] Pygmy cormorants live in small pockets around lakes, rivers, and marshes in places as diverse as Romania, Italy, Iraq, and Turkmenistan. They eat fish such as perch, roach, and carp.[7] I was pretty disappointed not to find a pygmy cormorant in the trays at Tring.

I learned later that I had just gotten confused. Robert Prŷs-Jones is one of those people who are so smart and organized that I not only feel foolish around them, but I inevitably manage to do even stupider, sloppier things than I normally would. The Bird Group does have every cormorant species. I somehow had just not read all of the labels on the drawers carefully enough to compare the different scientific names.

"It's right here," he says.

And it was, next to all the others. Predictably, in fact, the pygmy specimens were right next to a tray of several specimens of another tiny cormorant, the Javanese cormorant (*P. niger*). At different times some of these individuals have actually been considered in the collection as pygmy cormorants themselves.[8] And as early as 1898 biologists and collectors had already published the Javanese cormorants under almost twenty different scientific names, including as *pygmaeus*, within a few different genera.[9] So I could claim that this was why I got confused, but Prŷs-Jones would have seen through that.

And to be fair, cormorant taxonomy, whether you're reading labels of scientific names or sorting through common names in different field guides, is a thicket. Discussions of whether or not the pygmy cormorant is just a subspecies of Javanese, for example, make the job of Prŷs-Jones all that more complicated as he catalogs and organizes his collections. If he made a real attempt to stay updated he would have to do so endlessly, since new studies and examinations are conducted year after year.

"We have *never* aspired to keep up to date with the latest taxonomic changes," he says to me, "but rather to ensure that the collection is stored according to a standard system that should make any taxon easy to locate by visitors — or ourselves. Though clearly that didn't work for you in the case of pygmaeus."

The truth is that you can only be comfortable saying vaguely that there are somewhere between twenty-four and thirty-nine extant cormorant species across the globe. English ornithologist Bryan "Lord" Nelson in his *Pelicans, Cormorants and Their Relatives* (2005) went with thirty-nine. Throughout this book I use his taxonomy and common names because it is up to date at the time of writing and relatively clean and simple for our purposes. (See appendix 1.) Paul Johnsgard in his 1993 monograph settled on thirty-three living species, while Peter Harrison in his seminal *Seabirds* (1985) would commit to only twenty-seven or twenty-eight cormorants.[10] In order to assemble their work, Nelson, Johnsgard, and Harrison each spent a great deal of time with the collections held here at the Bird Group. Whether or not the pygmy cormorant is separate from the Javanese was actually much less of a hurdle for these authors than was sorting out the Antarctic blue-eyed shags. These are the cormorants that really shift the numbers and account for the major differences in total cormorant species.

Part of the complication with cormorants, which is true for the ever-plastic taxonomy of all animals, is that the early classifications were based on some type of morphology or behavior. Aristotle likely gave the first rough attempt that included cormorants in his *History of Animals* (c. 350 BCE):

> The whole genus of birds may be pretty well divided into such as procure their food on dry land, such as frequent rivers and lakes, and such as live on or by the sea. Of water-birds such as are web-footed live actually on the water, while such as are split-footed live by the edge of it. . . .
>
> Of web-footed birds, the larger species live on the banks of rivers and lakes; as the swan, the duck, the coot, the grebe, and the teal — a bird resembling the duck but less in size — and the water-raven or cormorant. This bird is the size of a stork, only that its legs are shorter; it is web-footed and is a good swimmer; its plumage is black. It roosts on trees, and is the only one of all such birds as these that is found to build its nest in a tree.[11]

The nineteenth-century German naturalist Johann Friedrich von Brandt died before he could finish his opus on the cormorants, so the

British naturalists were the first to publish comprehensive classifications of these birds and their relatives — using in large part the skins and skeletons of the Natural History Museum.[12] The first thorough attempt was conducted by a man named St. George Jackson Mivart in 1878. He chose subtle differences in the skulls as the primary way of classifying cormorant species.[13] As the decades progressed, other scientists took up cormorant classification as a project. Each in large part placed in front of him all the trays of dead cormorants and similar birds, and decided that it should really be the structure of, say, the plumage that is *the* way to classify the cormorants and their relatives. Notable among several that followed was the English biologist Jerry van Tets, who in the 1970s added behavior as a factor — such as the ways that various species dance to attract mates or how exactly they dive under the water.[14] Van Tets placed the pygmy and Javanese cormorants, for instance, separately under a subgenus that he called *Microcarbo*, which I like to call the "micro cormorants." Theoretically, recent DNA technologies and the analysis of egg-white proteins should put an end to much of this taxonomic subjectivity. Yet debate continues.[15]

Even the genus for all the cormorants, *Phalacrocorax* (from the Greek for "bald" and "raven"), after shifting from the wonderful title of *Graculus*, is not universally agreed upon today. Often the cormorants are divided into multiple genera, sometimes along the somewhat arbitrary line of cormorant and shag, meaning primarily their dimensions, selected behavioral traits, and their primary habitat of fresh or salt water. Some cormorant taxonomists also use sub-genera, of which Jerry van Tets used five. Most biologists also use subspecies for cormorants, as is the case with the double-crested in North America. "Lord" Nelson conceded at least two subspecies for about a quarter of the thirty-nine cormorants that he cataloged.

It is tempting to nod off over these sorts of classification debates — perhaps you already have. As one ornithologist in Oregon told me, and she's the author of several chapters and papers on the seabirds of this coast: "It doesn't matter to the cormorants whether they are in the Phalacrocoracidae, which is most closely related to the du-du-da, or maybe better related to the blah-blah-blah. Who the hell cares? Now, someone does. Which is good. Because I do see the utility. And everyone must have their passion in life!"[16]

In some ways all of the shifting taxonomies seem only to keep Robert Prŷs-Jones and his staff hurrying along the aisles and constantly feeling like they wish they had time to make new labels. In 1964 the British ornithologist Sir Arthur Landsborough Thomson slipped in this editorial in his definitive bird dictionary: "To regard nomenclature as more than a means to an end is pedantry, and to take a minority course in a matter of convention is merely a nuisance."[17]

In short, it can turn nearly everyone off if various biologists are constantly using new names.

There are some circumstances, however, where a careful consideration of cormorant classification has broad implications. For example, delineation of subspecies affects management policy at the mouth of the Columbia River. If the cormorants on East Sand Island did not make up 40 percent of one of the world's subspecies of double-crested cormorant, dissuading or even culling would be less controversial. Or if, for example, the Macquarie shag (*P. purpurascens*) is indeed its own separate species, then if its breeding population were to slip below a few hundred pairs, we should respond quite seriously and quickly before this bird is lost from the planet forever. Recognition of species can determine official classification of threatened or endangered status.

For researchers with more of a focus on "knowledge for knowledge's sake," the careful classification of birds — or any animals — with the absolute best information possible gets us closer to understanding the evolution of species, their past movement around the continents, and their anatomical and ecological progression. For all the work of biologists and paleontologists over the centuries, even the very basic arguments over how feathers evolved and from where and why has yet to be definitively settled. Did wing-driven autonomous flight, for example, evolve from the gliding behavior of ancient tree-living dinosaurs, or, as a majority of evolutionary biologists now believe, did solitary flight evolve from little ground dinosaurs who already had developed feathers for other reasons and began to jump and get some air (while trying to keep up with an ancestor of the Prŷs-Joneses)?[18]

Defining taxonomical relationships between cormorants helps us to come incrementally closer to understanding the relationships of behaviors and physical traits between birds. Does the absence of the wing-spreading behavior in certain cormorants or the difference in how deeply one species of cormorant dives teach us something about their historic evolutionary movements on earth? Or let's say two swimming birds have sharp hooks at the end of their beaks. They might be closely related and evolved from the same ancestors. Or maybe they are derived from birds that evolved on opposite sides of the globe and progressed through totally different lineages: they both survived and reproduced by convergently developing that pointy thing to hold on to the available slippery fish. Understanding the genetic relationships of species can help assemble larger stories. It might even provide clues as to the movement of tectonic plates or the ability of animals to swim underwater.

Jerry van Tets took up these broad questions in regard to the cormorants. Van Tets had left England as a young man and moved to Australia. He suggested in one study, presumably without the bias of an expat, that based on fossil evidence, global patterns in plumage and behavior in all cormorants, and the diversity of species in the southwestern Pacific, the greater Australia region just may perhaps be from where the first cormorant (or cormorant egg), the progenitor of every cormorant alive today, first evolved. Van Tets believed that all the cormorants spread westward from Australia in an arc around the Indian Ocean and over to Africa.[19]

While waiting for Robert Prŷs-Jones to bring out the prize pair of stuffed cormorants from the Bird Group collection, I stood before an open drawer of cormorant skins. The most striking aspect in many ways is the brittle, often crumpled feet of the birds. The dead blank eyes filled with cotton wool draw the eye, but there is something about the cormorant feet— perhaps because of the historic yellowed labels attached—that renders them at once creepy, portentous, and genuinely beautiful, somewhat like the skinny, arthritic hands of the very old. The cormorant feet in the trays seem clenched, gnarled, frozen, and even wise somehow. (See figure 11.)

"Lord" Nelson sticks with convention and puts the cormorants within one of the six families of the order Pelecaniformes. (This is in the kingdom Animalia, phylum Chordata, and class Aves.) Cormorants are the sole occupants of the family Phalacrocoracidae within this order. The other five families in this order are the pelicans, the gannets and boobies, the frigatebirds, the tropicbirds, and the anhingids. The combined total of all the species in these other families does not make up the diversity of the cormorant family.

All the seabirds of the Pelecaniformes share a few physical traits, most conspicuously their four-webbed toes, long thin beaks, tiny tongues, lack of brood patch, and a mostly featherless gular pouch. This pouch is huge, of course, on pelicans. It's that same big red balloon that male frigatebirds puff up under their beaks.[20] All the birds within the Pelecaniformes family have their salt-excretion gland fully enclosed within the part of the skull that houses the eyes. Their chicks are born blind and bald. In terms of behavior, Pelecaniformes are colonial nesters, fish eaters, and they have comparable mate-attraction rituals.

Recent DNA evidence and other advanced methods seem to show that pelicans might not be as closely related to the others in the Pelecaniformes as once thought—perhaps they are closer to storks and vultures—and the frigatebirds and tropicbirds might find themselves in a different order someday soon, too. On the other hand, those scientists who have conducted electron microscope examination of eggshell structure are sticking with these six bird families as a natural grouping.[21] Thus the taxonomy of Pelecaniformes, like the cormorants within, is still very much in flux.[22]

Cormorants or any of these families, however, are *not* closely related to penguins, loons, geese, mergansers, terns, or gulls—even though they have several convergent traits.

The anhingid family represents the birds most often connected and confused with cormorants. There are, arguably, two anhingids: the anhinga and the darter. The anhinga lives in the Western Hemisphere. The darter lives in the Eastern. Like cormorants, anhingids are diving black birds with long beaks that spread their wings after swimming. In several parts of the world, cormorants and anhingids live together in close quarters, such as in the Florida Everglades. When a cormorant swims you can see its back, while an anhingid more often swims much lower, sticking only its neck out of the water, which is longer and skinnier than a cormorant's. Thus anhingids have been given the common name "snake bird." Anhingid plumage is also quite different from a cormorant's. Anhingids have a much longer tail than most cormorant species, they don't have a hooked tip at the end of their beak, and anhingids are more vocal. So anhingids are not cormorants, but they are the most closely related to them.

So then what makes a cormorant a cormorant? What connects the pygmy cormorant to the Japanese cormorant to the South Georgian blue-eyed shag? Almost all ornithologists agree that cormorants are colonial coastal or inland birds that nest and sleep primarily, if not always, on land.

Cormorants have the traits of their larger family outlined above, most notably: totipalmate feet, altricial chicks, the salt gland within the skull, long thin beaks, long thin necks, and a mostly featherless gular pouch. Most cormorants have dark plumage on all or the majority of their body. They have twelve or fourteen tail feathers. Cormorants all share very similar mating display behaviors. Cormorants almost always dive from the surface to pursue their live prey underwater, using primarily their feet to propel them. Cormorants have wettable feathers, aiding their ability to dive deeply. And they all have sharp hooks at the end of their thin beak to catch the live fish and invertebrates that make up their exceptionally varied diet.

The Bird Group is connected to a Victorian redbrick building that was begun in 1889 to house what is now called the Natural History Museum at Tring. Until recently it was the Walter Rothschild Zoological Museum because of its founder and patron. The museum still maintains Rothschild's Victorian presentation of the world's biodiversity. Within glass cases are hundreds of stuffed birds, including a full line of hummingbirds and a few cassowaries from northern Australia and New Guinea — some of Rothschild's favorites. The bird taxidermy makes up only a fraction of the experience, however. Behind glass are preserved and mounted polar bears, tigers, and two rhinos. There is a large collection of mounted insects and lizards and fish. You can look at stuffed bats, a stuffed orangutan, and stuffed giraffes. There are sea otters, camels, and a hippopotamus. Added after Rothschild's death, showing diversity at the species level, is a display of stuffed dogs, some of which were once prized as competitors or were companions to famous Englishmen.

The founder and creator of most of this, Lord Lionel Walter Rothschild, was born to a wealthy banking family. He participated in the finance business somewhat, but his soul was with his research museum collection. He started this from scratch as a boy, then opened parts of it to the public and the rest to researchers. Rothschild rarely did the collecting himself, but was more interested in the taxonomy and the scientific study of the animals, so he often financed expeditions. Among others, the type specimen of the Chatham shag (*P. onslowi*) was once originally named after him in 1893: *Phalacrocorax rothschildi*.[23]

Rothschild wrote a careful book on extinct birds. He kept several exotic animals alive within the grounds, including flightless cassowaries and giant tortoises. He wanted to observe their habits and colors in life. Rothschild also kept zebras. He used them to pull his carriage around town and once rode them at Buckingham Palace.[24]

When Rothschild's money began to run dry, he was forced to sell almost his entire collection of bird skins. He did so to the American Museum of Natural History in New York City.

Robert Prŷs-Jones says with a pained look: "The Rothschild Collection ended up in the AMNH. It represented over a quarter of a million bird skins that actually should have come to us."

Prŷs-Jones can take pride in the fact, however, that the American Museum of Natural History does not hold one of the rarest bird specimens on earth. In fact, there are only six of these cormorant skins that remain. Two are held at Tring. One is in a museum at Leiden in the Netherlands, one at the Zoological Museum of Helsingfors University in Helsinki, and the final two are at the Imperial Academy of Sciences in St. Petersburg.[25]

Tring is the high point of the Grand Union Canal that connects London to Birmingham. To keep the locks of the canal flowing during the ascent up to Tring, four reservoirs were carved out of the ground in the early nineteenth century. Since the 1950s the reservoirs have been designated nature preserves and remain a popular place for bird-watching, hiking, and recreational fishing.

According to the BBC, the most popular sport in England is fishing, and the most popular hobby is bird-watching (although the difference in activity and fitness of these two is worth debate).[26] Correspondingly, two organized groups in particular make use of these man-made lakes: the Friends of Tring Reservoirs and the Tring Anglers. The Friends are mostly bird-watchers who keep regular tabs on the migratory waterbirds and passerines, including the great cormorants (*P. carbo*) who live in the area during the winter. The Tring Anglers have fishing competitions, social gatherings, and help facilitate the sport for species such as bream and perch, and for catfish — which was introduced intentionally in the nineteenth century.

"Tight Lines," the pseudonym of the editor of the fishing club newsletter, acknowledges the area suffers illegal fishing and threats from human-influenced natural imbalances, which include increases in minks, the unintentionally introduced American crayfish, and cormorants. One of the reservoirs is stocked frequently with trout for the sport fishery.[27]

The editor writes that the cormorants "do not seem to be eaten by anything and the European variety which dominates the winter migratory birds is now of plague proportions. We are not allowed to cull on Natural England Sites of Special Scientific Interests, however once away from the reservoirs . . ."[28] That is his ellipsis. It serves as a malevolent, if humorous, suggestion.

Fishermen taking action against the perceived threat of cormorants in England goes back for centuries.[29] In Cornwall in 1911, for example, the birds were removed from a listing on the Wild Birds Protection Acts (1880–1908), and residents of the county were given one shilling (about £3 or $5 in today's currency) for every head of a cormorant or shag they brought in. Cormorants and shags were shot on their breeding grounds, and men destroyed their chicks and nests. In less than two decades, almost 11,000 heads were delivered. One environmental historian estimates the number actually killed in Cornwall and the Isles of Scilly during this time to be closer to 20,000 birds. The bounty program was eventually abandoned because the effect on fish stocks seemed negligible and a new threat of seals garnered more attention.[30]

Today cormorants along with all other wild birds are protected in Britain and throughout Europe. The license to shoot cormorants in Britain can be granted by Natural England within the Department for the Environment, Food, and Rural Affairs (DEFRA). This is only if a real need is demonstrated regarding an effect on a fishery after other deterrents have been tried. The upper limit is set each year. The most shot thus far under this license was 1,603 cormorants killed from August 2005 to August 2006.[31]

Robert Prŷs-Jones says he knows of no real conflicts between the bird-watchers and the anglers in Tring, who both enjoy a reason to get out for a bit of fresh air. At least he knows of nothing that has made the newspapers. The Tring Anglers do represent, however, the anxiety that has been growing among certain Brits since at least the 1980s about these birds, evidenced by particularly nutty websites such as "Cormorant Busters."[32] (Click here for an example of a personal, semi-psychotic, raging diatribe against the evil and gluttonous cormorant.)

"Cormorants are not greatly loved birds," Prŷs-Jones says. "And fishermen get uptight about them."

He explains further: "I very much doubt cormorants eat more than other birds. It's just that they are fairly large birds that catch a lot of largish fish, which they bring to the surface before they swallow. They're just very obvious. It doesn't help that they're not particularly cuddly birds, as it were."

As the wintering populations of inland great cormorants grew in England, the conflict started to make national news in 1996 when the editor of the newspaper *Angling Times* published on the cover a masked man, looking like a terrorist, holding a gun standing above four dead cormorants. The bold, all-capitals headline shouted: "These Birds Must Be Killed." Inside this and a following issue were ideas about how to poison and drown the birds, which the authors called the "Black Plague."[33] The police brought the editor to court for inciting his readers to kill the birds as a violation of the Wildlife and Countryside Act. The case was thrown out.[34] I was able to get hold of an image of the cover and the article, but could not, after much effort, get the legal rights to print it here.

Of more or less six subspecies worldwide, two subspecies of great cormorant live in Europe.[35] *Phalacrocorax carbo carbo* stays on the coast of the British Isles most of the year beside the shags, with perhaps some time spent inland during the winter months. *P. carbo sinensis* only lives inland and is joined during the winter months with more *sinensis* who fly in from continental Europe.[36] The two subspecies are very difficult to tell apart without a bird in your arms and a set of calipers to measure the shape of the gular pouch.

In England, as in North America with its double-crested, the inland, more-migratory population of cormorant species has been the source of the controversy.

Recently a lobbying organization called the Angling Trust created a website called Cormorant Watch. This is an interactive Google map where anyone who sees a cormorant on any body of water, such as one of the reservoirs of Tring, can enter the location, adding to a database. The iconography of this website, including the poster the organization provides, is similar to that of neighborhood policing programs. The image on the poster is a blue box containing the shape of the view from within binoculars: a black, Miltonesque silhouette of a cormorant with its wings spread, standing among tree limbs. A second postcard/poster that the Angling Trust created to be sent to one's respective representative in government is still more fascinating. (See figure 13.)

The Cormorant Watch campaign also produced and posted a video of the chief executive of the Angling Trust, Mark Lloyd. Wearing a tie and jacket, the clean-cut, well-spoken Lloyd, seemingly of similar age and breeding to Prime Minister David Cameron himself, speaks to the camera while standing beside a lake with an island in the background on which several cormorants waddle on the dirt or perch in the limbs of bare trees.

He explains: "Over the last few years cormorants have become a major problem in our waters in the UK and have become a massive threat to native fish stocks."

The camera pans away from Lloyd, views the birds, cuts to a few cormorant chicks feeding, and then returns to him.

"We need your help to help make this work," he says. "To prove to the government that there is a problem with cormorants."

He explains how to use the Cormorant Watch website. Then he concludes: "There are millions of anglers out there in the UK. We need to stand up and be counted and make our voice heard. Please use this site and get everyone you know to record any sightings of birds or colonies they can. If you love your fishing, help us protect it."[37]

In February 2012, a dozen men from Angling Trust and the Salmon and Trout Association delivered to the door of the DEFRA offices in London a petition with sixteen thousand signatures, requesting faster action on controlling "non-native invasive" cormorant populations migrating from Europe.[38] They posed for a photo with a sign that said "Biodiversity in Danger," showing a cormorant eating a huge fish.

Both the great cormorant and the European shag are native throughout the British Isles. Records of cormorants are common in fossil beds and archaeological sites, such as in Norfolk, Kent, Essex, and the Orkneys. Stone Age peoples in Yorkshire ate cormorant. Cormorants and shags are found in trash middens of the Bronze Age, Iron Age, and of the more recent early peoples throughout the British Isles.[39] Likely the birds' numbers have been so low in inland waters because of centuries of dense human presence. The birds were shot, or potential rookeries were just too disturbed. Echoing the situation in North America in many respects, their inland population has only begun to rise substantially since the 1970s as a result of national protection, reduction of pesticide use and industrial chemicals, and the increased construction of reservoirs, gravel quarries, and the subsequent stocking of these waters for recreational fisheries.

The number of cormorants in England, including breeding and wintering birds, doubled from 1987 to 2003, peaking at about 30,000 individuals.[40]

The Royal Society for the Protection of Birds (RSPB) saw this increase in the cormorant population as an encouraging sign of the country's improving environmental health and legal protections. The leader of the RSPB's policy position on cormorants, Sarah Dove (appropriately named), explained to me that since 2004, when the evidence needed for DEFRA licenses to kill cormorants was eased, the breeding and wintering populations of cormorants have decreased. She said the very population models DEFRA uses are scientifically suspect, and research has found that killing cormorants seems no more effective on fish stocks than just scaring away the birds. Furthermore, she added, "The Angling Trust's call for the killing of an unlimited number of birds is unnecessary because sustainable, nonlethal approaches, such as fish refuges [man-made underwater structures], can reduce the damage caused to fish stocks by cormorants. Government research shows fish refuges can reduce cormorant presence by 77 percent, and the fish predated by 67 percent."[41] She says the RSPB does not oppose killing birds when there is a genuine problem. But this is not the case here, in their view.

Countries throughout Europe have had similar conflicts with cormorants because of recent population increases and much larger colonies. Norway now has twice as many cormorants as Britain, and there are over 160,000 *pairs* in over five hundred colonies around the Baltic Sea. The European Parliament began developing a cormorant management strategy in 2008, but progress has been slow and almost hit a stalemate.[42] A group of researchers in the Netherlands and the UK was set up to create a future nonbinding EU Cormorant Platform, organize continent-wide counts, and establish a website and a new organization, "Sustainable Management of Cormorant Populations," with the acronym CorMan.[43] CorMan has partnered with an older organization, the Wetlands International Cormorant Research Group, which meets annually and publishes a bulletin of papers. In 2011, for example, the Cormorant Research Group conference in Medemblik, the Netherlands, attracted representatives from twenty-two countries — all devoted to talking about the science and management policy of cormorants. (I wasn't able to make it to the conference, but from what I heard about this flock of colleagues, *What happens in Medemblik, stays in Medemblik!*)

Current public estimates of the numbers of great cormorants throughout Europe, Scandinavia, and western Asia have varied enormously, but the Cormorant Research Group arrived at a rough total of about 1.2 million birds, of both subspecies, in 2007. They hope to complete a new total in 2012.[44] Likely fewer than 10,000 great cormorant *pairs* now nest in the UK, and of these about 2,000 pairs nest inland, mostly in England—but this has expanded from one single inland nest identified in 1981. Only in a handful of cases in England are cormorants in colonies of more than a few hundred birds, and never more than 600 pairs. The trouble comes more in the wintertime, when tens of thousands of additional great cormorants fly inland for a few months. They come from the coast and from across the North Sea and the English Channel, from Scandinavia and mainland Europe. Once in Britain, the birds winter beside inland ponds, lakes, and rivers.[45] A sizable portion of these are the subspecies *P.c. sinensis*.[46]

Cormorants, at least of this inland subspecies *sinensis*, were once exceptionally rare in Europe in the nineteenth and twentieth centuries, likely because of persecution and loss of mammal-free rookeries. Great cormorants in Europe have since recovered after protective legislation in several countries, as a result of the same factors at play in North America: these protective measures, the reduction of man-made toxic chemicals, and an increase in their food supply thanks to the construction of aquaculture facilities in several parts of Europe.[47]

If a group of cormorants stays only a few months in Britain and then spends the rest of the year in another part of Europe, who, if anyone, has the right to manage these birds? The Dutch, who have one of the largest populations of cormorants, are completely opposed to culling the birds. Yet the French have been killing about 40,000 cormorants a year.[48]

Robert Prŷs-Jones returns at last. He has presumably completed a few dozen errands while I've been waiting. Yet while he carries the birds over he is not sprinting. He is walking slowly. He looks down at the stuffed specimens. On a desk he places the tray and the two extinct spectacled cormorants (*P. perspicillatus*). He closes the entire row of window blinds, darkening the room and casting wire-thin stripes of daylight obliquely, poetically, across the two dusky birds. For at least 150 years, no living spectacled cormorant has walked on earth.

As my eyes readjust to see these cormorants—these giants compared to the pygmy — I notice that one of the specimens of the extinct cormorant is missing a tail, and at one point had the top of its skull coarsely lopped off and filled with cotton — maybe for the breeding crest or to remove part of the contents of the skull. One of the necks is much thicker, but this is perhaps due to amateurish taxidermy. The two birds are large: longer and wider than any cormorant that now lives. The long outer toe is as long as a postcard.

The hairlike filoplumes coming from their necks are yellowed, as are the once-white flank patches. The rest of the feathers are dark — the whole reminiscent of a massive eggplant, an aubergine — but this is partly because of the lumpy shape of the birds on the white tray. The feathers are dark purple, with iridescent greens, visible only at an angle, within the neck and breast. The birds look in many ways like huge pelagic cormorants or like Mr. Yamashita's Japanese cormorants, both of which still today spend time in the Bering Sea. The larger, intact spectacled specimen measures well over three feet (95 cm) from its head to the end of its tail. It is lying on its back with its beak pointed upward and to one side. One stiff foot overlaps the other daintily.

The specimens have had over seven different scientific names since they were first brought to the Natural History Museum. The label on the ankle of the full specimen says "Bering Strait" and "John Gould (purch[ased])," with the registration number and year filled in on the back by hand: "1858–2–3–1." There's a much older, brittle label tied around the same ankle, written with a fountain pen in a rough script.

Beside this corpse and laid in the opposite direction is the scalped specimen. Its feet are placed so the claws are just touching, as if ready to hold a bowl. The impression is accentuated by the lack of tail underneath. Around its ankle is one old label that has been repaired with tape. The name *Phalacrocorax urile*, written in a lovely hand, has been crossed out in pencil, replaced by *Perspicil*. The location is "N. W. America," given by a "Capt. Belcher."

Robert Prŷs-Jones grins proudly down at the two extinct cormorants. He backs away a step. Then just before hurrying off, he pauses and says looking over his glasses: "Please be careful."

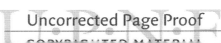

A tropical storm, nearly a hurricane, barrels over the Mystic River, screeching enormous southeasterly winds for almost two days. Gates Island is flooded and beaten. The wind and seas blast the guano off the boulders and wash away nesting material, corpses, shells, everything. Some of the cormorants, gulls, and other coastal birds of the estuary fly upriver and hide in various leeward refuges, such as behind the support to the highway bridge or behind a soil ledge in one of the upper marshes. A few birds are killed by one merciless gust of wind that funnels and flings them against a set of rocks while they are trying to land.

The four juveniles born first by the boulders on the island try to weather the storm with their parents and dozens of others by floating just behind a stone breakwater. Along with several gulls the cormorants keep their heads tucked low while they paddle in the seas to ride the waves and keep position. Dozens of the recently fledged birds perish from exhaustion after over thirty hours paddling into the blast. One of those to die is one of the juveniles from that first clutch. Her body floats in the marsh.

The next day a cold front rushes in after the hurricane, dropping the evening temperature to an early, unseasonable frost. The island is barely recognizable. Several huge rocks have been lifted and rolled. An immense tree limb and part of the cockpit from a smashed fiberglass boat have been left on the windward side. The gull beach is nearly gone.

So a few days before the equinox, without any visible preparation, most of the adult cormorants begin to fly away, including the first breeding pair. Their three surviving juveniles remain on the island for now.

7

Bering Island

RUSSIA

About forty years ago the Great Auk of the Northern
Atlantic became exterminated. A vigorous search has been
made for it and its remains; fabulous sums have been paid
for skins and eggs. . . . Within the same period another large
water bird has become extinct in the North Pacific, without
having as yet attracted the attention. . . . Yet, this bird
was the largest and handsomest of its tribe.

LEONHARD STEJNEGER, 1890

The most compelling and creative work of art I've seen that features a cormorant is a table. Audubon's paintings of North American cormorants are extraordinary and often narrative. Maine artist
Andrew Wyeth painted a dynamic watercolor of a soaring cormorant in
1942.[1] Thanks to a friend, I own a lovely painting in several shades of blue
of a pied cormorant, by a talented local artist in New Zealand. Several
ancient Chinese and Japanese paintings capture the essence of cormorants
with just a few melancholic brushstrokes. Many countries have issued
indigenous-bird stamps with cormorants: the cormorant stamps from
Serbia and Palau are especially appealing. It is easy to find stunning photographs of cormorants in books or online, but they usually lack a time-
less quality. I've seen bronze cormorant sculptures at tourist sights and
public parks in Laguna Beach, California, and Port Angeles, Washington.[2]
Among others in England, there is a five-foot cormorant sculpted with a
chainsaw in a public park in Gloucestershire, another crafted of salvaged
scrap metal that has been installed at a park by the River Thames, and
there is a cormorant sculpture that a woman welded and assembled in a
park in Peterborough.[3] Most of these sculptures evoke a single cormorant

134

standing with its wings spread. For me, none of these pieces of cormorant art, in any media, holds a feather to that created by the English artist Anna Kirk-Smith.

Kirk-Smith crafted her piece in 2011 as part of a large London-based exhibition involving over 120 artists. The "Ghosts of Gone Birds" project was produced to bring attention to extinct birds and those birds currently endangered. Kirk-Smith's table is titled *The Unfortunate Repercussions of Discovery and Survival (Spectacled Cormorant)*.

When you approach the sculpture it appears to be just an old piece of furniture. And it is. Except that she has varnished it to match the color of the body plumage of the spectacled cormorant (*Phalacrocorax perspicillatus*). She embedded the table with several glass-covered boxes in which the lining is painted iridescent green or deep blue to match the wing and tail feathers of this largest of cormorants. In one of the boxes is placed an antique sounding lead and a scale set to what would have been the weight of the bird. One end of the table has a historic map of the Bering Sea varnished into the surface. In the middle of the map she embedded the largest of the sunken boxes. In it is a book open to pages created from her sketches during her visit to the extinct cormorant specimens at Tring.

I contacted Anna Kirk-Smith after I saw the table. We've kept in touch since. Looking back on "Ghosts of Gone Birds," Kirk-Smith says: "I had a choice of which metaphorical corpses to resurrect. When I first learned about this cormorant it was variously described as a clumsy, stupid and ludicrous bird so, being English, I guess that cemented my decision to support the underdog. What I didn't expect however was how involved I would become with this bird and its history, what journeys it would take me on."[4]

"The Spectacled Cormorant introduced me to an intriguing array of characters, a gripping plot line," she says. "And I became more delighted with my choice as time progressed, and strangely possessive of this bird's memory."

It is September 1741. Vitus Bering's ship the *St. Peter* gropes somewhere among the long desolate string of Aleutian Islands. No one is admitting this aloud, but nearly all aboard recognize it is unlikely they will ever make it home. Several of the crew are flattened with scurvy. Two men have already died. Bering himself is also terribly ill, perhaps with scurvy and heart disease.[5] Much of their diminishing water supply is foul. The storm-force winds and seas are constantly in their face. They are not even clear as to where they are.

Sailing aboard the *St. Peter*, sharing the cabin with the sickly, aging Bering, is a German physician and naturalist named Georg Wilhelm Steller. In his early thirties and on his first voyage, Steller is a fiercely ambitious man fresh from university life and research. While the ship and her officers grope westward, Steller is doing a great deal of questioning of the mariners and still more praying to God for divine deliverance. He is certain they are near to land and to the south of the Aleutians, since he sees floating seaweed and various birds such as owls and gulls, which he knows to be strictly coastal. But no one listens to him, in part because he has not been shy in showing that he thinks them all buffoons, and also perhaps because his assessment of where they are, though confident, is not quite correct either.

Bering's expedition continues, struggling to claw its way west toward Siberia's Kamchatka Peninsula and home.

"We heard the wind charge periodically as if out of a channel with such terrible whistling, rage, and frenzy that we were every moment in danger of losing a mast," Steller writes in his journal. "The waves struck like out of a cannon, and we expected the final blow and death every instant."[6]

The veteran sailors shout to him that they have never seen storms as horrible.

Steller explains: "Let no one think that the dangers of this situation are exaggerated, but believe rather that even the cleverest pen would find itself incapable of describing our misery sufficiently."[7]

The officers begin to discuss the idea of finding an island to spend the winter because they cannot last at sea any longer. They watch sharks circle their ship. Their food is low. Their brandy runs out. Bering prays and leads a collection of money for a donation to the church in case they survive. The *St. Peter* weathers storm after storm and somehow keeps afloat. Two more men die.

Steller writes one morning that God "pulled up the fog."[8] They can see shorelines, and the crew take a sounding. It turns out they are cutting north through the chain of islands, and if it were not for the break in the weather and the timing of daylight, they would have wrecked.

Steller meanwhile watches for birds. The night before, a little black

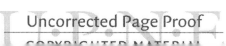

murrelet had flown aboard. He writes: "It passes the night in cliffs and, like an owl, flies during the day against everything it sees indistinctly close by. That is why they are caught alive by hand in large numbers around Avacha [the port of Kamchatka from where they departed]. A person supplied with a coat simply sits down close to one; they are accustomed to assemble under the coat as if it were a prepared nest."[9]

As the ship coasts safely through the channel, surrounded by sea otters, only about ten of the men remain in any condition to work. The officers decide nonetheless to continue sailing north in the hopes of making the mainland.

In the first week of November, with Captain-Commander Vitus Bering sick in his bunk and the officers and Georg Steller on opposite sides of the quarterdeck, the *St. Peter* limps into what is now named the Bering Sea. One of the officers would write later: "Our ship was like a piece of dead wood, with none to direct it."[10]

Steller scrawls in his journal that the sick are "suddenly dying off very quickly."[11]

Robert Prŷs-Jones told me that at most one or two people come each year to the Natural History Museum's Bird Group in Tring to see the spectacled cormorant specimens.

To be honest, when Prŷs-Jones placed them in front of me, I was initially paralyzed. I had traveled far just to have them in front of me. Then I didn't know what to do once they actually were. So I drew pictures. I took photographs from a dozen different angles. I made a lot of measurements of all the parts of the stuffed birds in case Prŷs-Jones hustled by — so I might look somehow slightly legitimate. Most of the time I just sort of looked at the skins. Silenced and moved, I had never before seen the body of an extinct animal, a species that had not breathed for over 150 years and would never again. I put my fingers on each chest as if I could feel a heartbeat.

When Anna Kirk-Smith made her own pilgrimage to the Bird Group, she wrote beside one of her pencil sketches, now shown inlaid into the table: "It's humbling to behold."

At one point I heard a person walking down the aisle. The steps were too sloppy for Prŷs-Jones, but I picked up my tape measure anyway, just in case. I began recording the length of the upper mandibles. A bald man with a rusty beard paused to chat, a pastime of which he was clearly fond. He chewed gum.

He looked at the spectacled cormorants and said: "Beautiful specimens, eh?"

The man told me about work he was doing that includes painting extinct birds. He said the spectacled cormorant specimens in St. Petersburg are "simply marvelous." The double crests are still intact.

He said: "It's harder to paint with the double crests, however." His fingers felt the beaks of the cormorants as he chatted. "I'm a paleontologist mostly, bird bones. But any time I can get a bit of painting in, as a sideline, I'm pleased as punch."[12]

Miraculously in early November of 1741 the *St. Peter* finds land and a natural harbor. The officers believe it to be connected to the Kamchatka Peninsula from where they departed. Bering is still too ill to come on deck. The officers drive the ship unintentionally, chaotically over a reef, but eventually in the middle of the night they settle into a relatively calm anchorage. The naturalist Georg Steller is shocked to walk on deck and witness a few of the sailors heaving some of the dead men overboard without ceremony — believing the corpses to be the cause of their trials.

Steller, his servant, and several sick men are the first to row ashore the next day. The landscape looks similar to Siberia. The plant life does too, but Steller thinks the clouds suggest this is an island. He is surprised by how the sea otters come swimming over to the boat as it approaches the beach. Once ashore he notices a huge unknown manatee swimming along the coast. Several fat Arctic foxes approach them without fear. The men easily shoot and eat several ptarmigans.

He tells Bering, when the captain-commander is brought to the island: "The abundance and tame self-possession of the animals alone clearly indicate that it must be a sparsely inhabited or entirely uninhabited land."[13]

It turns out that though they are only a few days' sail from Kamchatka, this island is desolate and indeed without a single human settlement. What they would name Bering Island is over fifty miles (80 km) long, however, so it takes them months of hiking around and exploring to confirm this to be surrounded by water.

Steller's status as a Jonah likely fades as he leads the establishment of a makeshift camp on shore and does much to save the lives of his shipmates by crafting a sense of order and a social structure that divides the labor as they prepare to winter over. He organizes the hunting of animals and collecting plants to heal those with scurvy. Steller leads the digging out of rude shelters in the hard sand.

Another storm shoves the *St. Peter* onto the beach. The vessel is damaged but safe, and the shipwrecked men begin to settle into a sort of routine in order to survive. Once ashore the sailors and soldiers begin to revive, but the severest cases among them die in the first few weeks. Despite Steller's efforts, Bering himself doesn't make it. The captain-commander probably dies of heart disease, and likely exposure, age, infection, and exhaustion. When he breathes his last he is in a miserable condition, with his joints swollen and pus discharging near his anus. He has half-buried himself under the sand to stay warm. He is infested with lice.[14] Bering dies in December after a full third of the seventy-eight men aboard the *St. Peter* have already perished. A few more will not make it through the winter.

While organizing rations and food and the daily chores, Steller steadily follows his passion as a field naturalist. He observes, collects, and describes all that he can. He writes down how they caught food and preserved pelts. His account of combating the foxes, for example, is at times gruesome. The foxes attack the sickest of the men and eat and steal seemingly everything. In one day Steller and another man kill sixty foxes with an ax and a Iakut hunting knife.[15] In their ineffectual attempts to scare these animals away from camp, the men dismember a few foxes, even half-skin them alive or burn their feet. The shipwrecked men eat the foxes, too, and they club sea otters, fur seals, and sea lions for food.

Steller and the officers ration the flour from the ship's stores to always have some sort of biscuit, however meager. Finding any wood to make a fire becomes more and more difficult, however, particularly when all the ground is covered in snow. Their shoes and clothes are rags. Sometimes a blizzard entirely buries their shelters.

After a seemingly endless winter, spring finally arrives in May. The rains flood out their shelters at times, but the castaways now have access to driftwood and a few sprouting plants. Despite the fact that all of the ship's carpenters have died, the men begin to build a smaller boat out of the wreckage of the *St. Peter*.[16] Stomachs continue to grumble, however, since they must travel farther and farther away from camp in order to find meat.

Before wrecking on Bering Island, the *St. Peter* had achieved its goal of doing a preliminary survey of the North Pacific and touching mainland America not far from today's Kenai Peninsula and the town of Valdez. Yet Georg Steller's landing in Alaska was for him infuriatingly brief. Once on Bering he took extreme care of all his journals and notes. He intended to bring back the first descriptions of the native peoples of the Alaska-Aleutian region, as well as his observations of dozens of animals unknown to Europeans. Several of the species still bear his name, notably the Steller's jay, the Steller sea lion, and a type of sea duck, the Steller's eider. He also observed a dramatic, gorgeous predator now known as Steller's sea eagle, which is found today only in this corner of the North Pacific.[17]

Most famously Steller gave the first and only description of living cold-water manatees that grew to be some thirty feet long. The animals had, as he put it, an "extraordinary love for one another."[18] Though the shipwrecked men saw the creatures regularly, it was not until the end of their stay that they were forced to figure out how to row out and harpoon them because of the scarcity of any other remaining meat. Steller's sea cow would be extinct in less than a century.

The spectacled cormorant is the second, much lesser-known species for which Georg Wilhelm Steller provided the only known living description left. These were huge cormorants, numerous, and easy to catch. The men ate them with the same understandable relish as they did everything else.

"The flesh of one would easily satisfy three hungry men," Steller wrote. "They were a great comfort."[19] This would be the thematic principle for Anna Kirk-Smith's sculpture: she carved this line into the top of the table.

Steller may have first discovered the huge cormorants during one spring excursion for food. He, his servant, and a couple of the other men were far from their camp when a blizzard overtook them. After a grueling night, near frostbitten, they found an ideal cave in which to shelter themselves. They later named this Steller's Cave.

Steller explained: "The sea birds and the migratory birds that I had opportunity to observe on Bering Island are almost the same as those one meets with on the eastern coast of Kamchatka. However, a special kind of large sea raven with a callow white ring about the eyes and red skin about the beak, which is never seen in Kamchatka, occurs there but only on the rocks near Steller's Cave."[20]

This "sea raven" had a thick, powerful bill. It had two large crests while

THE DEVIL'S CORMORANT

in breeding plumage, oriented one behind the other, like a Mohawk. Steller wrote in one translation: "From the ring around the eyes, and the clown-like twistings of the neck and head, it appears quite a ludicrous bird."[21]

Another description, translated in part from a version of Steller's descriptions, was later published in Sir Walter Rothschild's book of extinct species: "Of the size of a very large goose. Of the shape of the former [another cormorant], which it also resembles in the white patches on the flanks. The body is entirely black. A few long, white, narrow pendant plumes round the neck, as in Herons. Occiput [head] with a huge tuft, doubly crested. Skin round the base of the bill bare, red, blue and white, mixed, as in a turkey. Round the eyes a thick, bare white patch of skin, about six lines wide, like a pair of spectacles. Weight 12 to 14 pounds [5.4–6.4 kg]."[22]

Steller wrote that the cormorants had a certain *stoliditas*, a stupidity, but this might be more owing to their lack of fear of man and perhaps a sluggish, awkward flight. If they flew at all.[23] Because if the spectacled was indeed that heavy, then it was almost twice the weight of the largest of the airborne cormorants today, the great cormorant, whose wings are about the same length.[24] So if the spectacled could fly, it surely wasn't that efficient.

It might be that the cormorants on Bering Island, which Steller observed, did not breed on the rocks near Steller's Cave at all. Their nests would need to be entirely inaccessible to the foxes. Perhaps they had colonies at the time only on the tiny islets just off Bering and then swam and flew, in whatever fashion, back and forth to fish.[25]

In their efforts to survive, the men surely ate as many of the spectacled cormorants as they could acquire, presumably by wringing their necks or clubbing them. According to the most rigorous of Steller's biographers, the shipwrecked men likely prepared this bird the way that the natives of Kamchatka did other cormorant species: "namely by burying it encased—feathers and all—in a big lump of clay, and baking it in a heated pit." Steller thought it was quite tasty.[26]

This is about all the information that is known about this cormorant alive, none of which we have from Steller's original hand. Most of Steller's descriptions and his journal entries are usually from the translations by Peter Simon Pallas, a natural history professor in St. Petersburg who in the late eighteenth century was the first major scientist to work from

Steller's manuscript material. But it seems Pallas was at times a liberal, even creative translator. Steller's original journal manuscript is now lost. There might as yet be more materials about the cormorants from Steller in archives kept in St. Petersburg, but no one has done the full work yet. Pallas never met Steller, nor did he ever see the bird himself, dead or alive.[27]

Since Pallas brought this animal to the attention of the scientific community in 1793, the bird is often known as the Pallas's cormorant. Pallas christened the bird with the scientific name *perspicillatus*, which means, in a rough Latin, "wearing spectacles."

At last, in August of 1742, the survivors of the *St. Peter* cram into their new boat to try to sail back to the mainland. The cobbled, untested craft is loaded down. The officers permit Steller more weight than almost everyone else, but this allocation is still next to nothing: not nearly enough to bring the majority of his skins, skeletons, and plants. He does not have room for his skin of a young sea cow, which he had stuffed with sea grass. Steller will have to abandon practically everything to the foxes. He negotiates, cajoles, pleads with his shipmates and the officers. Eventually Steller steps off the island with only his notes, journal, a nearly complete manuscript describing the marine mammals, a few seeds, and a couple of sea otter skins. He brings home nothing of the sea cow or the gigantic sea raven.[28]

Of the fewer than ten thousand known bird species on earth, conservation biologists believe that at least 130 have gone extinct since Columbus sparked the European age of ocean-crossing exploration.[29] These extinctions include a few species of owls, sandpipers, finches, grebes, doves (e.g., the dodo), emus, macaws, herons, starlings, wrens, and parrots. The list of extinct birds also includes one gone species each of oystercatcher, ibis, quail, and petrel. Living out only perhaps a couple more generations in some sort of conservation facility are four more birds that are believed to be extinct in the wild: the Hawaiian crow, the Guam rail, the Alagoas curassow, and the Socorro dove.[30]

In addition to these 134 extinct wild birds are populations of nearly 200 other avian species whose numbers are so low or have not been credibly witnessed alive in so many decades that they might be considered functionally extinct. These have been designated as "critically endangered."[31]

The International Union for Conservation of Nature and Natural Resources (IUCN) is the central body for keeping track of endangered species, which are published as the "Red List." A significant percentage of the bird species that are in real danger are birds of the open ocean. Over a dozen species of albatross and petrels are listed as critically endangered. This is due mostly to human disturbance, introduced species at the rookeries, the prevalence of ocean plastics, and the efficiency and practices of deep-sea fisheries.[32]

Of the thirty-three extant cormorant species recognized by the IUCN Red List, nineteen species, including all six in North America and the two in the British Isles, are "of least concern." (See appendix.)

Four cormorant species are considered "near threatened." This includes the guanay cormorant and the red-legged "chuita" cormorant, who live off the coast of Peru and to the south. The other two are the Cape cormorant and the crowned cormorant, who both live in a small set of colonies off the coast of South Africa and Namibia.

Seven of the cormorants on the Red List are considered in worse shape, "vulnerable," largely because they have only a limited number of small rookeries: these are the flightless cormorant of the Galápagos and the Socotra cormorant of the Middle East, and five of the Southern Ocean species, such as the Stewart shag and New Zealand's king shag.

The IUCN labels two cormorants of the thirty-three living species with the bell-ringing alarm of "endangered," meaning that they are at a very high risk of becoming extinct. The first is the bank cormorant, long ago given the portentous species name of *neglectus*. This bird lives along the South Atlantic coast of Namibia and South Africa, alongside the two species of "near threatened" cormorants. The second officially endangered cormorant is the Pitt shag—a true cormorant dandy with orange feet, lime-green facial skin, and a tall, jaunty crest tufting both forward and aft off the head. This endangered shag lives on a set of isolated islands far to the east of New Zealand in a total population sputtering at around 500 pairs.[33]

The IUCN scientists believe that one cormorant is "critically endangered," meaning in genuine risk of becoming extinct. This is the Chatham shag, which lives in the same set of desolate islands as the Pitt shag, yet in still fewer numbers. This shag has caruncles by the beak, red facial skin, and a green metallic sheen to the feathers. Other than fencing to keep out introduced mammals from human settlements on the islands, there is not much more that can be done.

The IUCN does not make any predictions on the future of these endangered and critically endangered cormorants—or on any endangered animals. Its aim is primarily to provide some centralized information and to promote awareness of the frightening, spiraling loss of global animal diversity. In this way the goals of the IUCN are the same as the art of Anna Kirk-Smith and the "Ghosts of Gone Birds" project.

There is still more to be concerned about, unfortunately. An attempt to rethink the IUCN Red List for birds—its historic and current extinction totals—was published recently by an American group led by two men out of an Edgar Allan Poe bibliography: Stuart Pimm and Peter Raven. The scientists calculated that rather than the over 130 extinct wild birds that the IUCN recognizes, probably closer to 500 species have actually gone extinct since the turn of the sixteenth century. In other words, about 5 percent of all the known birds of modern times will never again be seen alive. Pimm, Raven, and their colleagues contend that the Red List underestimates because the IUCN is too cautious, too reluctant to include those "critically endangered" as actually irrevocably extinct or doomed within generations. New extinct species are also continually being found and identified—birds that were lost before Western scientists got a chance to describe them. Dozens of birds were eradicated by the exploration and overexploitation by Polynesian and other native peoples. When Steller was paging through his books aboard the *St. Peter* in 1741, his fellow naturalists knew of only a tiny fraction of all the birds that lived around the globe. Linneaus in 1758 listed 446. Rothschild in 1900 would have known of only about 80 percent of the birds science now describes. The more we learn, the more species we find both alive and that have gone extinct.[34]

Early European exploration and commerce led to quick, thoughtless destruction of species. Several groups of native peoples did the same, such as the Polynesians settling throughout the Pacific.

Pimm and Raven's team believes that if the current trends of rapid habitat destruction, introduced species, and global warming continue, then the near future will be absolutely precipitous for bird life on earth. Quoth Raven: "Some 1,200 more species are likely to disappear during the twenty-first century. An equal number are so rare that they will need special protection or likely will go extinct, too."[35]

That would mean by 2099, about one in six bird species that were alive

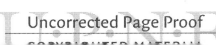

in the time of Columbus will either be extinct or facing this eradication imminently — to be seen nevermore.

The largest of Anna Kirk-Smith's sketches inlaid into the table is the head of the bird with a tuft of cotton wool in the cavity of the eye. No one today knows the actual color of the spectacled cormorant's eye. Steller provided no description of the iris itself, inside the white ring. An animal's eye does not survive preservation. Any biologist or artist who goes to an unaltered skin finds only a hole stuffed with cotton.

Cormorant eyes throughout the world span nearly the entire color wheel: dark blue, turquoise, gray green, emerald green, olive, light brown, dark brown, and even off-white. The eyes of the long-tailed "reed" cormorant are bright crimson. In a few species there can be a large range of color between individuals, and a few cormorant species have orbital rings — the fleshy skin around the eye — of varying widths and a variety of colors, such as yellow, red, and, of course, blue. Some orbital rings even have polka dots, as do Mr. Yamashita's Japanese cormorants. Sometimes cormorants have a third color on a thin inner rim between the iris and the orbital ring. All this in mind, know that the person who prepared the best existing mounted specimen of the spectacled cormorant, held at a museum in the Netherlands, inserted green glass eyes. This is a logical guess.

The spectacled cormorant living off Bering must have survived on a variety of fish, including the salmon species that migrate through the area and were supporting all those fur seals, sea lions, and Steller's sea eagles. Many of the salmon — perhaps even some individuals that once hatched in the Columbia River — would be quite large at the stage they were swimming through the Aleutians — too large for the most part for current cormorants, but perhaps not so for the huge spectacled. Bering Island and the other islands of the Komandorski group host hundreds of other fish species in enough quantity today to still attract millions of sea and shorebirds a year, including murres, puffins, several types of ducks, and two other species of cormorants.[36] (Steller also seems to have left a tantalizing note about a "white sea raven," as an additional species, but there isn't enough to go on. Perhaps it was a juvenile or an albino — some sort of Moby Shag.[37]) The spectacled cormorants surely would also have eaten some of the invertebrates, such as shrimps and crabs, that lived off the coast of Bering, hunting for them among the vast and tall kelp beds that had once also sustained the vegetarian sea cow.

Biologists believe the eyes are the primary tool for a cormorant to catch fish. Vision is its most developed sense. As evident when looking at the great cormorant skull inlaid into Kirk-Smith's table, the eye socket makes up most of the bird's head. Vision is the primary sense for most shorebirds and seabirds, except for those that in part find food in the open ocean with their remarkable smell — such as the albatross and petrels — snuffling up food in an immense ocean haystack.

Cormorants, like most birds and fish, have monocular vision: their eyes are on both sides of their head. They reconcile these separate images in their brain. This gives them a sliver of a blind spot dead ahead and immediately astern, but overall, they have an almost 360-degree field of vision. I love how when you slowly approach a cormorant it will turn the side of its face to you — as if it is being nervously nonchalant about your approach. Actually, it is getting you directly in its field of vision. It is very hard to sneak up on almost any species of wild cormorant today. I have tried. The spectacled cormorant wasn't stupid or blind, of course. It just didn't know it was supposed to be afraid of a person.

Cormorants, like most if not all birds, probably see a particularly vivid color picture, much better than humans. To another cormorant, the black plumage must look startlingly different from how it does to us, particularly considering the luminescent traces within the black feathers. Cormorants, like other birds, might be able to see into the ultraviolet spectrum or perhaps see polarized sunlight.[38] Birds are also better at seeing movement than we are, but not as good at detecting a predator or prey that is motionless or not visible: hence why bird observation "blinds" and the East Sand Island tunnels on the Columbia work so well.[39]

The anatomy of a cormorant's eye is fairly similar to that of a human's but with significantly more color receptor cells and a flatter retina, which seems to give them a wider in-focus view. Like other birds, cormorants have oil droplets in their color receptor cells. Biologists are still not entirely sure of the reason, but they speculate that it might help with the perception of so many colors and the reduction of glare — the latter being especially important for a seabird.[40] Birds have a third eyelid, the nictitating membrane, which comes across the eye vertically to help lubricate and protect. For cormorants, this membrane serves like goggles, replete with a high-refraction lens to help them see better underwater.[41] Cormorants in particular have a thick, flat cornea on the outside of their eyes, but a more flexible lens within the eye, adjusted with exceptionally strong muscles associated with the iris and pupil. All of this likely helps cormorants focus when submerged and adjusting to light refraction when under over one hundred feet (30 m) of water.[42]

Ornithologists have long marveled how the cormorant eye, functional also in flight at well over thirty-two hundred feet (1,000 m) above sea level, then allows the birds to see darting fish when the birds are swimming underwater, often in murky, dark, cold conditions.[43] The great cormorants in the Arctic today, for example, feed primarily on sculpin, a fish that has a variable, camouflage patterning to it scales, helping it blend into the bottom and among the rocks. Sculpins can't be easy to see underwater. Yet cormorants in this region will even feed at night.[44]

Scientists have been studying this vision in cormorants. At the University of Birmingham in England, ornithologists built a high-tech aviary specifically for these birds, with multiple bays and swimming tanks. They trained a half dozen cormorants to swim inside a tank in which there was a "swimway" with a fork at the end and an image of a fish that they could vary in terms of contrast, amount of light, and so on. Other studies have attached little cameras to the heads of cormorants.[45] In this way researchers have found that the cormorants might not actually have superb vision after all, at least in terms of identification underwater. Maybe, the researchers speculated, that cormorants do not really zip all around to pursue fish underwater but need to get within a few feet to allow their quick neck to really do the work, more like a heron. Or perhaps when in high-density schools of fish the cormorants use the sense of touch from their beak.[46] Maybe their skill is really about the extraordinarily quick perception of movement rather than the object itself. I wonder if this is related to how the researchers on East Sand Island say cormorants go right for people's eyes: wet, delicate, quick movement. Or maybe the cormorant's specialized vision is why it is so opportunistic, often eating whichever species is easiest to catch.

Thirteenth of August, 1742. The forty-six survivors of over eight months shipwrecked on Bering Island lie crammed shoulder to shoulder as one of them steers. Another man bails, since the boat has a leak. They sail past the southern point of the island and call it Cape Manati because of the number of sea cows.

They do not have a favorable wind, so most of their passage is spent rowing. But eventually, after almost two weeks, they pull into Avacha Bay. Everyone in the small village had thought they were long dead.[47]

Probably the most significant tangible result of the expedition of the *St. Peter* and Steller's work was his careful, thorough descriptions of the marine mammals on Bering and throughout the Aleutians. The sea otters in particular were already of enormous value in China. From Bering Island the crew brought back almost nine hundred pelts to sell back home. Even though Steller had to leave nearly all his samples, he seems to have ended up with about three hundred of the sea otter skins — some of which he purchased from his shipmates back in Kamchatka, while many others he received as gestures of gratitude.[48] Steller wrote: "This animal deserves the greatest respect from us all, because for more than six months it served us almost solely as our food and at the same time as medicine for the sick."[49] At one point Steller brought over to the dying Vitus Bering a baby sea otter still suckling on its mother, asking if he might cook it for him. Bering refused to try it.

In his first letters and reports after he arrives back safely in Kamchatka, Steller speaks of the potential profit from these furs. Soon the quality and abundance of sea otter and fur seals along the Aleutians ignites a Russian westbound trade and an arc of exploration, cultural impact, and commerce that will be devastating to both the marine mammals themselves and in many ways the coastal indigenous peoples as far south as California.[50] Near East Sand Island in 1805, Lewis and Clark saw sea otter pelts being traded and worn by the Columbia River First Peoples.[51] As the years progressed, the Russians, Americans, Spanish, and British all had to sail farther and farther to find these furs.

Scott O'Dell's young adult novel *Island of the Blue Dolphins* (1960) is set on San Nicholas Island off the coast of southern California. In this story a young girl named Karana, probably of the Gabrielino Indians, is orphaned in the mid-nineteenth century after a Russian vessel manned with Aleutians plunders the island for sea otters and then kills her father and most of the Native American men. The Russians had come this far south in search of these furs. In the novel, the girl ends up alone on the island. With the skins and feathers of the cormorants that she eats to survive, she works on sewing a skirt in the evening hours in her cave. When finished, Karana explains: "I had never seen the skirt in the sunlight. It was black, but underneath were green and gold colors, and all feathers shimmered as though they were on fire. It was more beautiful than I had thought it would be."[52]

The novel is based on a true story. The young woman eventually ended up at a mission on the California mainland. The cormorant garment was supposedly sent to Rome after she died, but it no longer seems to exist.[53] Toward the end of *Island of the Blue Dolphins* — a novel written just as the American environmental movement was about to really take off and cormorant populations were beginning to recover — Karana decides that she will not kill animals any longer, for any purpose.

She explains: "Nor did I ever kill another cormorant for its beautiful feathers, though they have long, thin necks and make ugly sounds when they talk to each other."[54]

Georg Wilhelm Steller is dead at the age of thirty-seven. He never makes it back to St. Petersburg. He sees none of his discoveries published or any of his ambitions realized. Steller had imagined a triumphant return to the capital and the academies, but he gets held up in Siberia twice for what seem to be planted charges regarding his dealings with the native Kamchadals. He dies of a fever in the middle of winter trying to get home. Desparately poor people dig up his body to steal the red cloak in which he was buried. A pack of dogs ravages his corpse. Decades later the river slowly washed away the gravesite and the stone marking where his remains were buried.

It is likely that Steller's legacy would never have been established, regardless of Pallas, if it were not for a man born about a century after Steller died in Siberia. Leonhard Stejneger, a Norwegian, would be as close to a reincarnation of Steller as even the German naturalist could have wished for. Bering Island would be Stejneger's professional and spiritual high point, too, and the remains of the spectacled cormorant was one of his most compelling finds.

Leonhard Stejneger arrived in Washington, D.C., in 1881. Though he was not proficient in English, he knew several other languages and had already published a few scientific papers. He walked up the steps of the Smithsonian Institution to ask for a job.[55] They put him to work helping to prepare and catalog specimens of birds from the Caribbean. The next spring he was asked on short notice if he would be willing to take biological and meteorological observations way out at Bering Island. He would serve in part as the keeper of a U.S. Signal Service station. Leonhard Stejneger packed his bags.

He lived for nearly two years on the island. Once he almost drowned when he became trapped underwater after he flipped over in a skin canoe. Stejneger returned to Washington as the expert on the fur seals of the North Pacific. From Bering Island and the surrounding region, he brought back to the Smithsonian a staggering collection of skulls, skeletons, fossils, preserved specimens (including three full Kamchatkan mountain sheep), and about seven hundred bird skins.[56]

Stejneger had been intrigued by Steller's writings of the rare cormorant. He interviewed some of the local people as he traveled, to find out if he might see some of these birds. He collected the very first skeletal remains of the extinct spectacled cormorant when he was excavating some buried middens in a cliff at the north end of Bering.[57]

The young biologist concluded by the end of his stay in 1883 that the spectacled cormorant was extinct. The animal had been wiped out by the 1850s. On returning to Washington from Bering, Stejneger wrote: "You will not be more disappointed than I am in learning that there is no hope whatever of getting a specimen."[58]

If the birds lived anywhere besides on Bering, Stejneger found no records. After the *St. Peter* and the discoveries of Steller and other contemporary Russian explorers, seal hunters had begun visiting the island. They ate the cormorants while camping there. In the 1820s the Russian-American Company brought Aleutian natives to Bering Island to hunt for the mammals full time, to settle. These Aleutian natives also ate the spectacled cormorants. During his travels, Stejneger was able to speak to some of the Aleuts, who remembered these cormorants as being plentiful, especially on Aij Kamen, a rocky islet just off the northwest of Bering Island.

Stejneger wrote: "The reason they give why this bird has become exterminated is that it was killed in great numbers for food. They unanimously assert that it has not been seen since, and they only laughed when I offered a very high reward for a specimen."[59]

The Aleuts that had been brought to Bering by the Russians found the spectacled cormorant better eating than the other cormorants they had already hunted farther to the east off the Alaska region. Apparently the Aleuts had a long history of eating cormorants, mostly the pelagic and red-faced varieties. The native people seem to have collected shag pelts as a valuable trade item and even told of the month of January as the one best for hunting cormorants.[60]

"The natives of Bering Island inform me that the meat of this species was particularly palatable compared with that of its congeners," Stejneger wrote. "And that consequently, during the long winter, when other fresh meat than that of the cormorants was unobtainable, it was used as food in preference to any other."[61]

Perhaps it was the foxes that eradicated the last of the spectacled cormorants, or even volcanic activity, as one theory goes. But surely humans — Russian and Aleut hunters — sealed the extinction of this bird.[62]

Very little evidence exists to put together any kind of past distribution of the spectacled cormorant. According to one set of researchers, Pallas and the Russian explorer Otto von Kotzebue recorded anecdotal stories of natives collecting and eating spectacled cormorants on the Siberian mainland.[63] A nineteenth-century naturalist named Lucien Thompson wrote of Aleutian natives of Attu Island describing a big cormorant only recently extinct as being "fully twice as large" as the other cormorants. The species was apparently quite abundant in the area, and, Thompson writes, it could have "been none other than the greatly desired Pallas's Cormorant."[64] And a recent archaeological dig on Amchitka Island, also several hundred miles to the east of Bering, found among the remains of hundreds of other cormorants one single indisputable two-thousand-year-old wing bone of a spectacled. This could be identified thanks to Stejneger's collections.[65] But these few tidbits hardly assemble a former range.

Stejneger and other scholars of extinct birds, including Lord Rothschild at Tring, believed the only known specimens still in existence were collected by a Governor Kuprianof, once placed in charge of the region. One of Kuprianof's directives was to supply samples of all known animals to the museum at St. Petersburg.[66] It seems he had a few extra spectacled pelts, one of which he gave to an expedition of the British navy under the command of Captain Edward Belcher, sometime between 1836 and 1842. This is one of the specimens that is now at the Bird Group in England.

In contrast to the quiet demise of the spectacled cormorant, as Stejneger wrote in this chapter's epigraph, it is hard to underestimate the

Victorian fervor ignited about the loss of other extinct birds, such as the great auk. About eighty skins of great auks are still around, and nearly as many of their eggs. An entire, gorgeous, painstaking volume is devoted to describing the great auk and the craze of final collections, art, and stories concerning this bird.[67] The spectacled cormorant, meanwhile, never received any of this sort of attention, even as a few new specimens were identified in dusty European collections. Perhaps this lack of fervor was because of this bird's lonely location in the far northwestern Pacific. Or perhaps it is reflective of a general cultural lack of interest in cormorants.

Unlike Steller, Leonhard Stejneger was able to return to Bering Island. The U.S. government sent him back a couple more times as an expert on the region. Stejneger found more bones of the spectacled cormorant and gathered enough remains of a sea cow to craft an entire skeleton. Stejneger's final professional act in 1936, at the age of eighty-four after a lifetime's worth of study and experience, was to complete his magnum opus, his definitive biography titled *Georg Wilhelm Steller*.

Leonhard Stejneger, like most of his fellow naturalists before the age of easy field photography, was a careful and gifted artist. His idol Georg Steller was not. Stejneger wrote: "It may be asserted with the greatest confidence that Steller was utterly incapable of making recognizable drawings of the natural objects, plants and animals which he observed or described."[68]

Thus there are no remaining illustrations of the bird alive, unless somewhere yet to be identified there is a preserved work of Aleutian or Kamchatkan art or a rendering in a journal by an early nineteenth-century sealer. The early paintings of this bird — all done from preserved dead skins — have been the only lasting images.

The first known depiction was published by John Gould. He is the same man who purchased the second skin for the Bird Group — the scalped one without a tail. Gould painted two birds in an island cliff scene to accompany Captain Belcher's *Zoology of the Voyage of H.M.S Sulphur . . . 1836–42*.[69] Though normally he was a deft and elegant bird artist, Gould's spectacled cormorants look sickly and immobile, like crosses between penguins and rooted zucchini. One of these birds looks as if it has a rodent in its neck, which it is still digesting snake style.

Since then, a few other painters have taken on the task of trying to give the bird life when only looking at a couple of musty pelts with no eyes and faded colors. The most-often reproduced work is a hand-colored lithograph by Joseph Wolf in 1869. This is static but elegant, and is mostly faithful to the bird, aside from the facial skin as Steller described it.[70] (See figure 14.) For Lord Rothschild's *Extinct Birds* in 1907, John G. Keulemans painted another attractive image, but he also represents the facial skin incorrectly.[71]

"I considered the early illustrations, their differences, their potential discrepancies from the actual specimens at Tring," Anna Kirk-Smith tells me. On the top of the table, on a ceramic plate, she engraved her own version of a profile of the cormorant.

"I wanted on my plate to have an amalgamation of these references at once," she says. "Hence the nod to Wolf's and Keulemans's with the pose, but with variations in the anatomy that I'd noticed from the skins at Tring. In this way I hoped to reinterpret the historical exploration and identification of this species again. The illustrations intrigued me more as a group — rather than singly."

Kirk-Smith was obsessive about research. She got especially excited about this project. She was already fascinated with seabirds and their conservation — she once ended up keeping a small collection of common scoters in her kitchen as part of an RSPCA rescue after the oil spill of the *Sea Empress*. To prepare for her sculpture, Kirk-Smith went out and observed cormorants and shags on the English coast. She recorded their sounds and tried to play them back at a slower pitch to perhaps match the croak of the larger bird. She read enormously about the animal's biology and about the history of Steller and Bering's expedition. She read Stejneger's accounts. She corresponded with dealers of antique maps in Alaska. All this in an effort to "try to re-create a gone bird, to conjure up a character."

The emotion and history oozes out of Kirk-Smith's sculpture, *The Unfortunate Repercussions of Discovery and Survival (Spectacled Cormorant)*, whether you know the story of the Bering expedition or not. Kirk-Smith calls her table an altar, an *in memoriam*. There are hours worth of details. One leg of the chair has collaged pictures of currently endangered Bering Island birds. A portentous quotation from Bering's lieutenant is scratched into the wood of another leg. She painted white wax rings on the table, what would be water stains left by a glass, to evoke the "spectacles" of the extinct bird.

For Anna Kirk-Smith the loss of the spectacled cormorant is about food and the mariner's survival. The initial focus of the sculpture, when you first approach the table, is three seemingly formal place settings for the ghosts of three men. The men feast on the ghost of one gone bird, represented by the cormorant skull with the huge eye sockets. At the setting of the first man, Vitus Bering, is a map of the island etched on a ceramic plate. Bering's gravesite is marked on the map. Kirk-Smith inlaid in resin a pair of navigation dividers to serve as Bering's utensil. For the second place setting, that for Peter Simon Pallas — the man who introduced this bird to the scientific community — there is a plate decorated with Steller's handwritten notes and a taxonomic tree. A quill pen is his utensil. The third and final setting, at the head of the table, is for Georg Steller. On this plate is Kirk-Smith's painting of a spectacled cormorant. To the right of Steller's plate is a replica of his Iakut hunting knife.

October

A flock of about one hundred cormorants flies just below a layer of gray stratus clouds. Southbound before a stiff cold northwesterly wind, they are making good time. The birds form a sort of loose V, an uneven echelon with only a quarter of the individuals on the windward side. All the cormorant wings beating the air create both a steady shushing sound and a thin squeaking, as if their joints need lubrication.

The migrating cormorants look down on the coast of New Jersey. They look down on casinos and boardwalks and the long straight string of Atlantic beach. The roofs of small homes crowd toward the shore, separating slightly as the human developments move inland. One cormorant looks down on a fishing boat with its wide-mouthed trawl stretched behind. Another cormorant sees beneath her dozens of gulls circling toward the beach in pursuit of a run of menhaden.

Flying together in the middle of the longer arm of the echelon are the three surviving juveniles from the first clutch of Gates Island: two males and one female. Like the rest in the flock, they fly with their necks stretched forward but with a slight kink in the neck toward the torso. Each parts his or her mouth slightly to make the breathing easier.

Small groups within the flock occasionally pause from flapping and soar in unison—expanding their tail feathers to their widest and extending their outermost primary feathers like outstretched fingers. Then they resume flapping. The echelon shifts occasionally toward the east and back toward the west. Individual birds drift away and the echelon opens up, then rejoins again, reorganized. The flock remains at the same altitude for the moment, optimizing the tailwind just beneath the cloud layer.

There is no clear leader of this flock of adults and juveniles. Maybe the birds are watching for landmarks. Maybe they are responding to the movements of the sun, or the earth's magnetic field.

The juveniles from the island had not only never seen this part of the world; they had never seen the ocean proper before yesterday. The juveniles are hungry, tired, and light-headed. Their breast muscles ache. This is only their second morning of the long flight south.

FIGURE 1. *Gōdo, Cormorant Fishing Boats on the Nearby Nagara River*, by Ando Hiroshige and Keisai Eisen, c. 1840. Courtesy Williams College

FIGURE 2. Junji Yamashita fishing with his cormorants on the Nagara River, Japan. Courtesy Kazuto Hino, Gifu Convention and Visitors Bureau

FIGURE 3. A young Frenchman holds a trained cormorant, named Tobie, in a similar fashion to a falcon. This is the frontispiece to *La Pêche au Cormoran* by E. Le Couteulx de Canteleu (1870). Courtesy of Williams College

FIGURE 4. Irene Mazzocchi of the NYSDEC collects a pellet between double-crested cormorant nests on Little Galloo Island, New York.

FIGURE 5. Sport fisherman Ron Ditch, one of the leaders of the men who slaughtered thousands of cormorants on Little Galloo Island, New York, holds a wood carving of a double-crested cormorant that he made over a decade afterward.

FIGURE 6.
A raven-like cormorant is Satan perched on the Tree of Life to illustrate the first letter of book 4 in John Milton's *Paradise Lost*, 1720, designed by Louis Cheron. Courtesy of Connecticut College

FIGURE 7. In a fable by Jean de la Fontaine, illustrated by Gustave Doré in 1868, an old cormorant deceives all the fish in a pond in order to eat them one by one.
Courtesy Connecticut College

FIGURE 8. Blue-eyed shags building nests near Port Lockroy, Antarctica.
Ralph Lee Hopkins, National Geographic

FIGURE 9. The crests on the double-crested cormorant appear a few weeks
a year while the bird is breeding. West of the Rockies the crests can be all white or,
as on this bird on East Sand Island on the Columbia River, a mixture of
black and white feathers. Bird Research Northwest

FIGURE 10. Research technicians band double-crested cormorant chicks after capturing them at night from the tunnels on East Sand Island at the mouth of the Columbia River. Adam Peck-Richardson, Bird Research Northwest

FIGURE 11. Specimens of guanay cormorants at the Bird Group in Tring, England. Note the middle label from a bird collected by Captain Robert FitzRoy in Valparaiso, Chile. The label on the right is a specimen from the Chinchas collected in 1910 by the Scottish naturalist H. O. Forbes, who recorded five million birds on Isla Centro.

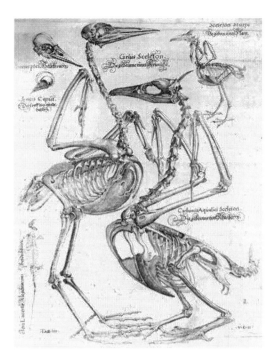

FIGURE 12.
An engraving of the full skeletons of a crane (left), a starling (upper right), and a great cormorant (right) by Volcher Coeiter, 1575. Courtesy of Connecticut College

FIGURE 13. In England the Angling Trust printed these postcards to put in local clubs and tackle shops to encourage sport fishermen to mail them in to their representatives in Parliament. The organization is concerned about the great cormorant's impact on recreational fish stocks.

FIGURE 15. A pair of flightless cormorants on Fernandina Island, Galápagos, Ecuador. Note the difference in size between the much larger male (left) and the female—this extent of sexual dimorphism is unique among cormorants.
Ingo Arndt, Minden Pictures, National Geographic

opposite:
FIGURE 14.
The extinct spectacled cormorant as painted
in this lithograph by Joseph Wolf, 1869. There is
no known illustration of this bird alive.

left: FIGURE 17. Brady Thompson uses a shovel to pick up a double-crested cormorant
that he just shot flying above this catfish aquaculture pond near Belzoni, Mississippi.
right: FIGURE 18. Tens of thousands of guanay cormorants pack a slope of
Isla San Lorenzo, Peru; photographer and date unknown.

opposite:
FIGURE 16.
Flightless Cormorant, by Heather Carr. Original is in color, acrylic
paint and mixed media, 24 inches by 36 inches, 2012. Note the two
chicks stenciled to the lower right, beneath "I'm not going anywhere."
Heather Carr; heatherunderground.com

FIGURE 19. In this Moche image drawn from a ceramic design, the god Quismique traps one of the guanay cormorants to help him travel under the sea. From Gerdt Kutscher, *Nordperuanische Keramik: figürlich verzierte Gefässe der Früh-Chimu* (Berlin: Gebr. Mann Verlag, 1954), plate 62.

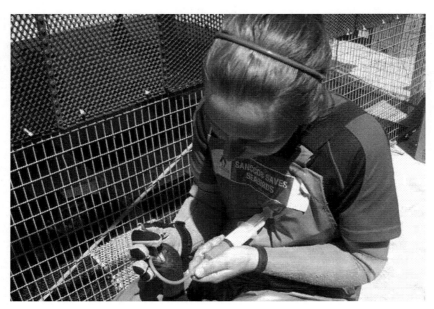

FIGURE 20. Staff member Marguerite du Preez cares for a Cape cormorant by giving the chick a fish formula with a syringe at SANCCOB in Cape Town, South Africa. Courtesy of Nola Parsons and SANCCOB

FIGURE 21. A crèche of double-crested cormorants
on Gates Island, Connecticut.

FIGURE 22. A cormorant using its feet and tail to steer in response to a school of fish.
Chris Ross / Aurora Collection / Getty Images

Galápagos Islands

ECUADOR

Flightless cormorants on the Galápagos Islands
decided, "fuck this, we're going to swim."

KURT VONNEGUT, 1987

ary Hepburn, a high school science teacher from upstate New York, is the character critical to saving the human race in Kurt Vonnegut's novel *Galápagos*. In this story, published in 1985, Vonnegut places on the wall of a travel agency a large photograph of a flightless cormorant (*Phalacrocorax harrisi*) just so that Mary would be in the position to play this most pivotal of roles after she is shipwrecked on one of the islands of the Galápagos. All the seal-like, cormorant-like humans living one million years from now have her to thank for their being around at all. These future people, however, are not capable of this sort of gratitude, let alone of understanding the process of evolution that brought them to this point. Humans are far less complicated than they are today. Their brains have shrunk. They have beaks and flippers.

Vonnegut, godlike, crafts a series of extraordinary events throughout *Galápagos* to get Mary Hepburn in the one location and position to save *Homo sapiens*. He has her do so on the hallowed ground where Charles Darwin began as a young man, with his own big brain, to percolate his theory of evolution by natural selection.

Darwin's experience with and study of cormorants are spare. He was never that excited about seabirds. The Angling Trust in England will be pleased to learn that Darwin did once write in his journal about shooting a great cormorant when he was seventeen. "The capacity of the stomach was very great," Darwin wrote after the dissection. "There being four sole about half a foot long [15 cm] in it."[1]

While aboard the HMS *Beagle* several years later, Darwin shot and collected the blue-eyed shags of the Falkland Islands. Darwin wrote of the "extreme wildness of [the] shags," and that he "saw a cormorant catch a fish & let it go 8 times successively like a cat does a mouse or otter a fish."[2]

With Captain Robert FitzRoy, Darwin sailed up to the Galápagos Islands in mid-September of 1835. There is no record of his or anyone else's aboard seeing a flightless cormorant during the six-week visit. These are the only cormorant species that live in this isolated archipelago. They nest today in only several small colonies on Isla Fernandina and Isla Isabela, which were known then as Narborough and Albermarle. Darwin must have barely missed seeing these birds, because the *Beagle* was becalmed and at anchor for a few days in the small channel between the two islands, right near where these cormorants are known to nest. Darwin went ashore on the northern part of Isabela, too, walking around on a day that was "overpoweringly hot." He observed the marine iguanas, the "great black lizards" that can swim and grow to four feet (1.2 m) long. He made notes on the geology of the island but said nothing about the bird life.

After leaving the anchorage Darwin wrote on October 3, 1835: "We sailed round the northern end of Albermarle [Isabela] Island. Nearly the whole of this side is covered with recent streams of dark-coloured lavas, and is studded with craters. I should think it would be difficult to find in any other part of the world, an island situated within the tropics, and of such considerable size (namely 75 miles [120 km] long), so sterile and incapable of supporting life."[3] There were quite likely several colonies of flightless cormorants on this northern coast, almost as they are today. Maybe there were fewer than normal if, as evidence suggests, Darwin's visit occurred during an El Niño year.

In the decades around the historic exploration of the *Beagle*, the most active years of Pacific whaling in the era of sail, literally dozens of ships stopped at the Galápagos each year for water, food, and an anchorage.[4] For example, a young Herman Melville was becalmed aboard a whaleship for over a day just north of Isabela in 1841. During about three weeks sailing in the vicinity of the island, he saw at least eight other whaling vessels.[5] Later in 1854, in his fictional "The Encantadas," Melville's narrator describes Isabela and Fernandina. He writes of the seabirds on the tall, isolated Rock

Redondo just to the north. The epigraph of Melville's "Sketch Third" in "The Encantadas" (which is pulled from Edmund Spenser's sixteenth-century *The Faerie Queene*) has this verse:

> And cormoyrants, with birds of ravenous race,
> Which still sit waiting on that dreadfull clift.[6]

But Melville does not write about the flightless cormorant, although if he saw one I'm certain he would have seized on its metaphoric possibilities. In these "Encantadas" Melville parodies Darwin's *Voyage of the Beagle* and other science narratives of his time, especially the desire to collect and quantify. He also conveys the dark, devilish mystery of the Galápagos and its unknown inhabitants. The flightless cormorant's habitat appeared to Melville, Darwin, FitzRoy, and other nineteenth-century visitors as hellish, bewitched, and antediluvian. The arid archipelago is the last place anyone would want to visit for pleasure.

After Darwin and Melville, even expeditions with staff devoted to finding birds and other animals somehow missed the flightless cormorant or failed to mention them, even though the birds seem to have nested as they do now, near regular anchorages such as Tagus Cove and Banks Bay on the long, mountainous island of Isabela.[7] If they spotted the flightless cormorants from afar, perhaps early visitors mistook them for the neotropic cormorant (*P. brasilianus*), another dark cormorant that is found commonly on the coast of South America. Or without any flashy facial skin or feet, the cormorants may have just blended too well into the black rock.

The only account of this bird by an early mariner of which I'm aware is by Captain David Porter of the U.S. Navy in 1813.[8] Porter landed his sailors in a small cove to kill fur seals and turtles: "We also found plenty of birds called shags, which do not appear alarmed in the slightest degree at our approach, and numbers of them were knocked down by our people with clubs, and taken on board."[9] Porter did not know that he had in hand—and presumably then in his belly—a previously unrecorded species.

The first official record of the flightless cormorant did not come back to the Western scientific community until 1898. Lord Walter Rothschild from Tring backed the expedition that observed this first known specimen. He named the flightless cormorant with the species title of *harrisi*, after the lead collector, Charles Harris, whose commitment to the mission was undeniable. Three of the five men in Harris's expedition died of yellow fever in Central America, including his intended captain. The fourth man called it quits and went home. Harris continued on to San Francisco alone, found three more collectors, and chartered a new captain and two-masted schooner to sail south to the Galápagos.[10]

Upon Charles Harris's return, Rothschild wrote of the flightless cormorant to the British Ornithologists' Club: "This is the most remarkable discovery made during the expedition."[11] Rothschild continued: "This bird is the largest known Cormorant, being if anything bigger than the extinct *Ph. perspicillatus* [spectacled], and its wings are quite soft and incapable of flight, and of about the same size as the wings of the Great Auk."[12] Though incorrect regarding the size of the bird in comparison to the much larger spectacled, Lord Rothschild helped ignite a twentieth-century fascination with scientific adventures to the Galápagos and for capturing this flightless cormorant.

Several new expeditions were launched in the twentieth century, some of which mention the cormorant as one of their prize captures. Others caught the animal as merely an incidental trophy, as did an exploration under sail in 1906. "We collected a number of these," wrote one of the naturalists about the flightless cormorants, "putting them in sacks with holes cut in, so that they could stick their heads out."[13]

In 1923 William Beebe undertook one of the more famous expeditions to the Galápagos. Beebe, best known for his later deep-sea dives inside his steel "bathysphere," led this trip by steamship with a dozen biologists, artists, photographers, taxidermists, and a "Chief Hunter." They returned with live animals, and thousands of insects, bird skins, eggs, nests, and preserved specimens of plants, plankton, reptiles, and fish.

An encounter with a flightless cormorant was one of Beebe's holy grails. Ruth Rose, the project's historian and "Curator of Live Animals," wrote about their first sighting of these birds. "We secured the two rare creatures that we had come to Tagus Cove to see. On a shoulder of a rock above the ravine sat a large, dull-coloured bird; it turned its head and a spark of clear greenish blue glinted from its eye and a long, goose-like neck stretched out. Only the sight of a great auk could have been more thrilling, for here was a flightless cormorant, a bird probably doomed in a few years to an extinction as complete."[14]

Beebe wrote of the "drab brown" plumage and "dusky" bill, suggesting that it was this "inconspicuousness" that must have helped it escape the notice of Darwin and so many previous visits. "The eye is the only exception," Beebe wrote, "being of a clear, glittering Italian blue."[15]

Beebe then moved in to catch one of the cormorants: "Quick as a flash she dodged, leaped off her nest and when I caught her body, she raked my hand fore and aft with the cruel curved tip of the beak. Three times she went to the bone before I could secure her."[16]

Beebe's Galápagos expedition returned to New York City and delivered to the Zoological Park and Aquarium two living female flightless cormorants, the first live specimens ever seen away from the islands. The *New York Times* ran a story with the headline "Flightless Birds in Cargo for Zoo." The article announced the expedition's return, the tameness of the creatures on the island, and Beebe's future lecture series. With a Barnum-like flare Beebe told the newspaper: "I have already arrived at some theories regarding the mysterious origin of these birds, which penned up in these tiny islands through the ages, have forgotten how to fly, but until I have gone through the mass of data we have collected I must defer any announcements of these conclusions."[17]

One of the cormorants seems to have not survived long enough to arrive at the aquarium. The other died a few months later in captivity, but not before she had been so aggressive to a Galápagos penguin that the penguin had to be moved because it was feared the cormorant would kill it.[18]

The Zoological Park and Aquarium was not long without these exotic cormorants for display and observation, however. A couple of years later a private expedition to the Galápagos supplied three more. This made the *New York Times* with the headline "Vanderbilt Back with Sea Trophies." The reporter wrote that the cormorants were the rarest of this collection: "savage" toward one another, but "tall, gracefully shaped birds" with the eyes of a "strange sapphire."[19]

Natural history expeditions to the islands continued. So many were launched that by 1930 the *New York Times* reported the return of another voyage with a weary page-seventeen headline: "Again the Galapagos."[20]

Scientific and popular interest in the Galápagos has not waned, due mostly to the gregariousness and exotic nature of the animals and to the continued aura and significance of Charles Darwin. On the centennial in 1959 of the publication of *On the Origin of Species*, Ecuador declared all of the island territory that had not yet been settled to be national park land, which included most of Isabela and Fernandina. By the time Vonnegut was writing his novel *Galápagos*, UNESCO had named the islands both a Biosphere Reserve and a World Heritage Site. At this time in the mid-1980s there was a resident population of over six thousand people on four of the islands, living off tourism and fishing. About twenty thousand tourists a year were visiting.[21] Among those were Kurt Vonnegut and his second wife. By 2011 the number of tourists had skyrocketed to over 185,000 people. These days about half the visitors to the Galápagos go on some sort of nature cruise.[22]

I visited the Galápagos about 10 years after Kurt Vonnegut and about 160 years after Darwin. I was just out of college, sailing as a teacher on a school ship. We anchored off one of the busier, settled islands. I had never, and have not since, been in a place of such fervent, prolific, outlandish wildlife. I scuba dived with hammerhead sharks and sea turtles. During one of my dives I was completely enveloped in a vertiginous school of glittering pompano. While I bubbled underwater during another dive, dozens of sea lions zoomed around me and peered straight into my mask. Ashore, at the Darwin Research Center, I put my hand on the shell of Lonesome George, then the last living representative of one of the island's species of Galápagos tortoise. I walked on the sharp black rocks among the marine iguanas and beside the blue-footed boobies. As we departed the islands, with a perfect breeze to fill our sails, huge schools of dolphins played in our wake and dozens of frigatebirds rested on our stays and yards. Amid all this natural wonder, I also saw for the first time a starving sea lion wrapped in a noose of plastic.

In Vonnegut's novel *Galápagos*, an entrepreneur and promoter in the 1980s builds a new passenger ship named the *Bahía de Darwin*. He manages to publicize a "Nature Cruise of the Century" for its maiden voyage out to the islands. The cruise is scheduled to depart from the port of Guayaquil, Ecuador. Thanks to the fortuitous influence of Jacqueline Onassis, the passenger list grows into a who's who of celebrities of the day, including Mick Jagger, Henry Kissinger, and Walter Cronkite.

Vonnegut was well aware of the elevation of the archipelago in the cultural imagination. He writes of his fictional promoter: "He had done as much as Charles Darwin to make the Galápagos Islands famous — with a ten-month campaign of publicity and advertising which had persuaded millions of people all over the planet that the maiden voyage of the *Bahía de Darwin* would indeed be 'the Nature Cruise of the Century.' In the process, he had made celebrities of many of the islands' creatures, the flightless cormorants, the blue-footed boobies, the larcenous frigate birds, and on and on."[23]

The only reason a public school teacher such as Mary Hepburn is on this passenger list is that she is the very first person who signed up—before the snowballing of celebrity interest. Her sick husband, Roy, purchases the tickets when he sees the travel agent taping up a poster of a flightless cormorant to advertise the cruise.

In several ways and on a few levels, the flightless cormorant on the advertisement serves as the running metaphor for the novel. Vonnegut writes:

> The travel poster in Ilium depicted a very strange bird standing on the edge of a volcanic island, looking out at a beautiful white motor ship churning by. This bird was black and appeared to be the size of a large duck, but it had a neck as long and supple as a snake. The queerest thing about it, though, was that it seemed to have no wings, which was almost the truth. . . . Somewhere along the line of evolution, the ancestors of such a bird must have begun to doubt the value of their wings, just as, in 1986, human beings were beginning to question seriously the desirability of big brains. If Darwin was right about the Law of Natural Selection, cormorants with small wings, just shoving off from shore like fishing boats, must have caught more fish than the greatest of their aviators. So they mated with each other, and those children of theirs who had the smallest wings became even better fisherpeople, and so on.[24]

The flightless cormorant's life is simple in the novel: dive into the water when you want fish. Then pretty much stand around for the rest of the time until mating. Vonnegut uses the cormorant as an example of the evolutionary path humans might take if stranded on these islands for a million years. The "use it or lose it" concept might just work for brains as it might have done for the wings. The cormorants, and subsequently Mary Hepburn and the first "settlers," have no predators. Aside from an occasional shark. Which is how Mary eventually dies.

As comic and outlandish as Vonnegut's scenario might sound, the novel and his depiction of the cormorant and the other animal life of the Galápagos serve to teach some of the major tenets of Darwin's theories. He did a great deal of research to try to get these premises correct. It was a point of pride with Vonnegut that he received an approving letter after publication from the prominent zoologist and author Stephen Jay Gould.[25] Gould began to teach the novel to his undergraduates in science at Harvard.[26] Vonnegut emphasizes reproductive success as the major driver of evolution, not the survival-of-the-fittest model that is often misapplied to human society, known as social Darwinism — the philosophy that Vonnegut most deplored.[27]

Vonnegut revolves Mary's story and the entire plotline around "contingency," the concept of luck and chance that is often overlooked in explanations of evolutionary theory and studies of historical progression on the earth, whether from a biological or geological lens. Vonnegut demonstrates that evolution does not necessarily progress in a linear fashion toward some sort of improvement. Sometimes evolutionary change is just a confluence of absurd, unpredictable, often unrelated events. On the smallest scale, for example, a shag living on Inishmore simply dies and does not pass on its genes because it is standing in the wrong place when a goat kicks a stone off a cliff.

Stephen Jay Gould wrote: "In Vonnegut's novel, the pathways of history may be broadly constrained by such general principles as natural selection, but contingency has so much maneuvering room within these boundaries that any particular outcome owes more to a quirky series of antecedent events than to channels set by nature's laws. *Galápagos*, in fact, is a novel about the nature of history in Darwin's [theoretical] world."[28]

Because of a series of events, the passenger list of the *Bahía de Darwin* when it grounds on one of the islands of the archipelago turns out not to include any celebrities after all. It is not the fittest, fiercest, smartest, or sexiest selection of people. The list is instead composed of a small group of girls from a nearly extinct tribe up in the rain forest, the pregnant wife of a Japanese computer genius, the blind daughter of a millionaire businessman, a bumbling German captain, and the schoolteacher Mary Hepburn. In the twisted unlikely plot that gets these people onto the ship and these islands, Vonnegut discusses and introduces the roles of hereditary diseases, trait mutations, massive moments of natural or human-caused upheaval, and the regular miracles of split-second chance. Vonnegut is also eerily prescient in 1985. His motley group of passengers end up on this ship as a result of, among other reasons, a worldwide economic crisis and a small handheld device that can translate languages, keep exact time, and search all the world's literature.

The normal story biologists tell about the evolution of the flightless cormorant is that their wings atrophied over time after the birds flew to the archipelago. The cormorants found a lot of food and no predators, so they stayed and thrived and shed their ability to fly. It was a disadvantage to invest the energy and calories to maintain these hefty flight muscles, or the thick keel-like bone to support them. It was a misuse of resources for cormorants in the Galápagos to continue growing the feathers necessary for flight when they were no longer crucial to the primary, already exhausting job of surviving long enough to reproduce and raise young.

No one knows definitively how, why, or when the ancestors of the flightless cormorant flew to the Galápagos. To try to answer the "when," a coalition of researchers from New Zealand and Ecuador put together a DNA analysis. They confirmed what ornithologists had suspected as early as the 1930s, that the lineage of the flightless cormorant traces to the double-crested cormorant and the neotropic cormorant.[29] This makes the most sense, at least superficially, since these species look the most similar to the flightless cormorant and are geographically the closest today. The researchers believe that the flightless cormorant arrived on the islands somewhere around two million years ago. The cormorants settled about the same time as Darwin's finches, but a couple of million years after the Galápagos penguins.[30] What is especially fascinating about this dating is that the cormorants' present islands, Isabela and Fernandina, did not even exist at the time of the presumed cormorants' arrival. These two islands would not emerge from volcanic activity for more than another million years.[31] The researchers suggest that the volcanism in the region would have shifted the currents and the movement of the local nutrient-rich upwelling, upon which this species of coastal bird relies.

Evolutionary biologist Richard Dawkins, who writes adoringly of one day when he snorkeled with flightless cormorants hunting underwater, has used this species as an example for his own teaching. Though Charles Darwin never saw a flightless cormorant, Dawkins points out that vestigial limbs on other flightless birds were one of Darwin's "key arguments" for evolution by natural selection. How could a creationist explain why God would give these birds wings that seem to not have much function?[32] Perhaps we do not yet understand how these sparse wings are indeed useful to the bird's ability to survive and reproduce — not unlike the caruncles on blue-eyed shags. The flightless cormorants still seem to use their wings for balance when hopping around the rocks. And they still spread them out after fishing, which may indeed be unnecessary: a vestigial behavior. If we had Vonnegut's narrator to look ahead a few million years, perhaps we would see the flightless cormorants with even smaller wings and with the wing-spreading behavior entirely gone.

Dozens of flightless bird species have either recently gone extinct or still live in a variety of environments, including penguins, ostriches, emus, moas, kiwis, grebes, wrens, and even a huge, flightless parrot. In New Zealand and Hawaii about one in four of *all* terrestrial and water birds were flightless before the arrival of Pacific peoples, who rendered many of them extinct.[33] Evolutionary biologists believe that the ancestors of all flightless birds once took to the air.[34]

In *Galápagos* Vonnegut reminds us metaphorically that given enough time and a series of contingent events, even the flightless cormorant could fly again. In the novel, that group of girls from the nearly extinct tribe not only reproduces but also is the foundational gene pool of the new *Homo sapiens* after a devastating and tight bottleneck both for their cultural group and their entire species. Those on William Beebe's expedition in the 1920s presumed that the flightless cormorant would inevitably be extinct soon — a common perception then and now: these cormorants are not equipped to survive in this rapidly changing, human-dominated world, living within such a small radius. They are indeed barely hanging on today, although the IUCN recently improved their status from "endangered" to "vulnerable."[35]

These days about 1,300 individual flightless cormorants remain, living along a stretch of narrow coastline on two islands.[36] These numbers are actually a dramatic improvement in recent years from as low as 409 cormorants in 1983 after an especially severe El Niño year. During an El Niño year, the water warms, upwelling is reduced, primary productivity plummets, fish starve, seabirds starve, and so on. In 1982–83, half the cormorants died, as did 80 percent of the Galápagos penguins.[37]

I like how David Day in his *Doomsday Book of Animals* speaks to a public that might not be concerned about endangered species in tiny populations on the brink. Some might think that only the "fittest" will remain and that this is OK: flightless birds just weren't meant to survive, and this is the harsh, natural order of things. Vonnegut would appreciate that Day compares the human race's impact on various species to the bombing of Hiroshima. It would be absurd for us to expect that humans in this part of Japan could survive by developing immunity to radiation, and, if they did not, to suggest that they were just not equipped—not selected in any natural survival-of-the-fittest sort of way.[38] The natural evolution of traits can take, as it does in Vonnegut's novel, a million years, not one lifetime or a few generations.[39] In more specific terms, the flightless cormorant is not necessarily less equipped to survive, nor is it necessarily a weaker species. These cormorants are threatened because they have had no time to respond to the bomb blast of recent human invasion on their native archipelago.

Thus in our lifetime, because of their tiny isolated pocket of a population, the flightless cormorants might get wiped out forever by any or a combination of the following, many of which they have already encountered at some scale historically or recently: a human-introduced disease; a mosquito-carried disease; their own disease; a fish disease or octopus disease or spiny lobster disease or tick disease; or an introduced mammal, like a rat, that eats all the cormorants' eggs; an introduced marine species that eats all the fish, octopus, and lobster; or human overfishing that eliminates all of the cormorants' food. Perhaps a group of men will kill the birds off because they believe the cormorants are competing unfairly for commercially valuable fish. Or maybe the flightless cormorant will be eradicated by a series of bad El Niño years, an oil spill, a tsunami, a volcanic eruption, or even by a meteorite.[40] Maybe some absurd series of events in the future will bring the flightless cormorants back to the mainland where they will somehow be well equipped to survive in a specific niche. They will thrive and look totally different again in a million or a hundred million years. Maybe a few hundred of them, randomly chosen, will end up on a cruise ship and waddle onto an Indonesian island. And thrive.

Vonnegut provides a surprisingly happy ending to *Galápagos*—a new, more content *Homo sapiens* is living a peaceful, fish-eating, cormorant-like existence, albeit incapable of any larger sense of self or soul. They cannot start wars or even fabricate tools, similar to the birds at the close of Amy Clampitt's poem "The Cormorant in Its Element." For Vonnegut, the evolved humans do fortunately sustain the trait of laughter, especially when there is a good fresh-fish fart on the beach.

Master and Commander is the 2005 movie derived from Patrick O'Brian's series of novels about the Royal Navy, set a few decades before Darwin and FitzRoy's voyage. The film is loosely based on the novel *The Far Side of the World*, published the year before Vonnegut's *Galápagos*. If the story of Mary Hepburn is the high point thus far for literary recognition of the flightless cormorant, then *Master and Commander* is the film where this bird is most celebrated.

In *Master and Commander* Russell Crowe stars as Jack Aubrey, the captain who shares his cabin with the Darwin-inspired ship's surgeon and naturalist, Stephen Maturin, who is played by Paul Bettany.[41] In order to track down a French man-of-war, Aubrey sails to the Galápagos Islands. Aubrey predicts the French captain will sail there, too, in order to attack the whaling fleet.

Captain Aubrey promises Dr. Maturin a couple of days ashore to collect and explore, telling him that he'll be the first naturalist to set foot there. When their ship first arrives, sailing in the channel between Fernandina and Isabela, Maturin looks over the rail with his telescope. With raised eyebrows he tells a young midshipman: "How extraordinary. Those birds." The camera cuts to his view in the scope. "By all that is holy, I think that is unknown to science."

Before they can land him ashore, information compels Aubrey to sail off after the French ship. The doctor isn't able to see the wildlife, despite his Steller-like protestations. As they sail away, Maturin lies in his bunk pouting, gazing at an illustration of a cormorant in a book.

Later on deck, one of the marines lodges a bullet in Maturin's belly while trying to shoot an albatross flying around their ship (a nod to Coleridge's "Rime of the Ancient Mariner"). Aubrey decides he must make camp back at the Galápagos to save the life of his old friend. As Maturin heals, he is soon out exploring. One day he treks, still limping, more than ten miles across the island for the sole purpose of trying to capture a flightless cormorant. He tells a midshipman that a live specimen will grant them a dinner with the Royal Society once back home. Maturin, exhausted, almost catches one, but it flutters off, hopping out of reach. When the surgeon looks over the horizon from where the bird just was, he sees that very French ship for which they have been searching. In other words, without this naturalist so eager to collect one of these cormorants, they would not have been able to find the enemy — the crux of the story.

Captain Aubrey leads his crew in the successful capturing of the French ship. As *Master and Commander* winds down, Aubrey intends to sail back to the Galápagos to get food and water. And he wants to give his old friend another chance to explore. Yet then, once more, a final plot twist forces the captain to head back out to sea immediately.

Aubrey delivers the final lines of the movie to Stephen Maturin, in mock appeasement: "Well, the bird's flightless. It's not going anywhere."

Flightless cormorant in literature. Flightless cormorant in film. Flightless cormorant in paint. In the dark, rainy gloom of a winter in Seattle, an artist named Heather Carr sat listening to a friend just back from his trip to the Galápagos. She looked at his photographs, hearing all about his nature cruise of the century. The flightless cormorant caught her eye. She told me that here was the metaphor for how she was feeling with two little kids, with wintertime, and with her career stagnating. She painted herself in the nude with not an albatross, but the flightless brown bird around her neck: a noose. The chicks are nearby wanting and wanting, and across this graffiti-like, splashy big painting is dripping "I'm not going anywhere."[42] (See figure 16.)

Fishing off the coast of Isabela and Fernandina, flightless cormorants dive as deep as 240 feet (73 m), but more commonly swim in waters less than 50 feet (15 m) deep.[43] The cormorants stay extraordinarily close to shore, rarely ranging much more than 750 feet (230 m) from the coast to fish.[44] With an especially long and robust beak, flightless cormorants pluck and pry octopus from out of the rocks. Like other cormorants, they also catch a large variety of bottom-dwelling fish and crustaceans.[45]

Aside from its more powerful beak and skimpy wings, the flightless cormorant has four other traits that set it apart from the other cormorants, enabling it to survive in such a tight range with such a small population.

First, not only are these the only extant cormorants that are flightless, but they are also the heaviest and largest of living cormorants. It is reasonable that the flightless cormorant's wings atrophied because flight wasted energy no longer necessary, since there were no predators from which to fly away. Cormorant flightlessness might also be connected to advantages in how and where this bird finds food. A larger animal can stay warmer underwater because it has a relatively low ratio of its volume exposed to the cold. As for mass, wings and flight limit an animal's ability to get heavier, whereas extra weight helps the cormorant dive under the surface. Imagine, say, the wings that would be necessary to get an emperor penguin airborne. The spectacled cormorant that swam in the frigid Bering Sea presumably weighed twelve to fourteen pounds (5.4–6.4 kg). The average male flightless cormorant today weighs a little less than nine pounds (4 kg).[46]

A second unique characteristic of the flightless cormorant is that the males are significantly larger than the females. In the rest of the cormorants, the difference is only slight. One theory is that sexual dimorphism occurs naturally as animals get larger, while another theory suggests that the large disparity in these cormorants allows a diversity of foraging capacities in such a tight area, reducing competition within species, within the colonies. Male flightless cormorants regularly dive deeper and for longer than the smaller females.[47]

Third, flightless cormorants nest several times a year, with little to no fidelity to mates.[48] With the flightless cormorant population so small, the danger of inbreeding is real.[49] There are regularly, for example, an inordinate number of infertile eggs.[50] These cormorants probably breed often to continue to diversify the gene pool as much as possible. And in an alteration of traditional gender roles, the females move around and try to attract different males. Biologists believe the females are the aggressors, while the males eventually do the choosing. After hatching the young—usually just one chick will survive to fledge—the female deserts her male mate, most commonly just after the chick has left the nest but still relies on adults for food. The male will then take care of the fledgling for as long as another six months. A female cormorant, however, will not leave to go mate again unless she is sure there is enough food available and the male is committed and capable of caring for the fledgling.[51] This behavior is not only unknown in any other cormorants, but it is a rare adaptation for any bird.

Professor Carlos Valle, a native of the Galápagos who earned his PhD at Princeton by studying the reproduction of the flightless cormorant, con-

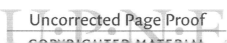

cluded after years of observation that the reason this system works is that the larger males are more efficient foragers. The males are able to take care of themselves and the young at the same time. If the male deserted, the female would likely not be able to raise her brood alone. His lineage would thus not be passed on. The female "exploits" this hard-wired need of the male to pass on his genes. She leaves her chick under the care of the male as soon as she can go mate again confidently.[52] One of the only reasons she sticks around in the first place after hatching is to help guard her newborn against the Galápagos hawk and even at times other adult cormorants.

In *Galápagos*, Vonnegut describes an educational documentary that Mary Hepburn regularly shows her high school students about the comic mating dance of the blue-footed boobies. The birds march, lifting up and down their wide feet. As the boobies get closer they "sky point": the male and female make "their sinuous necks as erect as flagpoles," aligning them together to look up and form a sort of tower with four blue webbed feet.

Flightless cormorants do not sky point, but all the members of their global family have an ornate mating dance, which generally involves some sort of throwing of the head back, some hunching or flipping of the wings, and in several species a weaving of the necks together.[53] Yet unique to all other cormorants — the fourth and final trait I'll discuss here — the flightless have part of their dance in the water.[54]

I've watched this myself. It is a sort of spinning waltz. During the event I watched, two other females swam around the pair spinning. Whenever one of them got close, the dancing female shot out and jabbed her beak at the intruder. The female continued to be the one to defend the pairing. After several minutes of twirling, the female and her male swam still closer and eventually entwined their necks. The two climbed on shore and walked over to their nest to continue a series of more postures and rituals — wing flapping, beak touching, gaping, throwing the head all the way backward — all common to the rest of the cormorants.

Because the flightless cormorant is unique among the rest of its family in having a more powerful beak, tiny wings and heavy body, much larger males, an adjustment in traditional gender roles when courting and raising young, and a mating dance in the water, several taxonomists have knighted this cormorant with its own separate genus. Instead of *Phalacrocorax*, the name sometimes ascribed is *Nannopterum*, meaning simply "small winged."[55]

If you've read any Vonnegut you know that he regularly inserts what feel to be quite autobiographical anecdotes, usually in the first person. In *Galápagos* the narrator is Leon Trout, the son of a science fiction writer. Leon is a ghost on the *Bahía de Darwin*, the ship that takes Mary Hepburn and the rest of the random assortment of people out to the islands. He tells the reader how Mary Hepburn was having sex with the bumbling captain Adolf von Kleist, seemingly out of boredom. One day she decides to walk his fresh semen over to the other side of the island and insert it into the fertile vaginas of the native teenagers. To her surprise, it works. And so reproduction begins on the island, eventually leading to the seal-like, cormorant-like humans.

Vonnegut was never accused of high lyrical art. He cared more for punch, for humor, and poignant storytelling. Yet at the same time he confided to a professor friend, after the reviews of *Galápagos*, that he was insulted when his work was pigeonholed as science fiction, that critics would not believe that serious literature could involve writing about science.[56]

One critic at the time described Vonnegut's narrative choices in *Galápagos* this way: "As for suspended concern for the well-made, big-brained novel, Mr. Vonnegut has opted to zoom in directly for the catch, the idea, the oracular bit. His books are not only like canaries in coal mines (his own analogy) but like the cormorants of the Galápagos Islands, who, in their idiosyncratic evolution, have sacrificed flight for the getting of fish."[57]

When Vonnegut and his second wife visited the island a few years before he wrote the novel, he went on his own nature cruise for two weeks. "We had advantages that Darwin didn't have," Vonnegut said, referring to motorboats and guides. "And, most important, we knew Darwin's theory of evolution, and Darwin didn't when he was there."[58]

While Vonnegut was touring the Galápagos, germinating ideas for this novel — as Darwin did for his foundation-shaking theories — he was acutely aware that his first wife of over twenty years was dying of cancer. During my visit, I was not aware at the time that my brother David had been growing his own vicious cancer. He died recently. I miss him terribly. When I was on that visit to the Galápagos I bought him a stuffed toy magnificent frigatebird, the black bird with the huge red ballooning gular pouch. These close relatives to cormorants can barely walk. They can barely swim. They're long and black and evolved only to soar and plummet-dive in the air. They gather some of their food from scaring other seabirds into coughing up their recent catch, hence their common name. This might be another advantage of flightlessness for cormorants, to avoid this sort of in-air predation, known as "kleptoparasitism."[59]

I had always thought my brother David was excited about collecting obscure stuffed animals. I learned in his final months that he only said he did so because it gave me an easy, cheap gift to buy him during my travels. I wasn't able to find a stuffed flightless cormorant at the tourist shops in the Galápagos. In fact, the only place on earth of which I'm aware that has anything like a cuddly souvenir cormorant of any species is in Gifu City, Japan. Believe me, I've looked. From Gifu I sent my brother a little stuffed cormorant dressed in usho's clothing. I think a cute, stuffed flightless cormorant toy would sell very well in the Galápagos tourist shops, in Guayaquil, and aboard all the nature cruises of the century. The flightless cormorant stuffed toy could have a sweet, lovable name, such as Mary.

November

Just before dawn. A pink blush where the sun will come up barely lights the dark turquoise water. Floating on the surface in a dim trough of wave is one of the three juveniles from the first clutch on Gates Island.

As the rim of the sun comes up over the Gulf Stream, the bird paddles toward a copse of mangroves. Nearby, two skimmers drop their long lower mandibles and trace the water's surface, seeking to scare up fish. Along this shoreline off a Florida key, beside a man-made beach of cement riprap, a heron looks down into the water. Its long white neck is bowed and perfectly still.

The cormorant's wet head reflects a tinge of orange from the sunrise. She holds her breath and dives again under the surface toward the dark sandy bottom. She swerves, her wings beside her body. Now she waves her torso and tail, swimming with both feet at once, like a seal. She swims back up toward midwater. She darts through a thin gap between two fan corals and rises quickly over a large lugubrious lemon shark. She paddles now with rotating feet toward the bottom. Air bubbles trail up from underneath her feathers.

The cormorant hears the crinkling sound of snapping shrimp. She hears a distant outboard engine starting up in the boat channel between this key and the next.

Then somehow through the dark salt water she sees a finger-size spiny lobster crawling on the tips of its feet. It is groping with its antennae within a patch of coral. With only a few more seconds of breath left, the cormorant lunges down. She flicks out her neck, nabbing the lobster. She swallows it before arriving back up at the surface.

9

Belzoni

UNITED STATES

An assemblage of cormorants is a "gulp."
An assemblage of waterfowl is a "plump."
An assemblage of seabirds is a "wreck."

DAVID M. BIRD,
The Bird Almanac, 1999

A white truck zips by and kicks up dust as it turns away. It races down one of the tractor paths between ponds.

"My man's out there seven days a week after these birds," says Chris Nerrin, a veteran of the catfish business who took over the farm from his dad. "It's just a pain in the ass, and I don't see it getting any better from the all the time I've been out here."[1]

As we are talking there is an occasional distant crack of a shotgun, and sometimes there is a scattering up of birds off a pond. A couple of cormorants fly overhead, but the more visible birds on the water this February afternoon are ducks and coots. On the edge of one distant pond beside a plowed field stand dozens of snow geese. The ducks, coots, and geese don't eat the catfish, though. So the guy in the white truck isn't after them.

"We've tried everything," Nerrin says. "The cormorants are smart. Adaptable. My man driving around is the best we can do. Sometimes we send out two or three trucks to run the birds. Whatever it takes to get them up."

Chris Nerrin used to have one of the larger spreads in the area, but in recent years he has reduced the number and acreage of his ponds by about half. He had to lay off a few workers. Some of his neighbors have got out of the aquaculture business completely, drying their ponds and plowing the fields for corn or soybeans. Recently farms throughout the catfish industry have had to downsize for several reasons. The recession has fewer people buying fish. Competition from farm-raised fish overseas has increased. The cost of feed and fuel has risen. And then there is this cormorant problem.

Nerrin still runs forty-four ponds of various sizes, totaling some five hundred acres of water. At one of the smaller fingerling ponds of about four acres, he has white twine stretched across the top. He tied the twine off with stakes sledged into the grass edges. This has helped keep the birds off this pond at least, which is important because the cormorants particularly like these small, young catfish.[2] In the past Nerrin has tied pie plates and ribbons on the twine. Because of the size of the other ponds and the fact that in those he and his workers regularly drag a seine net to harvest the marketable fish, it is just not practical to rig twine across them.

Nerrin says he has tried all the usual methods. He has tried "evil-eye" balloons, propane cannons set on timers, screamer shells, and a variety of other pyrotechnics. He put up a scarecrow. When that didn't work he bought an automated five-foot-tall orange figure that randomly rises up, blares a siren, and then deflates again. There's a company based up the road from Nerrin's farm that manufactures and sells products like this. Their logo looks like a frightened crow. The company markets this inflatable effigy as "Scarey Man®." For $1,200 you can get the digital timer.[3]

"If you don't stay on top of it, the cormorants can wipe a pond out," Nerrin says. "I shot two dead as a hammer this morning. Wounded one other. But it's hard to make much of a dent."

We thank Chris Nerrin for his time and get back in Dr. Jim Steeby's little suv. Steeby, an aquaculture researcher at Mississippi State University, explains more of the situation as we drive out a dirt road.

"You can tell as soon as you come into a farm whether they've been taking care of their birds," Steeby says. "If you drive in and the cormorants get up, you know the guy has been on his game. If the birds just stay and float and look at you . . . well."

The road is bumpy, so in order for me to hear him, Steeby holds on to the Mardi Gras beads hanging low from his rearview mirror.

Steeby continues: "You're not going to be able to kill that many of them, but you have to stay on it. I tell you what: you have to put the fear of God in the cormorants. The cormorants are smart, they'll figure out what's going to hurt them or not really fast. Screamers, bangers. We tried plastic alligator heads for a while. But cormorants figure it out quick if doesn't hurt them. You can't just leave a truck parked and expect that will scare the cormorants away."[4]

We drive down beside a few more ponds toward a brake of cypress where Chris Nerrin thinks the birds have been roosting at night. Steeby says: "It's important to know that a cormorant can live fifteen, twenty years. They're habitual. If they like a place, they're going to come back. That's a lot of years to be feeding a bird that can eat over half a pound of fish a day."

Jim Steeby went into the Peace Corps after college. It was there that he found his career and passion for aquaculture. In India he learned to raise carp. In Africa he helped raise tilapia. When he came home he got a job in catfish aquaculture in Mississippi at a time when the industry was just starting. He's been working with catfish ever since, for over three decades. He worked his way up in the business, then settled into a position with Mississippi State's extension service. Steeby conducts research and advises the men in the catfish industry. The secretaries in his office call him Doc. The farmers respect and like him. They know he's not a government man or an academic who hasn't done the farming work himself. He wears a belt buckle with a catfish design. Jim Steeby has poured his entire professional life into the catfish aquaculture business. He wants the men and the industry to succeed.

Steeby is professor emeritus these days, but he keeps his office across from the courthouse in Belzoni (pronounced "Belzona"). The town calls itself the Catfish Capital of the World. There is a Catfish Museum two blocks from Steeby's office. The World Catfish Festival each year brings several thousand people to this otherwise quiet, decaying downtown of but a few square blocks. For the festival, the chamber of commerce persuades Steeby to wear a large catfish suit: a furry mascot catfish wearing the hat and blazer of a traditional riverboat captain. I could not find a stuffed toy catfish, but Belzoni officials have commissioned dozens of brightly painted catfish sculptures at every possible corner or site of interest, in the way that other American towns and cities have done with artistic cows or whales or frogs. Belzoni is the government seat of Humphreys County, which until recently produced more catfish than anywhere else in the world. Farmers raise catfish in Arkansas, Alabama, Louisiana, and other states, but not at nearly at the level of catfish farmers in Mississippi in recent decades. Altogether this southeastern catfish industry has been raising more product than all other aquaculture industry in the United States combined — more than the total of all farm-raised salmon, trout, oysters, crawfish, tilapia, and other products. Catfish is by far the largest fully farm-raised seafood product in the United States, both by volume and total value.[5]

Steeby thinks that if the double-crested cormorants (*Phalacrocorax auritus*) had been around like they are today, catfish aquaculture would have never got a foothold. At least not around Belzoni and this farmland along the Mississippi Flyway.

The history of the catfish industry in the southeastern United States is not an old one. The first ponds were built on the Arkansas side of the Mississippi River in the early 1960s.[6] Soon farmers in the state of Mississippi got involved, particularly in the vicinity of Belzoni. This part of the state, composed of several northwestern counties, is known as the Mississippi Delta (not to be confused with the much larger area around the mouth of the river, often referred to as the Mississippi Delta region). In the 1960s and 1970s, prices for cotton, rice, and soybeans had dropped so low that families were looking around for something else. With lots of flat land and a mild climate, the Delta is well suited for aquaculture. The soil is mostly clay, which holds the water without leaking. The farmers also have access to a vast aquifer. They can pump water out of the ground relatively cheaply and easily to fill areas where they have plowed up levees to create the shallow rectangular ponds.

The early pioneers of the industry, such as Jim Steeby and Ed Nerrin, Chris's father, had some success through trial and error. Soon their neighbors built more ponds and got into the game, too. Learning to raise catfish is technical, and the investment is significant. A single pond can represent $50,000 worth of fish. Profit margins are small. The loss of one pond can be disastrous. If a man on one hot, summer night doesn't check the oxygen levels and doesn't flip the switch to start the aerator, all the fish in the pond could die in less than an hour. Diseases develop among the densely packed fish. Equipment fails. Reliable labor is hard to keep. And birds eat the fish.

The farmers primarily raise the channel catfish (*Ictalurus punctatus*), which has been caught wild along the Mississippi for centuries. It is a hardy fish that grows to marketable size in two to three years. Catfish do well in the man-made ponds, since they are adapted to living in the still water of river tributaries. Unlike farmed or hatchery-raised trout or salmon, catfish can grow without animal protein in their diet. This cuts down the price of their feed. Soybean and corn-based pellets also reduce risks of food contamination. Monterey Bay Aquarium's "Seafood Watch" program rates U.S. farmed catfish as a "best choice" for consumers who want to eat ecologically sustainable fish.[7]

As the catfish industry in the Mississippi Delta grew through the 1980s and 1990s, it found a good market, coinciding with developments in freezing and shipping technology. By the peak in 2003, the United States processed over 660 million pounds of farm-raised catfish (299 million kg), with the Delta being the biggest player in both acreage and dollar value.[8] In the Mississippi Delta alone, over eighty-five thousand open-air acres of ponds were devoted to raising catfish.[9] The fish farmers formed an association called the Catfish Farmers of America and a collaborative organization called the Catfish Institute, the logo of which Jim Steeby displays on the front license plate of his suv. Catfish farmers pay dues to help with national marketing and to lobby in Congress for initiatives like reducing competition from abroad. The Catfish Institute promoted federal legislation to halt imports of Asian basa from Vietnam being sold to U.S. grocery stores and restaurants as simply "catfish."[10] So now if you are in the business in Belzoni and the Delta, it is always called "U.S. farm-raised catfish," even in casual conversation.

Steeby takes me to visit a particular friend of his, a pond manager of Tackett Catfish Farms. By chance, the farm owner himself drives up and stops. A man in his eighties, he powers down the window of his brand-new truck and chats some with the pond manager. "Scare those birds now," he says before he drives off.

"That was Mr. Tackett," Steeby says. "You've just met the largest catfish farm owner in the business."

As the story goes, William Tackett walked barefoot across the bridge from Arkansas with a third-grade education. He couldn't make a living with traditional farming in Mississippi either, but then eventually in the 1970s he got himself a couple hundred acres of catfish ponds in Humphreys County. By the peak of the catfish industry three decades later, Tackett Farms worked over nine thousand acres of catfish ponds. Mr. Tackett built his own processing plant to get the fresh and frozen product to market. He built his own facility to hatch the eggs and to raise the fry.

One small part of the local catfish industry's success story is Stanley Marshall, the expert taster at Tackett's processing plant. Stanley has been tasting fish for decades. He supposedly smoked cigarettes for most of his years on the job, though I heard that he quit recently. Locals say he has the finest palate in Mississippi. Every farmer must bring in a sample of fish for tasting at several stages before the buyer at the Tackett plant will purchase a pond's worth. One story is that a man a few years back collected a sample of catfish from one of his particular ponds, but his Labrador retriever quickly put his mouth around one in the bucket. The man punished the dog, but saw no marks on the fish. He brought the samples in to Stanley Marshall. As he does pretty much all day long, Stanley cut out a piece of fish, put it in the microwave for ten minutes, and then had a bite. He said to the farmer: "This tastes like it's been in a dog's mouth."

Steeby does not doubt this story's veracity. He has had similar experiences with Stanley Marshall. Catfish's ability to absorb flavors has been one of the reasons it does well as a product. Catfish meat freezes cleanly and can be cut and prepared in dozens of different ways.

Yet despite all the advantages of the product, and the thrift and success of men like William Tackett, the profit margin for catfish farmers continues to narrow, while the risk of loss tied to several uncontrollable forces has not diminished. Like Nerrin and almost everyone in the business, even Tackett has cut his pond acreage down by more than half in recent years. Over half the growers and acres that existed ten years ago are gone, along with hundreds of jobs.[11]

The cormorants are not the most significant reason for the downturn. They are, however, a challenge that the catfish farmers can see in front of them each morning. So for several months each year in the Mississippi Delta, the cracks of shotguns from white trucks zipping around the ponds are ubiquitous. Yet when Steeby, William Tackett, and Chris Nerrin's father first started in the business they rarely saw more than a handful of these cormorants each winter.

In his *Ornithological Biography* (1831–39), the five-volume text accompanying his paintings, John James Audubon wrote eloquently of cormorants. He found the previous accounts in North America particularly lacking: "Few birds inhabiting the United States are so little known, or have been so incorrectly described, as the Cormorants."[12]

Audubon also knew the cormorant as an "Irish Goose." In the journal of his travels down the Mississippi River in the fall of 1820 Audubon wrote of seeing cormorants on the Ohio River south of Cincinnati, and at a couple other points on the way down to New Orleans.[13] On December 26, to the south of Belzoni, near Natchez, Mississippi, Audubon wrote: "Beautifull Morning, Light Frost — I began my drawing as soon as I could see — drawing all day — We saw to day probably *Millions* of those *Irish Geese* or Cormorants, flying Southwest — they flew in Single Lines for several Hours extremely high."[14]

Audubon, like all the naturalists of his time, shot his birds for collection, study, and illustration. He wrote how cormorants and anhingas tend to spread their wings when about to leave a perch or fly off the surface, "thus frequently affording the sportsman a good opportunity of shooting it."[15] Audubon found the cormorants difficult to hit, though. As the cormorants do around the farms of the Delta today, Audubon observed these animals returning to their same roosts at dusk, landing on the branches of trees to sleep for the night.[16]

During his travels to the Florida Keys, Audubon described the raiding of a cormorant colony for specimens. This was in April 1832:

There were many thousands of these birds, and each tree bore a greater or less number of their nests, some five or six, others perhaps as many as ten. The leaves, branches, and stems of the trees, were in a manner white-washed with their dung. The temperature in the shade was about 90° Fahr., and the effluvia which impregnated the air of the channels were extremely disagreeable. Still the mangroves were in full bloom, and the Cormorants in perfect vigour. Our boat being secured, the people scrambled through the bushes, in search of eggs. Many of the birds dropped into the water, dived, and came up at a safe distance; others in large groups flew away affrighted; while a great number stood on their nests and the branches, as if gazing upon beings strange to them. But alas! they soon became too well acquainted with us, for the discharges from our guns committed frightful havock among them. The dead were seen floating on the water, the crippled making towards the open sea, which here extended to the very Keys on which we were, while groups of a hundred or more swam about a little beyond reach of our shot, awaiting the event, and the air was filled with those whose anxiety to return to their eggs kept them hovering over us in silence. In a short time the bottom of our boat was covered with the slain, several hats and caps were filled with eggs; and we may now intermit the work of destruction. You must try to excuse these murders, which in truth might not have been nearly so numerous, had I not thought of you [my reader] quite as often while on the Florida Keys, with a burning sun over my head, and my body oozing at every pore, as I do now while peaceably scratching my paper with an iron-pen, in one of the comfortable and quite cool houses of the most beautiful of all the cities of old Scotland.[17]

Since these cormorants were nesting in the Florida Keys, we know they belong to what wildlife managers now call the Southern population of double-crested cormorants, a subspecies (*P. auritus floridanus*). Audubon thought of them as their own species, which he painted in an open-gaped, distressed posture, and called the "Florida Cormorant." (See cover.)

Like most naturalists up through the mid-twentieth-century era of Niall Rankin, Audubon sampled his birds not only with his gun, but also openly with his palate. Cormorants did not have a reputation as good food for European colonists or their descendants in North America. I've found no epicurean rhapsodies about cormorant dinners like those from the explorers and adventurers of the Antarctic. On the other hand, Native Americans on both coasts seem to have been quite fond of eating cormorant. In Monterey, California, for example, the Ohlone Indians particularly enjoyed cormorant eggs and chicks.[18]

Audubon mentioned the cormorant as food a few times, although he never tried it himself. He heard the meat was tough, dark, and fishy. The Floridian Native Americans and local African Americans killed young cormorants and preserved them with salt. In New Orleans Audubon observed young salted cormorant for sale in the market, bought by "the poorer people" to make "gombo soup."[19] When up in Nova Scotia he found that the men would not take the cormorant eggs because they were inedible, but they would eat young cormorant salted. "I have never eaten Cormorant's flesh," Audubon wrote. "And intend to refrain from tasting it until nothing better can be procured."[20]

A few decades later, ornithologist Oliver Austin wrote in his survey of

THE DEVIL'S CORMORANT

the birds of Labrador: "Not only is [double-crested cormorant] flesh too rank, tough, oily, and fishy for eating, but its filthy nesting habits and its unappetizing appearance have long eliminated it as a possible food bird, even among unparticular and half-starving peoples."[21]

Occasionally as I have listened to ideas on how to resolve some of the human-cormorant conflicts in the United States or Europe, I've thought that if some people ate cormorant, then the culling or proposed hunting programs would be easier to take. There could be similarities to the management strategies with Canada geese.[22]

During my visit to Belzoni I went out driving with Brady Thompson, a farmworker who was "running birds" in one of the white trucks. He said no one eats cormorant in the Delta. Steeby and all the catfish farmers scoffed at the idea of eating this bird.

One farmer named Larry Brown told me a story about one of his men a few years back: "He would eat anything. Coot. Raccoon. Anything. One time he took home a cormorant and cooked it up. He came back the next day and said 'Taste kind of strong.' Now if *he* said that, then cormorant must be *inedible*."[23]

Once I was teaching college students on a tall ship sailing out of Key West. On a late January afternoon we were out in the middle of the Florida Straits. I was writing up at the bow, minding my own business, when a cormorant landed on some rope right over my head. Honest truth. It was a small double-crested, probably a female. After some time, she flew around the ship and found a spot on a horizontal bar above the stern.

The cormorant sat there for several hours and well into the night, only a short distance from the person at the wheel. The bird would have stayed there longer, but we jibed the ship, and the huge main boom came across toward where she was sitting. Just before the boom was a hand's breadth away, she defecated on the deck and flew off.

The cormorant had first landed on the ship when we were southwest of Loggerhead Key. This was over thirty-five nautical miles from any land.[24] I missed from which direction she came, and it was too dark when she departed to see where she was going. I've seen cormorants on the rocks just outside Havana, but it is not common to see cormorants this far from land.[25] I wondered if she sought political asylum.

Over five centuries earlier on the south side of Cuba, Christopher Columbus once witnessed a huge flock of *cuerbos marinos*. This was the summer of 1494. According to one account of the voyage: "At sunrise, a cloud of cormorants, flying toward Cuba, obscured the sun's light."[26]

Audubon wrote: "Our cormorants are by no means great travelers, although they all migrate more or less at particular seasons."[27] Conventional wisdom has it that the double-crested cormorants of the Atlantic population, from New England and the Canadian Maritimes, generally migrate along the Atlantic Flyway—wintering along the coast of the southern states and as far south as the Florida Keys. The cormorants from the Interior population, from the Canadian Prairie provinces and the Great Lakes region, fly down the Mississippi River and winter on the Gulf of Mexico. The Appalachian Mountain range works as a natural ecological barrier in a way similar to how the Rockies isolate the Pacific Coast–Alaska populations of cormorants. The Southern population stays year-round in the Florida-Caribbean region, with some winter dispersal when not breeding.[28]

Biologists have been exploring the flexibility of these migrations. Are individual birds or small flocks or entire populations of cormorants hardwired to fly south to a specific spot? Or do these animals adapt to where the food is, to past experience, or to the given season's air temperature? How much do migrating cormorants respond to factors such as human disturbance? These are all difficult questions, particularly because biologists do not know much about how birds orient themselves in a local area or how they navigate to distant points. Cormorants, like other birds, might use the earth's magnetic field, prevailing wind direction, visual clues and landmarks, some sort of genetic memory, and/or the angle of the sun, moon, or even the stars if they are flying at night.[29]

Learning about the miracles of migration is fascinating on its own, but it also has significant management implications for cormorants in North America and around the world. A persistent position on cormorant management is that if we knew precisely how cormorant populations migrate, then managers could more comfortably manage cormorants within flyways, perhaps setting upper population limits this way, instead of using state or national borders.[30]

Along the Mississippi Flyway in the 1970s, very few if any cormorants paused to winter in the Delta for any significant amount of time. By 1995, counts suggest there were about 32,000 individual birds roosting from the Interior population. By 1999 there were about 64,000 cormorants.[31] Since then the wintering population in the Delta seems to have remained at about that level.[32] If you combine the roosting counts in the late 1990s throughout the Southeast, including in western Arkansas, the rest of Mississippi, and in Alabama, the catfish ponds attracted for a few months of the year something like one in ten of all the double-crested cormorants throughout the entirety of North America.[33]

The cormorants surely have been drawn to the plentiful, easy food supply of the catfish farms. They somehow communicated this to fellow birds. It is likely that the catfish aquaculture has been a major factor in the rapid growth of the Interior population of cormorants after the reduction of harmful pesticides. Biologists Jerome and Bette Jackson wrote in a paper about "the feathered pariah," as they termed the cormorant: "It is a most unfortunate coincidence that the very heart of the catfish-farming industry is located in the Mississippi Delta at the confluence of the Arkansas and Mississippi rivers. The birds are funneled in to the catfish farms!" The biologists continued: "Unfortunately, it was [at] the species' low point that the catfish industry in the mid-South underwent its major growth. The current cormorant 'crisis' might have been predicted by anyone familiar with cormorants, their winter range, and their behavior."[34]

The biology of cormorant migration and where these birds are traveling over the course of a year gets entangled in politics, similar to the issues with managing great cormorants in Europe. Which state has the right to manage a pocket of birds if they move across borders? Again, the United States Fish and Wildlife Service (USFWS) is in charge of the big picture of cormorant management. The United States Department of Agriculture (USDA) and its Animal and Plant Health Inspection Service / Wildlife Services program are also involved, particularly in the Delta region, because of this catfish aquaculture. Steeby told me there used to be two men who would monitor bird populations in the Delta and help the catfish farmers with cormorants. USDA staff are licensed to shoot cormorants off their roosts, or at least to try to dissuade them from spending the night in certain brakes of cypress trees that were close to catfish ponds. They have tried to keep the cormorants awake, to make the birds want to go elsewhere, by using a few strategies, including high-powered lasers. In recent years, however, funding for USDA staff in the Delta has declined.

Because of federal downsizing, the agency now supports just one part-time person. A farmer could call him for help with all manner of wildlife issues — including alligators, which are another relatively new problem for some catfish farmers.

It is worth a short digression to tell you that the Wildlife Services program of USDA is so sensitive about the cormorant issue that this one remaining part-time staff member who was going to meet with me during my visit to the Delta, even with Jim Steeby's blessing, was at the last minute told by his superiors that he should cancel our interview. As mentioned earlier, in 1998 the USFWS granted a "depredation order," which allowed fish farmers to freely kill unlimited numbers of cormorants off their ponds without a USFWS permit. The depredation order also allowed the USDA Wildlife Services staff to "dissuade" the cormorants from roosting locally. As of 2003, as part of the USFWS nationwide cormorant management plan, the USDA staff is now allowed to "lethally remove birds out of their roosts," in addition to helping out with killing on the ponds.[35] So it is pretty much all legal there in terms of killing cormorants. But the government agency does not seem to want to not publicize the matter.

The Mississippi office of the USDA Wildlife Services could not have known that one of my favorite artifacts from my cormorant studies over the years is a pamphlet they first published in 2001 titled *Living with Wildlife: Cormorants*. It is a "Children's Activity Sheet." Much of the basic information about cormorants and aquaculture is accurate and useful, except one unsubstantiated and value-shaping sentence: "Today, cormorant populations are at an all-time high."[36] The activity sheet has a section titled "Your turn — calculated costs," in which a child can presumably work on his or her math skills by figuring out the price of ammunition, labor, and "frightening programs" based on a 90-day or 150-day deterrence program. On the back page is a crossword puzzle. One down: "Firecrackers are an example of this." Answer: pyrotechnics. Seven across: "What IPM stands for." Answer: Integrated Pest Management.[37]

Anyway, several professional and very smart members of the Wildlife Services staff conduct research, often with scientists at Mississippi State University, on migration questions. They want to know how much effect the aquaculture ponds have actually had on the movement of cormorants as they mysteriously fly around the country.

One biologist who has been involved in this collaborative research is

Brian Dorr. In 2010 he was part of a study, led by Tommy King, that examined the recoveries of over 10,600 leg bands from 1920 to 2006. They had information from about 6 percent of all the double-crested cormorants ever banded officially, from all over the country. Biologists and amateur birders have banded and are still banding cormorants from places such as Little Galloo Island and East Sand Island, and colonies in Saskatchewan, Manitoba, and in the Gulf of Maine. In examining all the reported findings of these bands—by binoculars, or more often by natural death or shotgun—Dorr and his colleagues saw a dramatic trend after 1985. The recovery of leg bands increased by 450 percent in the southeastern aquaculture region, as opposed to an increase of just 55 percent in other areas around the country. Near the aquaculture ponds there was a 520 percent increase in the finds of those cormorants that had been banded in the colonies from the Great Lakes region and the Canadian interior provinces.[38] There are quite a few variables and admitted holes in this sort of study, but if you are thinking that all the men in the white trucks shooting cormorants after 1985 have been recording leg bands, I can tell you from experience this is not the case. Most farmers do not have the time or interest for this, although they arguably should. The original 1998 depredation order said that farmers had to at least log their kills monthly. The 2003 USFWS management plan actually wrote in a system to encourage aquaculture farmers to report how many cormorants they killed each winter. The USDA staff is mandated to shoot no more than 25 percent of the total number reported "removed" by the farmers.[39] A cynical way to look at this is that the more cormorants the farmers report shooting, the more the USDA can, too: a sort of matching-grant motivation plan. Yet the catfish farmers I spoke with did not seem to report the numbers of birds they killed.

Regardless, after the collection and analysis of the leg band data, Brian Dorr and his colleagues are confident enough to confirm what most everyone suspected: "The expansion of the aquaculture industry after 1985 has contributed to changes in cormorant migration and wintering patterns and has concentrated wintering populations in primary aquaculture areas of the southeastern United States."[40] In short: the cormorants have been flocking to the catfish ponds from all over the continent.

Brian Dorr was also a coauthor of another migration study, again thinking big about cormorant flight across North America. Dorr and his colleagues went onto fourteen breeding colonies across the Great Lakes, including with Russ McCullough at Little Galloo Island on Lake Ontario. With padded foothold traps, they captured a total of 145 birds and fitted them each with a little backpack harnesses equipped with a solar-rechargeable GPS transmitter accurate to within sixty feet (18 m) and which transmitted every couple of hours to a central computer.[41]

Dorr and his colleagues found that on the whole the Atlantic population cormorants did use the Atlantic Flyway, and the Interior population birds did use the Mississippi Flyway. But there was more to it. The cormorants seemed to observe a sort of longitudinal boundary somewhere between Lake Erie and Lake Ontario. Those to the east went along the Atlantic Flyway. Those to the west flew the Mississippi route.

The transmitters also revealed huge individual variation. One bird from Oneida Lake, New York, just southeast of Henderson Harbor, flew down the Mississippi Flyway and wintered in Louisiana. Two other cormorants from the Great Lakes did a loop migration, heading south down one flyway and then returning up the other. One cormorant soared right over the Appalachians. Several others migrated westbound over a particularly low spot in this mountain range. Other cormorants spent some time up north several hundred miles from their breeding grounds before eventually heading south. Dorr and his colleagues were not able to find any predictable trend in terms of gender or age or food supply or disturbance.[42]

In 2004, scientists estimated that from 100,000 to 200,000 cormorants were making their way down the Mississippi Flyway each year.[43] There has never been much nesting in the Delta region, and even at the height of the catfish industry in the late 1990s, only three nesting colonies of 133 pairs were known in Arkansas and Mississippi.[44] That number does not seem to have increased.[45] And several people are eager to make certain it does not.

In the American South the cormorant has been known with several different common names, including water turkey, water buzzard, crow-duck, and much worse.[46]

I once received a letter from a man in Broken Bow, Oklahoma. He had read a magazine article I had written about cormorants, so he sent me a photocopy of his old hunting license when he was nineteen. He had grown up in Calhoun, Louisiana, about 150 miles (240 km) to the southwest of

Belzoni, Mississippi. He wrote: "In 1958, nigger was just another word in the confederate dictionary and no one even thought about it. My years in the military with the best NCO I ever knew, a black Msgt. from Raleigh, NC, altered my thinking."[47]

The back of his old hunting license had a section under "Migratory Game Birds" that read:

> OUTLAW BIRDS—May be killed at any time between one-half hour before sunrise and sunset. Cormorants (nigger geese), vultures (buzzards), great horned owls, duck hawks, sharp-shinned and cooper hawks (blue darters), crows, red-winged black birds (rice birds), English sparrows, starlings, and, when destructive to crops, grackles (chocks) and bob-o-links (true rice birds.)[48]

Note that most of these "outlaw birds" are black.

To be fair, I've heard the same racial epithet used to describe cormorants in New England and in the Great Lakes—even recently.

I asked Jim Steeby and most of the fish farmers I met in the Mississippi Delta—including Brady Thompson, the African American man who took me out with him to shoot cormorants—if they thought they might feel differently about the birds if these animals were white, had a fancy song, and had some sort of frilly, pretty plumage. Granted, Dr. Steeby only took me around to see the more professional of the catfish farmers, but to a person they shrugged this off and answered similarly: "It doesn't matter what they look like if they're eating our U.S. farm-raised catfish."

These guys shoot snow geese and wood ducks and all sorts of other lovely animals for fun and food. In this case, in this part of the country— contrary to my expectations—I don't think the cormorant's color has much to do with demonizing this animal. The men in the trucks shoot just as earnestly the white pelicans and great egrets because these birds also feed on the catfish.

The southeastern states of North America are not the only place you will find inland freshwater aquaculture producers who are upset about cormorants eating their fish.

The great cormorant species is often involved. In aquaculture ponds in Germany, the Czech Republic, France, and other countries, fish such as carp, perch, and pike are raised in open ponds. As discussed earlier in

Tring, these ponds have provided a food supply to aid previously ailing great cormorant populations throughout Europe.[49] In Japan, wild great cormorants eat ayu and other desired species from inland waters, competing with commercial fisherman.[50] In Australia, great cormorants eat farm-raised silver perch and freshwater crayfish.[51]

Fish farmers in Israel, as another example, have been especially upset with pygmy cormorants — those micro cormorants I couldn't find in the drawers in Tring. This species, "Lord" Nelson writes, "has suffered more than most from man's activities."[52] Once common in Israel, the pygmy cormorant was eradicated by the 1960s. The pygmy cormorants then had a small revival in one part of the country in the 1990s, largely owing to aquaculture farmers who raise fish such as blue tilapia, carp, and bass.[53] In 1999, a year after the USFWS provided an open depredation order to catfish farmers and Ron Ditch and his crew attacked Little Galloo Island, an Israeli man in a tractor rolled over an entire cormorant nesting colony in the Bet She'an Valley.[54] The government responded by beginning a dissuasion program with cannons and pyrotechnics. The cormorants have since moved from this contested region, though more likely because of at least two unattributed fires that destroyed nesting sites. One in 2002 burned an entire nature reserve.[55]

In 2005, an Israeli fish farmer working with the Israel Nature and Parks Authority to help legally scare away the birds told a reporter: "People don't understand the long, hard work of fish farmers. They are people who love nature, but they also need to protect their interests. People think only of the birds, and they forget that fish are not grown in refrigerators."[56]

Just as it is a challenge to put a dollar value on a salmon smolt eaten by a cormorant on the Columbia River, it is difficult to quantify the economic impact of cormorants on the catfish industry in the Delta. One owner of a medium-size catfish operation to the north of Belzoni told me that the operations to kill and scatter birds — including labor, truck maintenance, fuel, ammunition, and any other hazing or deterrent strategies — cost him from $20,000 to $30,000 for the year.[57] This does not include the replacement cost of the fish and the loss of the investment of feed, time, and labor for raising these fish. Some farmers think cormorants injure the catfish they don't manage to eat, and that the birds disturb the feeding patterns of the schools in the pond.[58] Studies on the cost of cormorant predation to the entire industry in the Delta have ranged from $2 million to up to $25 million.[59] Brian Dorr led the most recent study, which totaled the annual economic loss of the region's raised fish to the cormorants at between $5.6 and $12 million.[60]

In terms of total number of fish actually eaten, one research model calculated that in the late 1980s cormorants ate about 4 percent of all the available fingerlings in the Mississippi Delta, as many as eighteen to twenty million little catfish in a winter season.[61] A 2002 study, which Brian Dorr coauthored, showed that cormorants in captivity, at specially designed research ponds, consumed between 1.14 and 1.34 pounds (5.16–6.08 kg) of catfish per day. This is the equivalent of ten seven-inch (18 cm) catfish fingerlings.[62]

In the Delta around Belzoni, the catfish farmers are pretty much out of ideas on how to stop the cormorant predation on their ponds. The farmers seem resigned to the white trucks zipping around in the winter months. The smart men string lines across their fingerling ponds and focus their deterrence strategies there, since the cormorants prefer these smaller fish. They know to double up the trucks around March, as the birds stock up on calories for their flight north.

A facet of all this worth considering is that these ponds, which have been in place for years, if not decades, in certain areas, are their own little ecosystems. The ponds do not just support catfish but also a range of turtles, amphibians, and insects. Other birds feed on the fish in the ponds, yet with a much smaller effect, including herons, egrets, pelicans, gulls, and kingfishers. Still other birds, such as ducks and coots, eat solely algae and small invertebrates in the ponds. White pelicans have been identified as a link in a complex trematode parasite cycle that infects the catfish and can kill their weaker young.[63]

Other fish species grow in the ponds, too—most notably the gizzard shad (*Dorosoma cepedianum*). The shad are transported from pond to pond by the occasional feeding bird, but more often by the farmers themselves during the regular fish transfers that move the catfish around from pond to pond. This shad is a smaller fish that does not have a food market. Farmers actually like them because shad keep the ponds cleaner and can provide a prey item for large catfish that might otherwise eat some of their own. Shad thrive in these ponds. Catfish fishermen agree that the cormorants might even prefer to eat these "trash" species to the catfish.

Studies from stomachs of cormorants hanging around the aquaculture ponds actually revealed that the birds' diet was only just over half catfish, with gizzard shad making up most of their remaining food. Cormorants roosting closer to the Mississippi River ate even fewer catfish as a percentage of their diet.[64] Thus there have even been proposals to fit out ponds *stocked* with gizzard shad or other "buffer prey" so that the cormorants would just chow down on these. However, most fish farmers shiver at the idea of paying for anything that might attract more cormorants to their ponds.

There are other strategies out there that men like Chris Nerrin have yet to try, such as floating ropes, underwater wire grids, and water dyes. Some of these are being researched on aquaculture ponds in the Netherlands and Germany. But this would all require more investment toward an unknown result. The Delta fish farmers could cut down the trees where the cormorants roost or organize their own roost dispersal teams, but, again, the perception is that this is just more hassle than it's worth. Jim Steeby and a few of the farmers approve of the government programs that oil eggs and shoot adults in the northern breeding colonies, such as at Little Galloo Island on Lake Ontario. The men I met in Belzoni were knowledgeable about cormorant migration and the broader story about pesticides and other factors that helped double-crested cormorants recover across the continent in recent decades. Though Steeby did not know him by name, he is grateful for the likes of Ron Ditch for helping to publicize the issue in the Great Lakes. Steeby thinks that once the people in the North got upset, this brought more help for the catfish industry in the South.

By the time of my visit, catfish farmers in the southeastern states had worked with a legal blank check to kill an unlimited number of cormorants on their ponds for almost fifteen years. Chris Nerrin thinks the cormorant problem on his ponds is getting worse, but his was not necessarily a representative opinion. Older men like William Tackett remember when the birds were far more populous.

Jim Steeby told me: "Each year you kept thinking it couldn't get any worse. But then there would be even more cormorants the following winter. I tell you what: they just kept coming."

Steeby thinks it is better now. But the birds are moving around. As the number of farms diminishes, one man might see a decrease while another sees more cormorants on his ponds than in previous years. Few if any of

192   THE DEVIL'S CORMORANT

the catfish farmers think shooting the cormorants does any real good on a larger scale. One study published in 2000 surveyed the region's farmers, who said they killed a total of 9,557 cormorants over the course of a winter under the new unlimited permit. (Fewer than half the farmers responded to this survey.) At least initially, bird counts showed that the increased culling had "no apparent impacts on wintering populations."[65]

On a farm to the north of Belzoni, Brady Thompson takes me out in a white truck to run cormorants. Before I get in the cab, Thompson says that he just saw a massive flock of cormorants flying off one of the ponds. "Shame you weren't here five minutes ago. I just saw for sure a couple thousand. It was just black up there with them."[66]

Thompson has been doing this for ten years. During the winter months, particularly January through March, but sometimes as early as September or October, he spends hours each day — sometimes the entire day — driving around in a truck shooting cormorants and trying to scare them off.

"Oh, they're smart. They're smart," Thompson says. "Not as dumb as you think they are. They recognize my truck. Especially when it's clean. I do a lot better when the truck is dirty."

We drive around with his twelve-gauge in the front seat, barrel down toward the floor mat at my feet. After a few turns around a couple of ponds, I've already lost my bearings. I can't tell one pond from another. While driving down the narrow tractor path between the ponds, approaching a pond with birds on the surface, he picks up the gun and swings it — "excuse me," he says — and lays it across his lap. He leans the gun out the window over his side mirror. He tells me how the herons have learned to dive and fish in the middle of the ponds like the gulls.

Thompson spits some chew into a cup. The cormorant juveniles are easier to shoot, he says. They haven't learned yet. Usually when he sees a cormorant come up on the surface with a fish in its mouth, it's actually a gizzard shad.

"They love them shad."

Thompson says the older cormorants will see him coming. If the birds are on one of the larger ponds, they'll just swim toward the middle, knowing they are safely out of range. Over the years, he has also realized that the flocks of cormorants have figured out how to fly behind his truck rather then in front, where he can get a clear shot.

On a day that he is running cormorants for eight hours, Thompson says his shoulder will get sore. If he runs out of shells, sometimes just honking the horn or spinning his wheels on the gravel are enough to get them up. On some days, Thompson will put forty or fifty miles on the truck.

"I actually don't mind this job. Whatever I do I try to be good at it. I'm a better shot now. I'm out here on my own. Listening to the radio. It's not bad."

Thompson says he can shoot maybe a dozen birds on a good day.

Steeby has told me that the idea is simply to deprive the cormorants of an opportunity to sit on one pond and eat all day long. Maybe you can scare them and hopefully make them want to move over to some other guy's farm or ideally away from the ponds altogether and toward the Mississippi River. That does not mean the farmers do not want their man to be a good shot, though. It's just hard to find that labor. I don't mention to Thompson the man I had been hearing about, the one who regularly shot over one hundred cormorants in a day. He is a sniper in the U.S. Army, and he worked on the fish farms between tours in Afghanistan. A couple of managers of different farms, including that of William Tackett, had hired this man to train other guys to shoot the birds more effectively.

But Thompson shoots accurately when I'm with him. On his first real attempt he slows down beside a pond with some fifty or so cormorants and dozens of various ducks, mostly shovelers. He coasts to a stop with his window already down. The shotgun rests on the side mirror.

The birds have to be much closer than I realized for him to have a chance. Thompson flicks off the safety. The cormorants get up. He lifts his gun. Traces them. Fires one shot. Over Thompson's shoulder I see one cormorant fold its wings and drop. It falls more slowly than I expect. The cormorant looks dead in the water at first, but then it lifts its head up.

"I better go finish him off," Thompson says, raising the barrel.

Then the bird begins to push its wings forward. This tells Thompson that the bird is going to die momentarily. It does.

We get out of the truck. The cormorant floats over to our side of the pond. Thompson scoops it up with a shovel. (See figure 17.) I can't find the wound. The cormorant's eyes are closed. It's a juvenile. No leg band.

"I just leave it here for the chicken hawks," Thompson says. "It will be gone by tomorrow."

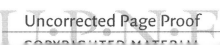

We get back in the truck.

"That wasn't even that great a shot," Thompson adds as we drive off toward another pond. "Sometimes I can shoot a cormorant three times before it hits the water."

December

The water around this key has a red-brown tinge. A bloom of poisonous plankton has left a line of dead crabs and fish on this stretch of Florida beach. Among the dead animals are redfish, lookdown fish, barracuda, and dozens of little sardines. Several large groupers have also washed up dead, as have dozens of small eels, now brittle and curled up in the wrackline.

A male from the first clutch of cormorants from Gates Island, Connecticut, staggers on the sand. It is blazing hot. There is no wind. He is dehydrated and trying to walk toward the shade, but to his confusion he is unable to walk along the unevenness of the beach. He recently ate a few pinfish that themselves had consumed many of the toxic dinoflagellates. The toxins have since accumulated in this cormorant's intestines. He drags his wings on the sand.

The bird eventually makes it up under a copse of scrub oak, but remains disoriented. He stumbles over a few sun-faded cans, then gets his feet tangled up in a ball of plastic fishing line. The cormorant sits down in all this and opens his mouth, fluttering his orange gular pouch.

Two turkey vultures that are tearing at a grouper on the beach now look over at the cormorant.

Isla Chincha Centro

PERU

Figuring in dollars and cents, and with reference
to effect on human life and human geography, I beg to
present my candidate for the post of king among avian
benefactors — the Peruvian cormorant or *guanay*.

ROBERT CUSHMAN MURPHY,
Bird Islands of Peru, 1925

The soft-spoken *guardias*, Hermilio Ipurre Choccña and Claudio Mallqui Padilla, live nearly year-round on the island. The creases around their eyes are perhaps from smiling — which they do often — or from age, or from squinting against the steady wind that blows the dust every afternoon. The two men live in a cement home by the dock. Painted on the side of the house is a huge cartoon of a cormorant and the word PROABONOS. Underneath it says: "Bienvenidos a Isla Chincha Centro." Then "Mejorando tu consecha," which means "Bettering your crop."

Choccña and Padilla are employed by the government company PROABONOS. The acronym translates to Special Project for the Promotion and Good Use of Fertilizer Supply from Marine Birds.[1] Their job on the middle of the three Chinchas is to protect their island from trespassers. The most common problem the men handle are fishermen who motor too close to the island or even try to land to shoot the birds to sell for food on the black-market docks of Callao. Living most of the year alone together on the island, Choccña and Padilla have no phone, e-mail, or television. They have no electricity. In the mornings before the wind it can get blistery hot. Year-round, it almost never rains. A bit of fog sometimes, and that's it. A boat from the mainland periodically brings water to pump into a concrete cistern.

197

This January morning, as they do each day before sunrise, Choccña and Padilla ease out of bed and pad toward the end of the dock. They had invited me to meet them there. We whisper as the sky begins to lighten. My Spanish is weak, so our communication is rough. They try to predict when exactly the cormorants will begin their departure. Both of these men once worked as miners on the guano islands, as did most of their friends and relatives. It had been brutal, back-bending work, but it paid three times as much as any job they could get at home. Years ago they were able to land this steady post on the island, a job especially coveted. If it were not for the isolation and living conditions, a guardia post is a sinecure. And it also carries a serious amount of responsibility. The two men sleep near their rifles, but they usually do not carry them around.[2]

Recently Padilla has been more successful predicting when exactly the birds will leave. He chides Choccña. They laugh quietly. On a beach around the bend, terns and gulls chirp and bark. An oystercatcher with its high-pitched shrill flies low across the sound between Centro and Sur, but it is too dark to see its long, crimson beak. Choccña asks me if I know where Elisa Goya and her assistant want to go on the island today. I say I don't. The men discuss where it will be best for the biólogos and me to go exactly. The guardias say it is toward the end of the peak of the local breeding on the Chinchas — tens of thousands of cormorant eggs have recently hatched — so we need to be especially vigilant.

Choccña and Padilla look over at the cliff to the east. Still no sign of any cormorants waddling out to the edge. Simultaneously the two men press the glow button on their watches to read the time.

Over the distant Andes the sun begins to rise. I can see that Chincha Centro, like Chincha Sur and Chincha Norte, is bare of vegetation. The islands are completely bald. A shin-high layer of light brown soil covers gray volcanic bedrock. Each island is less than one mile (1.6 km) in diameter; together they appear like three loaves of bread rising out of the ocean. The cliff edges are steep, pocked, and undercut with caverns and caves in which the howls and bellows of sea lions echo.

On Chincha Centro, the only bright colors that are on the island, and the only ones that catch the first sunlight, are on a house just up from where the guardias live. This is an aqua, maroon, and yellow building that has Victorian hints in its architecture because of a full, wide porch and wood-scroll details on the supports. It is the nicest of the shelters for the laborers who come for several weeks every seven years or so. The two scientists and I are staying there. We politely ignore the pornographic photographs pinned up in the bedrooms. Beside this brightly painted house with the porch is a neat row of over twenty anchors and a narrow railroad track that leads out to the wharf on which Choccña, Padilla, and I now watch for the birds' departure.

These three Chincha Islands were once the hub of a nineteenth-century boom in merchant shipping traffic — connecting clipper ships, ships of exploration, merchant ships, slave ships, and warships. Where we stand was once a focus of military tension and international political maneuvering. The Chinchas became a fatal and brutal destination for imported indentured servants, primarily from China. And then for the first half of the twentieth century, owing in part to the creation of this network of guardias, the islands were one of the most successful models of wildlife resource management anywhere in the world — a program benefiting both the economic good of a new nation and the conservation of the marine environment. The cause of all this activity on these three tiny islands was the European and American response to the discovery of massive deposits of an ignoble savior for starving soil: guano. The primary producer of this brown, dusty gold was the guanay cormorant (*Phalacrocorax bougainvillii*).

Guano is accumulated bird poop. Its smell is somewhere between ammonia, old fish, brass polish, and rotten eggs. Once, before I had the chance to visit the Chinchas, I was in a rare-artifacts research room in a maritime museum. I wanted to smell seabird guano in a historic container. No one was watching me. I unscrewed the cap. It looked like a pint of pale dirt. A few rocks and even a few tiny feathers were mixed in. In retrospect I don't know what I was thinking, but I just stuck my nose in there and breathed in as if I were inhaling the forest on the first morning of a camping trip. The back of my eyes went black, then red, and I nearly passed out. I had a headache for the rest of the day.

The Peru, or Humboldt, Current tumbles northward from the frigid Southern Ocean along the long west coast of South America. The consistent southerly winds along the coast create this cold current and an upwelling system, forcing nutrient-rich waters from the ocean bottom up to the surface, where they mix with waters lit by the sun.

The most common fish of the Humboldt Current are the *anchovetas,* the Peruvian anchovies, and the *sardinas,* the Peruvian Pacific sardines. Both swim at their highest concentration by central and northern Peru.[3] Stupendous schools of these glittering fish slalom along with the current around the Chinchas and the forty or so other guano islands.

Three major guano-producing species nest here and prey aggressively on the fish of the current. The cormorants have historically been by far the dominant bird.[4] The second, the Peruvian booby, is a white seabird that looks similar to the better-known blue-footed variety of the Galápagos. The third guano bird is the Peruvian pelican, a larger cousin of the brown pelican that roosts on East Sand Island in the Columbia River.

The guanay cormorants, known to Choccña and Padilla as simply *guanay* (pronounced gua-NIGH), are mostly black like the cormorants of North America and Japan, but they are more slender and have pink feet and a large oval red patch around the eyes, like a red costume mask. Most conspicuously they have a white breast, both as adults and as juveniles.

In his major reference work on cormorants and their relatives, Bryan "Lord" Nelson suddenly breaks from his normally sober collection of seabird data to write of the guanay: "a more gregarious bird is beyond imagination."[5] Guanay cormorants can live by the hundreds of thousands in one single colony, making the cormorants on Little Galloo Island look like a monastery of lonesome hermits. (See figure 18.) Guanays will cram three or four nests — that's two adults and a few chicks each — into one square yard (0.8 sq m).[6]

In the early twentieth century, Chincha Centro held over one million nests. In 1913 the Scottish naturalist H. O. Forbes calculated this to total over five million adult and juveniles birds all living at one time on this tiny island.[7] According to Forbes, the colony on Centro could eat a thousand tons of fish (907 metric tonnes) per day, which current research suggests might actually even be an underestimate, assuming Forbes's bird count was accurate.[8]

Robert Cushman Murphy, the first major ornithologist of Southern Hemisphere seabirds — the same guy who referenced Keats when observing the blue-eyed shags at South Georgia as a young man — visited the Chinchas and the nearby islands just a few years after Forbes. He described standing on that Victorian-style porch on Chincha Centro. I like

One line short

THE DEVIL'S CORMORANT

to think I slept in the same room as he did. I suspect the wall decorations were different. Murphy observed the cormorant densities this way:

> On such an ocean dark flocks of Guanayes form rafts that can be spied miles away. Slowly the dense masses of birds press along the sea, gobbling up fish in their path, the hinder margins of the rafts continually rising into the air and pouring over the van [the forwardmost group] in some such manner as the great flocks of passenger pigeons are said to have once rolled through open North American forests. ... At other times, when the Guanayes are moving toward distant feeding grounds, they travel not in broad flocks but rather as a solid river of birds, which streams in a sharply-marked unbroken column, close above the waves, until an amazed observer is actually wearied as a single formation takes four or five hours to pass a given point. Equally impressive are the homeward flights of these cormorants, after a day of gorging upon anchovies, when in late afternoon slender ribbons, wedges, and whiplashes of Guanayes in single file twist and flutter, high in air, toward the rounded plateaus of white islands which gradually turn black as the packed areas of birds swell out from clustered nuclei toward the borders of the available standing room.[9]

John Milton would have approved of Pablo Neruda's verse about the cormorant, written in 1966, almost exactly three centuries after his *Paradise Lost*:

GUANAY CORMORANT
Phalacrocorax bougainvillii

Crucified on the rock,
the motionless black-coated cross
stubbornly posed in twisted profile.
The sun fell on the coastal stones
like a galloping horse:
its shoes unleashed
a million furious sparks,
a million seadrops,
and the crucified steering wheel
did not blink on the cross:

the surf swelled up and gave birth:
the stone trembled in delivery:
the foam whispered softly:
and there, like a hanged Negro,
the cormorant remained dead,
the cormorant remained alive,
remained alive and dead and cross
with its stiff black wings
opened above the water:
remained like a cruel gaffing hook
plunged into the rock's salt
and from so many angry blows,
from so much green and fire and fury,
from the forces gathered
along the howling seacoast
it looked like a menace:
it was the cross and the gallows:
the night nailed to the cross,
the agony of darkness: but
suddenly it fled to the sky,
flew like a black arrow,
and climbed in cyclical form
with its snowy black suit,
with a star's or ship's repose,
And above the unruly ocean —
a gnashing of sea and cold —
it flew flew flew flew
its pure equation in space.[10]

The word "guano" derives from the South American Spanish word *huano*, which the first Spanish adopted from the native Quichua word *huanu*, which meant dung.[11] The Moche people of Peru mined the offshore islands for guano more than fifteen hundred years ago. The Moche discovered its attributes as a fertilizer and used it to help raise crops in their arid soil.[12]

Later the Incas got their hands dirty with guano, too. Sources, derived mostly from early Spanish accounts, claim that the Incas allocated islands to specific territories and groups of people.[13] The Incas believed guano was equal to gold as a gift to appease the deities. There was even an ancient Peruvian god of guano, named Huamancantac.[14] The Incas supposedly outlawed the disturbance of the seabirds' nesting grounds during breeding, posting their own guards on each island and punishing an unsolicited visit with death.[15] The Incas reportedly carried the pungent guano, hefted by long caravans of llamas, as high as two or three miles (3–5 km) above sea level into the Andes and some sixty miles (100 km) from the coast. They terraced hillsides and irrigated with canals and ditches to grow corn and potatoes.[16]

Seabirds and the use of guano appear in the indigenous pottery of mainland Peru. Images of cormorants are in early textiles and other art forms — perhaps even in one of the famous huge Nasca geoglyphs that might date as far back as 300 BCE.[17] Peruvian oral traditions and myths contain seabirds, such as the Moche legend of how the god Quismique trapped guanay cormorants. Quismique forced the *aves gauneras* to help him enter the sea.[18] (See figure 19.)

A Peruvian proverb translates: "Huano, though no saint, works miracles."[19]

Englishman George Shelvocke in 1726 left one of the earlier descriptions of European guano extraction along the coast. He described Iquique Island, which is well to the south of Isla Chincha Centro and of similar size:

> The Island is of moderate height, and the whole body of it consists of *Cormorant's* dung (a kind of sea bird very numerous on this coast.) Some will have it be a particular sort of earth; but the most probable and certain conjecture is, that it is the dung of birds; 'tis not in this place only that one sees large quantities of it; but also all along the coast of *Peru*, there are lofty precipices, and large rocks near the sea, cased over with it, which at a distance makes them appear like chalk-cliffs. That there should be a greater abundance of it here than on any other part of the coast, may be accounted for by observations made by the *Spaniards*, who agree, that these birds are more numerous in and about the latitude of this place than elsewhere; and to confirm the truth of it, they farther report, that after having dug to a considerable depth, they have found birds' feathers. As to a nice enquiry into this, our affairs would not permit it; all that I can affirm of it is, that the smell of it is very offensive.[20]

Shelvocke goes on to explain that the Spanish built a small town near the island and used the guano they dug up to fertilize their local plantations of peppers, worked by African slaves.

Although clearly the Spanish knew of the fertilizing qualities of guano, as did the Dutch from their colonial use of guano from the islands of South Africa, the German naturalist and diplomat Alexander von Humboldt is normally credited with bringing the fertilizing qualities of Peruvian guano to the larger attention of Europe. He and his botanist companion Aimé Bonpland arrived in Lima in 1802. They marveled at the deposits of bird droppings on the island of Mazorca. Humboldt supposedly caught a cormorant and measured its daily excrement: five fluid ounces (148 ml). He brought samples of the guano soil back to France and Germany.[21] At the time chemists were first beginning to understand how plants derived specific nutrients from the ground. Farmers throughout Europe and in the United States had recorded drops in soil productivity due to long usage, monocultures, and increasing human populations. To understand how important this was, consider that in the mid-nineteenth century approximately 80 percent of Americans lived on farms, and about 75 percent of the U.S. economy was involved in agriculture.[22] Farmers across Europe and then in the United States began to test Peruvian guano on small plots. It turned out to be much more effective than barnyard manure.

By 1825 Peru had won its independence from Spain, just as guano was gaining favor. To promote trade, the new government exempted this commodity from taxation and sent more samples to Europe. Then during the months of 1841, twenty-three ships — each absolutely reeking of cormorant guano from the Chinchas — sailed to the docks of England, most tying up in Liverpool. The ships delivered in their holds a total of over sixty-three hundred tons of dried cormorant feces from the Chinchas.[23]

The guano rush was on.

Chemists and farmers identified the deposits from the Chinchas as the best of the best. These islands had an especially large population of cormorants, who, because of their diet, numbers, and physiology, produced the greatest volume of guano of the finest quality. The Chinchas' accumulated feces, mixed with the remains of dead birds and feathers and all decomposed together over time, had a particular balance of nitrates, phosphates, and other nutrients. Rain and storms dilute and wash seabird droppings off rookeries in other parts of the world, but because there is almost no precipitation whatsoever on the Peruvian islands, the guano had piled and accumulated and piled higher still over the centuries. Guano fertilizer from the Chincha Islands represented the absolute premium product.[24] I like to imagine a lab-coated French chemist completing an analysis of a sample of Chincha guano. He would lean back after triple-checking the percentages of nitrates and phosphates, then theatrically kiss his thumb and ring finger: "*Oui, oui! Parfait!*"

The guano trade boomed from 1841 to the late 1870s. One historian estimated that Peru exported some 12.7 million tons of guano during this period, worth a sale value, figured in 2005 dollars, of $13 billion.[25] In 1857 roughly three-quarters of the entire income of Peru's government was from guano.[26] And the Chincha Islands was not just the hub because of the quality of the fertilizer. They were three accessible islands close together, and a short trip to Callao, the seaport of Lima. In the whirl of the guano rush, Chincha Norte had a town of between one thousand and three thousand residents, including a church, a graveyard, dogs, donkeys, and a three-story Chincha Hotel. The British consul, the Peruvian governor, and a few exporting companies all built offices on the island.

Though the United States did not purchase as much product from Peru as did European buyers, American farmers also sought guano fertilizer and pushed their legislators to make it affordable and available. I like to presume President Fillmore was thinking about cormorants when he declared in his first State of the Union address: "Peruvian guano has become so desirable an article to the agricultural interests in the United States that it is the duty of the Government to employ all means properly in its power for the purpose of causing the article to be imported into the country at a reasonable price."[27] Unable to lower prices, Congress in 1856 passed the Guano Island Act. This allowed mariners and entrepreneurs who were American citizens to claim uninhabited islands in the name of the United States in order to profit from guano. Among mariners trying to make money or plant the Stars and Stripes there was much confusion and even outright fraud in trying to match the ideal chemistry of that on the Peruvian islands. They shoveled and bagged inferior droppings of albatross, bats, and even turtles. One Captain Julius de Brosset from Louisiana took samples from the Chinchas and said they were from the Galápagos, a hoax that roped in a senator, a Harvard professor, and the U.S. secretary of state, who wanted to buy or lease the island in the Galápagos that supposedly had a huge deposit of high-grade guano.[28]

As a commodity, guano was worth the trip on its own, but the Chinchas also fit conveniently into the merchant routes of the more romantic California gold rush and its era of the clipper ship. These fast luxury ships often stopped to load guano on the return journey from California, sailing back to the U.S. East Coast or to Europe. Guano provided an alternative to a trans-Pacific return passage for Chinese tea. In 1853 a correspondent for the *New York Times*, who described the Chinchas as "hell on earth" and the birds as "too numerous to mention . . . like some great army," wrote about these clipper ships: "With their yards cockbilled, and rolling their royal masts almost against the face of the rock, all covered with guano, you would hardly recognize some of the finest clippers, that before they left New York or Boston were praised in the papers, visited by ladies, and, instead of guano, had their cabins perfumed by champagne."[29]

Nelson Crowell, an officer aboard the ship *John Q. Adams*, has left one of the best records of the guano trade, complete with some of his own illustrations. Crowell wrote in his journal that over 150 ships had dropped anchor off the Chinchas in the early months of 1853.[30] His ship was returning from San Francisco.

"In four days arrived at Chincha and immediately commenced discharging ballast and getting ready to take Guano," Crowell wrote. "The islands are not large but are quite high being about two hundred and fifty feet above the level of the sea. They are destitute of vegetation and water and covered with guano which in some places is One hundred & fifty feet deep."[31]

Crowell described the channels between the Chinchas as "bold" right up to steep cliffs, so vessels anchored and moored, when it was their turn, as close as possible for easier loading. While they were waiting, Crowell watched schools of whales, sharks, sea lions, and porpoises.

"Fish are very plentiful here," he wrote. "Often herring and sardines are bailed up with baskets and buckets and even with the hands."[32]

Captains' wives came to the Chinchas, too. In a later voyage aboard the *Scargo*, Crowell wrote in his journal that there were about thirty wives among the two hundred ships at anchor.[33] Less than a month later, on September 6, 1856, he wrote the following extraordinary entry about the wives aboard the vessels: "*Three* of their number have in one day given birth to *four* babies and I hope I shall not get intoxicated drinking their healths, God bless them."[34]

The Mystic-built clipper ship *B. F. Hoxie* anchored off the Chinchas in 1857 to load guano—a couple of years after sailing downriver and past Gates Island. One crew member aboard, Simeon G. Fish, wrote of the "millions of fish swimming about" and how he and his fellow sailors "had a fine time catching birds."[35]

Fish wrote, in another entry on a Saturday in April: "Pleasant as usual. Started off fishing again. Rather dull concluded we would explore South [Chincha] Island, the top of this island is covered with bones of birds fish and animals which the birds have carried there, the surface is perforated with holes I suppose the nests of birds. Lizards are also numerous. The rocks on which these islands rest are in some places very high one look over the edge, at the surf dashing below was enough to conclude we went back to fishing. Caught seven and started home."[36]

In May, Fish wrote in his journal that the captain of another ship had died and was preserved in the guano.[37] He then witnessed about a week later several ships coming to anchor, including the famous clipper *Great Republic*.[38] This was a massive ship, drawing twenty-five feet (7.6 m).[39] It was on the return home from its first and record-breaking trip to San Francisco. At the Chinchas, the *Great Republic* loaded forty-five hundred tons of guano. Later in its voyage, to the south of Cape Horn, massive seas smashed through the planking and the deck, and seawater poured into the hold. The water mixed with the guano, which sloshed around the hold and into the food stores, forcing an emergency stop for provisions.[40]

The *Great Republic* was not the only vessel that suffered from carrying dried cormorant feces. Mariners complained that the dust ruined the metal fasteners, cables, and anchors, and at least one physician thought a few shipwrecks might have been caused by the effect of the toxic smell on the sailors' health.[41]

Other accounts bemoan how the guano gases actually combusted spontaneously en route or were ignited by a sailor with an open-flame lantern. The word "shit" itself, some say, derives from this trade: when guano was sent aboard in sacks, they were stamped with the acronym S.H.I.T.—Store Higher in Transit—so the fumes could be ventilated better, and also so the bags would not expand if they sloshed in any bilgewater.

Farmers received the bags, and since they generally thought of guano as merely high-grade manure, the word spread.

This etymology is not true, of course. The word traces back much further, but perhaps you should still pass it around because it is good pub humor. I'm not lying about combustion on board, though. One event was reported in the journal the *Chemist* in 1845. On the return home to England from a South African guano island, the bark *Ann* grounded on a sandbank with a full cargo. Seawater splashed into the hold. Smoke billowed out. All hands quickly leaped off the ship. The author wrote: "And scarcely had they done so, when a tremendous explosion of the gas engendered by the partially fired guano, blew the stern out of the vessel, which then filled, and sunk in deep water."[42]

Regardless of the dangers of transport, nineteenth-century merchants were willing to risk the extremes of weather and a troublesome cargo. When these merchants heard "guano," they saw dollar signs, imagining the advertisement posters with the cartoon cormorants standing on burlap bags.[43]

The guanay cormorants of the Peruvian coast are different from most others in their family. The guanay have longer, thinner wings than most cormorants, enabling them to fly farther against stronger winds and to soar a bit more to save energy. Their wing and body shape inclines ever so slightly toward the shape of a shearwater or an albatross.

Guanay are also rare among cormorants in that they, as Robert Cushman Murphy observed, can "hawk" for their food. Murphy wrote that the guanay seem to find fish primarily from the air, by sight, and though they still do not regularly plummet dive as do the pelicans, boobies, and terns, their primary prey is the same: the anchovetas and sardinas that swim close to the surface.[44] Murphy wrote that a few birds head out first from the islands to act as "scouting parties" for the rivers of cormorants that will follow behind. It is "a system of efficient cooperation which almost suggests certain customs of ants or social insects."[45]

Though November to January is generally the most active time, guanay cormorants breed during all seasons, sometimes even laying two clutches of eggs per year.[46] On the dense cormorant colonies on the Chinchas and Ballestas, you can see at the same time nearly every behavior known in cormorants. During my few days on the island, I observed all at once the mutual preening, "necking," and mating gestures — as well as the behavior of chicks, fledglings, juveniles, and adults.

Another behavior I witnessed on Chincha Centro is the "running of the gantlet," as Murphy calls it: one bird runs through the crowd and is pecked and generally abused. Sometimes this happens as a bird lands into a large group. Murphy described it this way:

> Every now and then . . . some unfortunate Guanay, which seems to be the butt of all bystanders, will go dashing through the thong, holding its head as high as possible in order to avoid the jabs and bites which all others direct at it. If the victim would but stop fleeing, perhaps the blows would cease, but it keeps more and more desperately running the gantlet, flapping its wings, bumping into innumerable neighbors, until eventually it bursts from the vicious crowd into a clear space, shakes itself with an abused air, and opens and shuts its mouth many times with an expression of having just swallowed an unpleasant dose.[47]

Murphy surprisingly did not describe another behavior that I observed: stealing. Because of the density of the colony and the regular preening, guanay often walk around with guano splatter on their backs and with downy feathers stuck to their beaks or heads. Since there is little plant life in the area, guanay nests are mostly little craters, as the sailor Simeon Fish had suspected. The nests are carved, raised, and packed from older guano, a few sticks and seaweed, and the bones and decayed carcasses of dead birds and fish. As Herman Melville wrote of the value of wood on nineteenth-century Nantucket, the cormorants on Isla Centro seem to carry around sticks "like bits of the true cross in Rome."[48] What happens is that one cormorant will waddle several feet over to another area or perhaps fly a very short distance in order to pick up a stick or a few strings of dried seaweed from the nest or pile of another bird. This appeared to be done surreptitiously. If caught, the act is countered with the jabbing beak of another cormorant. If successful, the bird returns to its little pile, places the prize, and then goes out to pilfer another. This is common practice at cormorant colonies of all species, but I had never seen it at this scale. Because in this Peruvian colony, while one cormorant goes out for another stick, another two birds are moving in on his or her pile. Pan out and envision this behavior all happening at once: thousands of simultaneous, perpetual little burglaries.

Supposedly once the birds pair off and there is a female sitting on the nest, the thieving is less prevalent. But how a nest of anything but a dug crater ever gets built in a colony of kleptomaniacal guanays remains a mystery to me.

The harvesting of tremendous piles of seabird guano on the Chinchas did not take an extraordinary amount of capital or advanced technology. Workers with shovels and pickaxes chopped the guano out of the islands' hills. They chiseled out walking terraces and steep staircases. The Chincha Islands when the foreigners first arrived were somewhere between 100 to 330 feet (30–100 m) taller than they are now.[49] The men loaded the yellow-brown, noxious product into bags or carts. To deliver the guano to ships' holds, workers usually slid it down long canvas chutes to vessels moored at the foot of the cliffs.

Nelson Crowell explained that when his *John Q. Adams* arrived at the Chinchas in 1853, they put all their sails below. The carpenter and the rest of the crew caulked within the hold, trying to seal all crevices belowdecks in order to contain the dust. Crowell described how they worked to keep the guano out of the officers' quarters, battening all the cabin windows and doors with canvas. He explained how they wrapped up their watches, chronometers, barometers, and their personal trunks, but "in the end dust was found everywhere."[50]

When any of the ships loaded at the Chinchas, clouds of guano billowed up from the hatches. Sailors climbed up into the rigging to escape the dust. Crowell described how it blew to leeward a distance of some twenty-five nautical miles.[51] To avoid nosebleeds and loss of consciousness, the trimmers could not stay in the hold for more than minutes at a time. They wore masks of oakum and tar. When Crowell went belowdecks, blood burst out of his nose "in a torrent."[52]

If life was toxic and tedious for the sailors and officers, it was lethal to the island laborers. When the guano rush began in the 1840s, the first workers on the Chinchas were convicts, army deserters, African slaves, and some indigenous Peruvians and Chileans. By the height of the rush, the majority of laborers were indentured servants, "coolies" from China.

The vessels of several nations, including the United States, were involved in bringing these men. A young mariner named Henry Sargent Jr., in a letter from China aboard the American clipper *Winged Racer*, was disturbed by his vessel's new cargo. He wrote: "We are now fitting bunks in our between decks, which are calculated to accommodate one thousand Coolies who are to be transported to Callao where they are sold as slaves for life."[53] Sargent and the rest of the crew crammed 390 Chinese laborers into the *Winged Racer*.[54] He writes of the terrible smell on board, how the Chinese were forced to walk around naked, and how the crew whipped them with the cat-o'-nine-tails for stealing. Upon arriving in Callao, almost 250 of the Chinese had to stay on board the *Winged Racer* for over a month before being shipped to the guano islands.

Many Chinese did not survive the passage across the Pacific. They died of dysentery, scurvy, fever, lack of exercise, and rotten food. One historian calculated that from 1850 to 1865, on average, over one in four men died during the voyage.[55] In 1852, for example, the *Empresa* sailed with 323 indentured servants from China to Callao. The passage lasted 114 days, during which seventy-seven of the men perished.[56] That same year, aboard the *Robert Bowne*, the Chinese detested their unusually cruel and demeaning treatment. According to witnesses, they heard rumors that the ship was not bound for California as promised, but instead sailing for the Chinchas. The men rose up and killed the captain and five more officers and crew.[57]

The tens of thousands of Chinese men who did safely arrive on the Chinchas and the other guano islands discovered the conditions to be as brutal as rumored.[58] The guano dust seeped into their lungs. Their clothes were rags. Their food was sparse and rotten.[59] They ate the birds.[60] The Chinese men labored six days a week for more than twenty hours a day. They earned a pittance and lived in bamboo or reed shacks with grass mat beds and guano dirt for floors. They breathed in the toxic dust as they slept, as they pickaxed and shoveled and carted and bagged and trimmed. Though termed "colonists" or "under contract," they had no legal representation. Some men bought their freedom with saved money, but most of the Chinese workers spent any cash on food. Others spent their meager earnings on gambling, alcohol, and opium.[61]

Overseers and government representatives managed the labor on the Chincha Islands. The Peruvian government stationed soldiers there. Visitors reported harsh discipline at the hands of overseers, who whipped the laborers and punished them by flogging, by suspension in the sun, or by lashing them to a buoy in the sea.

A letter to the *Morning Chronicle* in 1855 explained that the conditions for a guano worker seem "to realise a state of torment which we could hardly have conceived it possible for man to enact against his fellow man."[62]

Most of the Chinese died on the Chinchas. If the workers did not suffer from nosebleeds, temporary blindness, and unconsciousness while trimming cargo, they suffered from other maladies like cramps, diarrhea, and scurvy. The men fell in trenches that caved in. Wounds became infected. Bones and muscles wore out from hard labor. Those who could not work as well as the others had to pick rocks out of the guano on their hands and knees, or they were tied to wheelbarrows to haul.[63] Multiple accounts recount how guano laborers when working shoveled up the bones of their dead countrymen.[64]

Author George W. Peck wrote of his 1853 visit to the Chinchas: "It is, moreover, the worst and most cruel slavery of all forms of slavery that exist among civilized nations. It was universally said to be so by captains who had visited every quarter of the globe; I am sure I never saw anything like it on the plantations."[65]

Peck continued: "Almost every week some of [the Chinese] commit suicide by throwing themselves from the cliff. . . . [The governor of Chincha Centro] told me that more than sixty had killed themselves in this way in the two years he has been there. One was driven over the cliff or jumped off, and was dashed to pieces to escape the lash of a black driver, who chased him to the verge in sight of a captain of an American ship, the week before we left. The cliff where he leaped is two hundred feet high, and almost perpendicular."[66]

Another traveler wrote: "I have heard fifty of the boldest of them joined hands and jumped from the precipice into the sea."[67]

The guano rush fizzled out a few decades after it blazed. Though as late as May 1869 shipmasters recorded 160 large ships with over four thousand men anchored off the Chinchas, the guano resource had largely been depleted, and only the saddest of ships attended the trade. The California gold rush was well past its peak. The clipper ship era and the need for a return cargo faded. The American Civil War stunted overseas commerce and even directly affected the guano fleet. In 1863 the css *Alabama* sank the *Golden Eagle*, a guano ship. Admiral Raphael Semmes wrote: "It seemed a pity, too, to destroy so large a cargo of fertilizer, that would else have made fields stagger under a wealth of grain. But those fields would have been the fields of the enemy; or if it did not fertilize his fields, its sale would pour a stream of gold into his coffers."[68] Labor for the Chinchas and other guano islands also became increasingly scarce because bringing in Chinese workers became too controversial and difficult.

Meanwhile continued desire for profits from cormorant and other fertilizer resources triggered two major military conflicts in the region. In 1864 a Spanish fleet seized the Chinchas, overwhelming the 160 soldiers and the one thousand workers on the island, as a reprisal for what they felt was mistreatment of Spanish emigrants into newly independent Peru.[69] Spanish leaders aimed to seize guano profits as compensation. A war followed, a conflict that brought Bolivia, Ecuador, and Chile in alliance with Peru.

During all of this, as the high-quality guano from the Chinchas and other islands diminished, inland rock nitrates — which have no connection to bird dung — and other competitive alternatives and mixes replaced cormorant guano on the international market. A French manufacturer of a competing fertilizer product claimed: "Guano is like champagne; light, effervescent and short-lived."[70] What rock nitrate deposits Peru had, it soon lost to Chile, now its enemy, in the War of the Pacific, also known as the Nitrate War.

The sale of guano, primarily from cormorants, had been the savior of Peru. Guano was the major force of its new economy, helping establish independence from Spain. But in the end Peru was left after the guano rush with only an exorbitant foreign debt to creditors in London.

Rock nitrates and guano mixtures with fish products continued to take the fertilizer market away from the Peruvian islands, leaving the three small Chinchas strip-mined, denuded of people and birds, and unnaturally quiet. When naturalist Frederic Lucas visited the Chinchas for a few months in 1869, he could not find any seabirds beyond a few pelicans, gulls, and Inca terns. He did not find any birds on the nearby islands, either.[71] The disturbance and destruction of habitat had been too extensive.

During the decades following the nineteenth-century guano rush, harvesters still dug the deposits on other Peruvian islands to fertilize local farms, getting what was thought to be the last of a finite resource. In an effort to conserve the guano, control the trade, and get the fertilizer to Peruvian farmers, Peru's finance minister Augusto Leguía created in 1909 the Guano Administration Company—the precursor to PROABONOS. Leguía hired an American scientist named Robert Coker to help. Coker gave a variety of suggestions. Most important, he recognized that the piles on islands such as the Chinchas actually did not require millennia to build. If there were enough birds they could produce harvestable guano in a matter of years. But the government had to protect these animals—the cormorants in particular.

So the Peruvian government nationalized the islands, unifying the responsibility and rotation of guano extraction. They left the Chinchas and the other guano islands alone for years at a time. They created the position of the guardia. The guardias counted birds and maintained nesting sites. Managers began thinking of the cormorants as a semidomesticated flock, like herds of free-range sheep.[72] They built walls for the purpose of retaining guano and to keep the birds from brushing off their own deposits when they took flight off a cliff. They restricted boat traffic to avoid scaring the birds. When the company did harvest guano on an island, the government hired hundreds of Peruvians from the Andes. On the islands they used some labor savers like bulldozers, railroad track, and gravity-driven conveyor belts, but it remained a low-tech operation. Men still shoveled, sewed up the bags, and hauled them around over their shoulders. Workers could sweep an island clean in a few weeks to months. Then they left it alone for years at a time so the birds could return.[73]

The conservation measures worked. The seabird populations doubled in twenty years to total about eight to ten million birds in the 1930s. The guano built up again, providing plenty of fertilizer for the domestic market.[74] Biologist David Duffy, who has worked on human-cormorant conflicts all over the world, called this early twentieth-century management "one of the first and most effective examples of sustainable exploitation of a natural population by a government."[75]

In the 1940s the Guano Administration Company and its leadership saw still more potential and moved beyond the islands. They constructed mammal-proof fences to allow birds to nest on coastal headlands. Guardias were encouraged to shoot other bird species, such as gulls, that ate cormorant eggs. Guardias shot condors and falcons, predators of adult guano birds.[76] The guanay population grew to four to five times the population at the start of the government action, back up to as many as twenty million birds, primarily cormorants, nesting in Peruvian waters.[77]

THE DEVIL'S CORMORANT

But then came a bizarre wave of events. Overfishing and canning technology crashed the sardine fishery off the coast of California, the fishery of John Steinbeck's 1945 novel *Cannery Row*. Demand, cheap boats, and gear were shipped down to the coast of Peru, where the schools of fish ran even larger and were yet untapped. The anchoveta fishery off Peru became the largest single-species fishery on the planet.[78] The seabirds and the guano islands continued to thrive, but then in a relatively short amount of time the effects of poor management, overfishing, and a powerful El Niño event in 1965 collapsed the fishery. Subsequently, about three-quarters of the guanay population—almost thirteen million cormorants—perished with the drop in fish populations.[79]

Throughout the 1990s and into the twenty-first century, fish and bird populations off the Peruvian coast have crawled back, but are still relatively low. Another steep El Niño event in the mid-1990s cut into the recovery. On the Chincha Islands today and all along the coast of Peru, the seabirds are a fraction of what they once were. Choccña and Padilla are not charged with protecting nearly as many seabirds as witnessed a century ago by Forbes and Murphy. In 2011, there were at most 44,000 cormorant individuals and as many as 576,000 boobies on Chincha Centro, but reproduction for both species was down steeply. The populations fluctuate from year to year, and even within a year. On other nearby islands, such as on the Islas Ballestas, biologists in 2011 counted colonies of about 50,000 *breeding pairs* of cormorants and 19,000 pairs of boobies.[80] A couple of years before, biologists counted on Isla Pescadores almost 200,000 cormorant individuals.[81] Government attention to seabird repopulation is ongoing, however, and unexpectedly the demand for organic, artisanal fertilizer for both home gardens and organic crops has continued to increase in recent years, both domestically and abroad.[82] PROABONOS now manages about twenty islands and continues its rotation of mining. These days it actually cannot keep up with the demand.

Elisa Goya, her assistant, and I prepare to go collect pellets. Although Choccña and Padilla insist on determining all her movements on the island, they both like Goya very much because she is knowledgeable, experienced, and, perhaps most important, she prepares wholesome food for them. She brought fresh fish to make ceviche for lunch today.

After the meal Choccña and Padilla decide that before they can allow

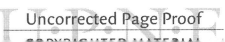

us into the colony, even though it is early afternoon, they want to be certain the area is clear of birds. From the dock they cannot see what is happening at the location where they want the scientists to go, so Choccña rows out in a small wood boat to view the colony from the water to make sure most of the cormorants are still away fishing.

The name of the guardias' boat is *Chuita*. This is the Spanish name for the red-legged cormorant (*P. gaimardi*), a second of three cormorant species that lives along the coast of Peru. The third is the neotropic cormorant (*P. brasilianis*), a widely distributed bird on both sides of the equator in the Western Hemisphere that ranges from about Texas to Buenos Aires. Both the red-legged and the neotropic occur in small numbers off Peru. The neotropic cormorants are mostly black or brown from afar — very similar to the double-crested or even the flightless cormorant. The red-legged, however, is a favorite sighting of the ecotourist boats that motor around the Islas Ballestas, because this is the most brilliantly colored of all the cormorants in the world. The chuita boasts warm gray feathers, a brilliant neon-orange beak, coral-red feet, and a black-striped ring around each of its green eyes. The red-legged cormorants prefer to nest on the steep cliff edges of islands and headlands. No one knows much about what they eat. "Lord" Nelson's research suggests eels, anchovetas, and minute invertebrates. The red-legged cormorant chirps like a songbird. It often makes its nest with worm tubes.[83] This cormorant, like the neotropic, does not occur in large numbers. In other words, it produces no guano deposits. No one has printed its picture on hundreds of thousands of burlap sacks. The early guardias in the area shot the red-legged cormorants for fresh meat.[84]

Rowing in the *Chuita*, Choccña waves to say it is safe to walk into the colony of guanay. Padilla leads Goya and the rest of us up a path. The wind spins up the dust forcefully, as it does most afternoons. Goya looks like a traveler in the Sahara. She wears sunglasses, a bandanna, and a tightly cinched broad hat. She has tucked her socks into her khaki trousers.

I am well prepared for the sun, wind, and dust — I wear a balaclava and sunglasses — but I scoff inwardly about her warning of the ticks. I live close to Lyme, Connecticut. I know about ticks. It turns out that one of the factors that might affect cormorant populations in Peru are the swarms of ticks and lice that feast within the colony. It's a slim possibility even that the very name of these islands is derived from the Spanish word *chinche*, which means bedbug or tick. The gray ticks are known as *garrapatas*, which translates literally to "claw feet."

Within a few minutes of walking in the colony, I learn that the ticks can leap olympically. They find my neck with outstanding speed and stealth. I realize a single Peruvian seabird tick would eat a bunch of American deer ticks like popcorn.

Elisa Goya is based in Lima at a government organization, the Instituto del Mar del Peru (IMARPE), similar to the U.S. Fish and Wildlife Service. Among her other seabird projects, Goya studies the diet of the guanay cormorant as it connects to the fisheries. She is rarely authorized to shoot birds to examine their stomach contents, so we walk out in the colony to search for pellets. For my part I am nearly useless and can't find a single pellet for some reason, even though they look fairly similar to those on Gates Island or on Little Galloo. Fortunately Goya and her assistant collect not only dozens of guanay pellets but also two warm regurgitants recently vomited from startled birds. One contains four whole anchovetas, each longer than a pencil.

For Goya and IMARPE the health of the guanay population is critical on its own, but also knowing which species of fish and the quantity consumed is of genuine national economic interest and an aid to the development and implementation of resource management — especially if her work can help estimate current fish stocks in the Humboldt Current and predict future changes in the schools of anchovetas. On the coast of Peru today both fisheries and guano are economic drivers. The two industries can conflict. Recently Goya and two of her colleagues proposed that the anchoveta fishery formally reduce a certain percentage of their catch to be left for the seabirds. They suggested that during certain parts of the year, fishermen should also observe "temporary closed areas" around the primary breeding islands.[85]

Elisa Goya has also been studying the islands to figure out why in recent decades the cormorant population has dramatically plummeted while the booby population has remained pretty much the same. After the nineteenth-century guano rush, the enacted bird conservation practices were largely successful, so, again, by the middle of the twentieth century there were about 21 million cormorants and about 2 million boobies on the once-decimated islands and coast. But now there are only about 2 million guanays and still 2 million boobies. The biologists knew that cormorants and boobies both prey here on the same species of fish, so they wondered if perhaps this dramatic drop in the cormorant population might have to do with *how* they go about finding fish.

Using GPS tagging and close, systematic observation, Goya and her colleagues, led by a French biologist named Henri Weimerskirch, found that the cormorants do not exactly scout as Murphy suggested, but form a system of what the researchers term "compass rafts." Nearly every cormorant that flies off to forage goes around the island once and then lands within a group of guanays, a raft of birds, sitting in the water a short distance away. Then they go off to feed. The cormorants who have returned from fishing also land in this raft before returning to the island. When these successful cormorants return they fly directly from where the fish were. Thus the compass raft incrementally shifts and changes throughout the day—as much as 180 degrees around the island—as the lines of birds come and go. A group of cormorants leaving the colony to go fish looks to the raft for the direction to fly.[86]

Biologists used to think this rafting behavior of seabirds was merely to clean themselves off after the mess of a colony. The concept of a raft serving as a range marker, a guiding compass for foraging, has ramifications for understanding social behavior, navigation, and intelligence in all sorts of animals.

Goya and her colleagues hypothesize that just maybe the reason the cormorants are not doing as well these days is that the anchoveta populations remain low and more dispersed, reducing the effectiveness of this compass raft method. The Peruvian boobies seem to fish more individually and rely on memory and other methods of transferring information to each other. Even so, there could be a dozen other variables at work to explain why the cormorant population has diminished so steeply. That is in part why Goya and other biologists are still visiting these islands.

I get up the next morning to watch the cormorants depart again off the cliffs of Isla Chincha Centro. Standing at the end of the dock, Hermilio Ipurre Choccña and Claudio Mallqui Padilla look at their watches. A few small groups of guanay leave, and then hundreds, thousands of birds—in waves, in lines—leap off the cliff. Some of the individual cormorants seem to get to the very edge and want to reconsider. Others leap off without a thought. Several free-fall precipitously for half the height of the cliff before

THE DEVIL'S CORMORANT

they take flight and soar forward. Once airborne and flapping they merge into long lines of other cormorants. The formations and "whiplashes" blend and split and scissor across each other. Sometimes during the cormorants' circles of the island on the way to the compass raft, a booby or a pelican will mix in. From Chincha Centro the full departure will take an hour or so — thousands of guanay outward bound to fish.

Later that day I tell Choccña in my broken Spanish that I covet his hat: a chocolate brown, trucker style baseball cap with the word PROABONOS in white under the logo of a cormorant. He is game for a trade, so I lay out what I have — some cash, a set of colored pencils, a fleece jacket, an audiocassette recorder, and a bird identification book in English. He looks over to Padilla, who nods. Choccña takes the bird book.

January

It is after midnight. The moon has yet to rise. The two surviving juveniles from the first clutch on Gates Island sleep beside one another on an unlit navigational buoy that bobs with the sea. The steel edge over which the birds wrap their feet is thin, but they remain balanced as they sleep. The female has bent her neck back to nestle her head under one wing. Behind them headlights whir across the bridge between two Florida keys. A wave washes against the buoy. A natural raft of palm leaves, sea grass, and coconut husks floats past with the current. The male's neck quivers as he dreams.

Cape Town

SOUTH AFRICA

She is the first bird I have cried over.

ELAINE HAYES,
director of Living Coasts, speaking about the
death of the first bank cormorant born in captivity

ola Parsons was born in Zimbabwe but raised in Johannesburg, South Africa. After earning a degree in veterinary medicine, young Dr. Parsons moved to London and worked for a few years in a small animal clinic, spending most of her time neutering dogs and cats. During her free time she backpacked around Europe. She spent a month on one of the islands of Tonga helping with research on fruit bats. She worked for a time on an organic farm in southern Spain. During her travels, she became a vegan, in part because of the ethics of the matter and her larger environmental concerns.

Eventually Parsons grew weary of the weather in England. Her wanderlust mostly satisfied, she decided to move back to South Africa. She wanted to explore projects in conservation biology rather than work in a private veterinary practice. So she found a position at the South African Foundation for the Conservation of Coastal Birds, known officially and colloquially as SANCCOB. This nonprofit organization is in a quiet northern suburb of Cape Town. Its mission is to conserve and protect South Africa's seabirds through education and awareness, training, seabird rehabilitation, and contributions to broader research.

Parsons had first learned about SANCCOB in 2000 because of the oil spill of the bulk ore carrier *Treasure* in Table Bay. The center treated and released an improbable 19,000 African penguins. Even in London, she

221

could not help but hear what an overwhelming yet enormously successful undertaking it all was.[1]

Parsons signed up as a volunteer at SANCCOB when she moved back to South Africa, which soon led to a full-time position as the rehabilitation manager. The job required that she help organize the progression of returning the birds to health. She kept track of all the paperwork and also did some of the initial hands-on care of the birds. She did this while working with a steady stream of eager but largely unskilled local and international volunteers, known as "vollies," who were generally surprised to learn that most of the job is trying to keep the facility clean.

For her first few months at SANCCOB she studied about seabird physiology and acclimated to how everything functioned. She learned the names of the various pools, the holding pens, the logistics and systems, and she picked up a few new unexpected skills as they came up, such as how to repair the aviary netting or fix the high-pressure water hose. She grew used to the pervasive smell of guano and fish. Her friends outside of work said she always smelled of it — even after she had showered and changed clothes.

When she first started, Parsons was amazed at just how busy the rehabilitation center was every day. Birds were constantly arriving, the phone always ringing. She would later calculate that for her first two years at SANCCOB, without any significant oil spills, the center took in an average of over four birds each day.[2]

Even though penguins have always composed the majority of the admitted birds at SANCCOB, as well as dominated the center's public mission and outreach, it was the cormorants that Nola Parsons first learned how to handle and rehabilitate. Four different cormorant species nest and fish off the Atlantic coast of South Africa and Namibia, and a fifth cormorant lives in the local marshes and inland ponds. These are the bank cormorant (*Phalacrocorax neglectus*), the Cape cormorant (*P. capensis*), the crowned cormorant (*P. coronatus*), the great "white-breasted" cormorant (*P. carbo lucidus*), and the more inland long-tailed "reed" cormorant (*P. africanus*).[3] On a few occasions all five of the cormorant species have been listed on the center's master whiteboard chart, which catalogs all the seabirds currently admitted.

Aside from the one penguin and five cormorant species, a handful of other common coastal seabirds breed in the region, including gannets, gulls, terns, and pelicans. In addition, because of Cape Town's geography and the wind patterns, all manner of straggler birds find themselves at SANCCOB, such as albatross and skuas. During the first year that Parsons was there, the crew of a deep-sea tugboat brought in a gravely ill red-billed tropicbird, a species rare to this coastline. The animal had landed on deck and never had the strength to fly away, even when the boat entered the harbor.

As time went on, Parsons earned a PhD in zoology at the University of Cape Town. She then worked for a year and a half out on the islands with South Africa's coastal management team and returned to SANCCOB in the position of veterinarian and researcher. Now a dozen years after first volunteering, she is in charge of several short-term and long-term initiatives, such as running the African Penguin Chick Bolstering Project. She coordinates the Seabird Health Survey, which includes a long-term research study on the parasites found in seabirds' blood.

These days at SANCCOB, Parsons spends less time directly with bird rehabilitation. She is quick to point out all the skills and hard work of all the other staff and vollies. Most of her days are spent in the lab or doing autopsies. And although she works primarily with penguins, and her PhD thesis was on African black oystercatchers, it is the cormorants, especially the endangered bank cormorants, that remain her favorites. Cormorants were the first birds she worked with, and since then she has been active in programs to assist this struggling local bank cormorant population.

Echoing the sentiment of artist Anna Kirk-Smith about the extinct spectacled cormorant, Nola Parsons tells me: "I love the bank cormorants because they are a little bit big and stupid and clumsy — and I always support the underdog!"[4]

The largest and most prestigious institutions for bird research in the Southern Hemisphere are a short distance from SANCCOB. Based at the University of Cape Town, the Percy Fitzpatrick Institute of African Ornithology, known for fun as "the Fitztitute," and the university's Animal Demography Unit both have active, rigorous slates of research projects, courses, and scholars studying all types of birds. Parsons earned her PhD through the Animal Demography Unit. Elisa Goya from Peru came here to UCT to study and collaborate with researchers. Robert Prŷs-Jones, before he took the position at Tring, zoomed along the Fitztitute's hallways for five years as a lecturer and researcher.

In the late 1980s, when Prŷs-Jones was here, biologists conducted some fascinating research looking at historic and current guano deposition and how it might reveal fish populations and the birds' effect on these stocks. The researchers compared the Humboldt Current off Peru to the Benguela Current off southwestern Africa.[5] This comparison makes a lot of sense because though the Benguela system is smaller geographically, the two upwelling regions are exceptionally similar. Strong, seasonally shifting southerly winds drive both currents, mixing productive cold waters from the ocean bottom past chains of small, coastal islands. In the Benguela Current, the upwelling sustains dizzying schools of surface-feeding anchovies and pilchards.

Fishermen in both regions have complained about seabirds damaging the fish populations.[6] In 1955, for example, a South African wildlife manager wrote in a government paper that the coastal cormorants "kill for the lust of killing even after their voracious appetites have been satisfied."[7] Yet the research from the Fitztitute suggests that seabirds in both currents have had little impact on stocks over the years, especially in comparison to the harvest by humans.[8] As on the coast of Peru, commercial overfishing in the region of the Benguela Current dramatically diminished these fish populations in the 1960s and 1970s. This in turn precipitously reduced the seabird colonies that had been feeding on these fish and that had once lived on the offshore islands in a population, for cormorants alone, of over a million.[9]

The guano islands of the Benguela system hosted in decreasing order of population the African penguin (*Spheniscus demersus*), the Cape cormorant, and the Cape gannet (*Morus capensis*).[10] This Cape cormorant is actually quite similar to the guanay cormorant physiologically and in general ecology. Cape cormorants are frequently admitted to SANCCOB, usually as weak juveniles.

In the mid-nineteenth century the Swedish naturalist and explorer Charles Andersson — after surviving a rhinoceros goring and having his kneecap blasted apart by native tribesmen — was one of the first Europeans to carefully observe and describe the Benguela seabirds. His account of the Cape cormorant in the early 1860s sounds like Audubon's description of the vast flights of double-crested cormorants on the Mississippi River and Robert Cushman Murphy's account of endless strings of guanay cormorants: "At some seasons of the year they may be counted not merely by tens or even by hundreds of thousands, but by millions: their numbers, in fact, exceed all computation; for it is no unusual thing to see a deep unbroken line of these birds winging their way for two, or even three, consecutive hours to or from their feeding grounds."[11]

Andersson observed: "During the nesting-season large numbers [of Cape cormorants] are to be found on almost every suitable rock and islet from the river Cunéné [Namibia's northern border] to Table Bay, in which situations, next to the Gannet and Penguin, this species is the principal depositor of guano."[12]

Dutch merchants seem to have mined the islands of the Benguela Current on a small scale for guano as early as the 1660s.[13] In the mid-nineteenth century, right after deliveries from the Chincha Islands sparked the guano rush in Europe and the United States, merchants quickly and aggressively mined the islands off southwest Africa. Elisa Goya feels right at home on this bald, dry African coast, though there is a bit more wind and rain here to erode and wash away the guano. In part because of this rainfall, the African deposits were never as large as on the Peruvian islands. The white-gloved European chemists determined that this mixture of penguin, cormorant, and gannet droppings was not quite as effective a fertilizer as that of the Chinchas. Nevertheless, British mariners secretively, then contentiously, claimed the African islands, trying to get a slice of the high demand sparked by the Peruvian product.[14]

Almost immediately after the guano rush began out of the Chincha Islands, mariners, primarily Brits, cleared Ichaboe Island of about two hundred thousand pounds (90,720 kg) of guano in less than a year and a half.[15] Stories go that there was such a rush on this island that sailors were murdering each other with pickaxes. In 1844 an American mariner described three hundred English ships and five American ships at anchor beside Ichaboe. In addition to all these sailors, two thousand more laborers from Cape Town had been brought over to work at the same time, all chopping at an island that is no more than eighteen hundred feet (550 m) long.[16] The ship *Ann*, mentioned earlier — the one that exploded after running aground because of the combustible properties of the stored feces — was in fact returning with a hold brimming over with Ichaboe guano.[17]

Though the guano resource off southwestern Africa was quickly exhausted, smaller-scale domestic mining has continued into the twentieth and twenty-first centuries. In the early 1930s, for example, an entrepreneurial carpenter built a guano platform off Namibia over partially submerged, and appropriately named, Bird Rock. After several years of work, his platform — a massive wood deck ingeniously built over the rocks — measured about 183,000 square feet (17,000 sq m). The operation connected with the platform eventually turned a handsome, consistent profit and supported each season hundreds of thousands of nesting birds, which included likely almost a quarter of a million Cape cormorants. The product, especially since it was not mixed with any sand or dirt, fetched an exceptionally high price.[18]

Bird Rock is still being harvested. Until quite recently a company named Guano Green Fertilizers continued to scrape small deposits off Ichaboe, other purpose-built coastal platforms, and perhaps another island or two. They ship the product to countries such as Germany and Belgium for the organic market.[19]

Although the Cape Cormorant has been by far the most abundant cormorant in the Benguela region, Nola Parson's favorite, the bank cormorant, also seems to have been historically an active breeder here, too.[20]

In 1854, just a few steps ahead of Charles Andersson, the German naturalist Johan Wahlberg traveled to the region. Between preparing ape skins and shooting flamingos for food, Wahlberg became the first European naturalist to describe the bank cormorant.[21]

On June 22, Wahlberg secured a trip aboard one of the most famous American clippers ships: "My trip to Walfisch [Walvis] Bay on board *The Witch of the Wave* was a slow business, because the vessel took on a cargo of guano on the way. While this was going on all my things had to remain on deck, and though they were covered they were completely drifted over by the fine guano dust. One compensation for this unpleasantness was that I had time to collect various things from the little islands along the coast."

Wahlberg and his team killed fifty fur seals. He recorded (and often collected) penguins, oystercatchers, gannets, grebes, skuas, and terns. And he found four different kinds of cormorants. "Two species of *Graculus* [*Phalacrocorax*] seemed to me possibly new," he scribbled. "But I have no time to describe it now."[22]

Later, Wahlberg explained that his newly observed bank cormorant is "tolerably frequent off the coast of Western South Africa, such as Possession, Halifax, Ichaboe, &c."[23] (His other new cormorant was the crowned cormorant, one of the "micro cormorants.") Unfortunately Professor Wahlberg did not make it back to his European halls of sciences to enjoy the ribbons of his discoveries. He was trampled by an angry elephant.[24] But the name that he prophetically assigned the new bank cormorant, *neglectus*, Latin for "neglected," has remained.[25]

The bank cormorant is a large, relatively stocky bird. It has an entirely black face — no colored orbital ring or yellow-orange gular pouch, just all black. Both the male and female when in breeding plumage have a few white filoplumes and a white rump patch, but otherwise they are entirely black from crest to toe. To compensate, the bank cormorant's bicolored eye is phenomenal: unique among the cormorants. It can only be appreciated up close or seen with a powerful lens. The iris is a lovely honey-orange color — Wahlberg described it as "ochre-yellow" — which then has an emerald-green lower crescent within the iris. The pupil is black.

The "bankies," as many at SANCCOB call them, rarely fish farther than five or six nautical miles out from the shoreline. Less interested in the large schools of anchovies and pilchards that dominate the offshore range and diet of the more prolific Cape cormorants, the bank cormorants feed more on gobies and prefer to hunt in the beds of kelp, eating local fish called blennies and klipfish. Studies have found that about a fifth of the bank cormorant's diet is also invertebrates, such as spiny lobster, crab, and octopus.[26] Similar to the Galápagos flightless in a few aspects, the bank cormorant is a heavier, weaker flier than other cormorants. The bank has evolved a more muscular lower jaw, presumably better adapted to probing around rocky bottoms for this type of prey. The bank cormorant can catch a wider diversity of species in a smaller range than the other four local cormorants — each of which has a slightly different underwater niche along this coast of southwestern Africa.[27]

Along with the Pitt shag and the Chatham shag, as mentioned earlier, the bank cormorants are at a genuine risk of going the way of the spectacled cormorant. The endangered bank cormorant nests at about forty-five rookeries.[28] In total today there are an estimated 3,000 breeding pairs of bank cormorants — down from a total of some 7,600 pairs in the early 1980s.[29] What is especially challenging is that 75 percent of the bank cormorants live on just two fragile island sites off Namibia: Ichaboe Island and Mercury Island.[30] When a rare bank cormorant is brought into

SANCCOB, it is usually from Robben Island, which — though home to only about an average of one hundred pairs — is one of the largest South African colonies of these birds.[31] Nola Parsons can see Robben Island in Table Bay from the beach nearest to the rehabilitation center. It is better known as the old prison island that held Nelson Mandela for almost two decades. It is an Alcatraz-like tourist attraction now. ("Alcatraz," you might like to know, meant pelican, gannet, or cormorant to the early Spanish explorers.)

A variety of factors probably have led to the demise of the bank cormorant. Their island rookeries were historically disturbed for the guano harvest, especially Ichaboe. Their populations have been plagued by human development, introduction of mammalian predators, oil spills, and, perhaps most significantly, the movements and diminishment of their sources of food likely due to overfishing and climate change. These include especially the kelp-dwelling fish species, the spiny lobster, and the pelagic goby.[32] To a lesser extent, predation on eggs and chicks by gulls and even pelicans has been a problem for breeding bank cormorants. And if that were not all enough, the adult and young bank cormorants are often chomped up by fur seals and the occasional great white shark.[33]

Usually it is a member of the public strolling the beach or one of the government or conservation workers on one of the islands who calls the rehabilitation center about a sick or injured bird. If the bird is not oiled, the most common observation is that the bird looks drunk. It isn't flying normally, it can't hold up its wings, or the bird is stumbling around or just sitting in a place or manner unusual for a wild animal. So a staff member drives out, catches the bird with a net, and transports it in a ventilated box, not unlike those in which dogs and cats are carried around. Once back at SANCCOB, the bird is weighed while in the box, and then a staff member gives the animal an introductory evaluation.

Nola Parsons says that when a bird is stressed and sick it gets sedate and silent. A healthy seabird, on the other hand, will not calmly let you handle it. Regardless, you must be confident and firm. The birds seem to know when someone experienced is caring for them. The standard uniform for the staff is a T-shirt, a pair of oilskin fisherman's style bib trousers, and waterproof boots. When handling birds, staff members wear neoprene gloves and arm guards up to the elbow.

A cormorant will quickly dart out its neck and peck with its sharply hooked beak. As Megan Gensler was taught on East Sand Island, cormorants go right for people's eyes, a unique behavior among the seabirds at SANCCOB. New staff and volunteers are often caught off-guard because they get accustomed to penguins, which do not have the ability to swiftly uncoil their necks or, as Parsons calls it, have "the strike distance." The only time Parsons has been injured by a cormorant at SANCCOB was when she did not properly instruct a vollie on how to hold the bird. She and the new person were working on the animal together, when the cormorant jabbed Parsons's eyelid. "A really good wound," Parsons tells me. "But luckily no damage to my eye itself."

(Parsons did get bit once by two cormorants at the same time when she was out at one of the colonies. She is embarrassed about this now, because it was improper form and judgment, but she was in a rush and carrying a bird under each arm. She tripped backward over a rock. Though she held on to the two birds, the cormorants pecked her face during the tumble.)

If Parsons is looking at a cormorant that has just been admitted at SANCCOB, she brings the bird to the intensive care unit. She learned long ago that cormorants are more likely to relax if they have their feet supported, so she keeps this in mind when she takes the cormorant out of the box. She holds the cormorant's beak firmly with one hand as she sits on the edge of a chair while supporting the bird's feet with the other hand. She puts the cormorant between her legs, transferring the bird's feet to rest on her calves. With her knees, she holds its wings firmly against its body.

Once the cormorant is secure, Parsons fingers through the feathers for any obvious wounds, broken bones, oil residue, or parasites. On one of the bird's legs, she secures a temporary tag of waterproof tape with a number. Then she leads a long tube attached to a syringe down into the cormorant's throat. She squeezes in a portion of electrolyte solution, similar to a sports drink. She opens the bird's beak again and inserts with her fingers half of a deworming pill.

That done, Parsons normally places the cormorant in a clean, towel-lined crate. She hopes the animal will relax while she fills out a card. Parsons writes down what she has given the bird and the reason for admittance. She grades its general initial health, what they call "habitus."

Typically, a cormorant will get fluids for the rest of the day and perhaps a mild "fish smoothie" enriched with vitamins. (See figure 20.) For the night, the cormorant will get an infrared lamp and a shade-cloth pen to keep it warm and help it feel safer.

The truth is that most cormorants of any species, by the time they arrive at SANCCOB, are not in a position to survive. If a cormorant does makes it through the night, however, Parsons will take a blood test for pathogens. She will conduct a more extensive evaluation to see how best to rehabilitate the bird and try to get it right back out there as soon as it's behaving normally and has passed a few basic health tests.

Over the decades, the long-term staff at SANCCOB has observed cormorants roosting and swimming at the rehabilitation center beside a variety of other seabird species native to the Benguela region. Some of these other birds are closely related to cormorants, such as gannets and pelicans, while others are evolutionary convergent in physiological and behavioral ways, such as penguins and terns. Staff regularly handle wild seabirds, examine them, and at times even hand-raise them from chicks. They observe them in captivity, clean wounds, take blood samples, and conduct autopsies. Thus the SANCCOB staff have a rare firsthand perspective on four traits I want to discuss here, which separate the cormorants from the other seabirds at the rehabilitation center.

First, and least conspicuously, cormorants lack external nostrils and excrete salt out the roof of their mouth, similar to the other Pelecaniformes. The closed nostrils surely restricts the cormorants' sense of smell, if they have any at all, but it likely helps them swim underwater more efficiently. In the 1970s an ornithologist set up a canvas blind on one of the man-made guano platforms on the Namibian coast. He observed that the cormorant chicks are born with open external nostrils, but as the chicks grow — and before they enter the water to learn how to swim — these nostrils actually close up.[34] The juvenile cormorants then breathe only through their mouths as they mature.

All cormorants have the ability to excrete salt quickly and substantially from their blood, as if they had a second pair of kidneys, enabling them to drink seawater and stay hydrated at proper salinity levels. Cormorants have two large internal glands to do this, which are near each eye. Albatross and petrels, truly pelagic seabirds, have tubes on their beaks to drip the salt out from their nostrils. For cormorants, the salt drains out from within the roof of the mouth, then down two thin channels on either side of their beak.[35]

A second attribute the SANCCOB staff notes is that the cormorants in particular rely on their tail feathers for flight: "Cormorants need these tail feathers to get liftoff," Parsons says. "And they struggle to get airborne if their tail feathers are damaged. It seems too that they use their tails for braking and changing directions in the air." In Japan, the usho used to clip a couple of feathers out of the tail to keep their birds from flying away. They now clip the feathers out of the wing instead, probably for aesthetic reasons.[36]

A third behavioral trait is what Junji Yamashita and his fellow usho also know very well: cormorants seem to be especially mischievous. Cormorants are curious. They engage in play. This has become evident in particular at SANCCOB, thanks to a few cormorants that have been permanently in residence because of injuries. These birds could no longer survive in the wild. They live in an area of SANCCOB called the "home pen."

Parsons says: "One of these crowned cormies was called Bobby. That was because she bobbed her head when asking for food. We had another tame one at the time who imprinted on penguins, but didn't like people. So he was called Evil Bobby."

Parsons immediately feels bad about this nickname, so she is quick to point out that it was a good bird: it just didn't care for humans. Evil Bobby even used to affectionately bring nesting materials to the penguins — who didn't seem to quite know how to respond.

"Cormorants are really inquisitive birds and will stick their heads and necks into any available opening," Parsons says. "They pick up any small object that they see lying around, so you have to be careful not to leave any cable ties or pens or elastic bands within their reach."

Bobby and Evil Bobby used to wander around parts of the facility and get underfoot. Bobby liked to sit on people's shoulders, like a pirate's parrot. This soon had to be stopped because she was too fond of earrings and occasionally stretched her neck around and struck at the person's eye — presumably wanting to complete the pirate effect with the need for a patch. These days there is another cormorant at the rehabilitation center who likes to swim underwater to nip at the feet of the penguins and gulls — and then dart away as if hoping to be chased.

I can tell you from personal observation all over the world that wild cormorant chicks at their breeding colonies practice what appears to be simple games, such as tugging a stick, throwing the stick to another, or even a sort of "King of the Hill" on a boulder or a tree limb. Sometimes in the stiff winds just off the Cape of Good Hope the cormorants will soar up and dip and weave in the air, more like gulls or petrels.

"I spent ages watching them fly," Parsons says. "Not very gracefully, but amazing in that strong wind, especially to be able to land on the cliff face. I do think it looked like these cormorants were having fun."

The fourth and most common behavior to cormorants that I'll discuss here is the spreading of their wings. At SANCCOB, the cormorants—whether resident or in the process of being rehabilitated—inevitably spend part of their day at the edge of one of the swimming pools standing still and silent with their wings spread out, at times fanning them slightly. No other bird at the center has this behavior, not even the closely related gannet.

This wing spreading is usually one of the first ways people identify cormorants almost anywhere in the world. The behavior is notorious because it is evocative of a bat or a vampire. Riverside signs and nature guides often explain that cormorants do this to dry their wings because they do not possess the proper oils to waterproof their feathers. All cormorants do, however, have near their tails an oil gland—a uropygial gland—which se-cretes a waxy substance similar to the material called sebum that mammals have in their hair follicles (and which gives humans acne). Cormorants nibble at the base of their backs to get the sebum and wipe it across their feathers, often spreading it around with the back of their necks. The cor-morant's uropygial gland is as functional as those of other diving birds.[37]

An important study in the 1960s by A. M. Rijke at the University of Cape Town showed that waterproofing in plumage is not all, if even pri-marily, about the oil. It is more about the structure of the feathers. Rijke found that the microscopic feather construction in cormorants was dif-ferent enough from that of other waterbirds, say from ducks, to render their feathers less water repellent.[38] Getting wet helps cormorants dive deeper more efficiently by reducing drag and the buoyancy of their hollow bones, their lungs, and that thin layer of trapped air close to their skin that enables them to stay warm in cold temperatures—as we discussed with the blue-eyed shags. In exchange for this adaptation to dive deeper more easily, cormorants must dry their wings occasionally after bouts in the water. Shedding the water weight helps them fly more easily and thermoregulate.

Author and biologist Thor Hansen put it this way: "In this context, the shags and cormorants come off looking rather smart. Rather than unkempt survivors with wet feathers and a crummy preen gland, ornithologists now view them as beautifully adapted to a diving lifestyle. They benefit from the negative buoyancy of soaking, while still keeping their skin and down feathers sealed inside a watertight blanket."[39]

This explanation alone, however, is not satisfying enough to rabid cormologists, because it is not complete regarding the wing-spreading behavior. As I mentioned earlier, a few cormorant species do *not* do the spread-wing thing, such as the blue-eyed shags in Antarctica and the red-legged "chuita" cormorants off Peru. Grebes and loons are deep-diving seabirds that also can fly, but they do not spread their wings. Some vultures, who do not even swim, will occasionally stand and spread out their wings. And cormorants sometimes stand there and spread their wings in the rain.

In an attempt to account for some of these inconsistencies, numerous alternative or additional explanations have been put forth, some of them by researchers at the University of Cape Town. Perhaps cormorants just spread their wings for balance. Maybe cormorants spread their wings to aid with swallowing and digestion. From observations near Cape Town, one ornithologist in the 1970s hypothesized that long-tailed cormorants might spread their wings to collaboratively advertise foraging success, a sort of flag to the rest of the birds in the flock to signal exactly where the food is.[40] Another intriguing study in the 1990s by a French ecologist who later became an honorary research associate at the Fitztitute suggested that cormorants spread their wings to aid thermoregulation, since, if the bird has just been fishing, it has filled its stomach with a significant amount of colder material. His study found that a group of cormorants at a German zoo always spread their wings when given cold fish, but only did so a quarter of the time when given warm fish. The energy to hold up the wings and flap them slightly might warm the body, not unlike how we shiver or run in place when cold.[41] (Despite this research, I've sometimes wondered if it is actually the other way around in the wild: feathers have insulating properties, and since the birds work so hard swimming underwater, perhaps they actually are spreading their wings to *cool* their bodies, similar to how an elephant spreads its ears.)

In 2008 a group of researchers at the University of Birmingham tested wing-spreading theories by using specially trained cormorants in a hermetically sealed aviary with high-tech scales and monitors for air and water temperature. They concluded: "Wing-spreading following aquatic feeding appears only to serve a wing-drying function. However, our observation that cormorants may commence wing-spreading when both dry and without having fed previously, suggests that there are further functions of wing-spreading behavior that require investigation."[42]

To sum up this behavior, cormorants probably spread their wings to dry off because getting waterlogged helps them fish better — for the same reasons a scuba diver wears a weight belt — but surely this getting wet isn't great for flying or staying warm afterward. It is not about the oil, but more about the structure of their feathers. Then again, there still must be something else going on with this behavior. No one has quite figured it all out.

"Unfortunately most of the volunteers come for the penguins," Parsons tells me. "They also like the gannets, because they are beautiful. The rest of the birds are less interesting. I think people don't react to cormorants in the same way because they're uniform black and brown. Not eye-catching. I also think the vollies are scared of handling the cormorants, because these birds are usually weak when they come in and they die so easily in rehab."

Parson continues: "I know the education officer always points out to visiting groups how the cormorants spread their wings. But I don't see too many people that interested in them. But I do think that when the vollies get to see the behavior and playfulness of the cormorants, then they grow to love them!"

Herman Melville in 1851 used the cormorants of South Africa to portend disastrous events. In *Moby-Dick*, he wrote of "inscrutable sea-ravens" perching on the stays of his death-bound ship *Pequod*. Melville's cormorants added to the dismal, dark tone of foreboding as the ship rounds the Cape of Good Hope, "as though they deemed our ship some drifting, uninhabited craft; a thing appointed to desolation."[43]

Over a century later, the Australian poet John Blight wrote the sonnet "Cormorants" (1963). He too perpetuates the evil — even fascist — image of the cormorant in literary imagery. Only in the last lines does Blight hint that there might be some beauty to this bird. Blight suggests that this elegance is not in our aesthetic appreciation but instead more in an understanding of the cormorant's utility: a Liam O'Flaherty–like recognition, acceptance, of this animal's dark suitability for its role in nature.

CORMORANTS

The sea has it this way: if you see
cormorants, they are the pattern for the eye.
In the sky, on the rocks, in the water — shags!
To think of them every way: I see them, oily rags
flung starboard from some tramp and washed
onto rocks, flung up by the waves, squashed
into sock-shapes with the foot up; sooty birds
wearing white, but not foam-white; swearing, not words,
but blaspheming with swastika-gesture, wing-hinge to nose:
ugly grotesqueries, all in a shag's pose.
And beautifully ugly for their being shags,
not partly swans. When the eye searches for rags,
it does not seek muslin, white satin; nor,
for its purpose, does the sea adorn shags more.[44]

The Southern Africa Foundation for the Conservation of Coastal Birds was founded in 1968 by Althea Louise Burman Westphal. She had been living in Cape Town when there was a large oil spill from the tanker *Esso Essen*, the first major spill to hit this region. Her daughter volunteered with the local SPCA to treat the oiled penguins, but their facilities were no match. Westphal took about sixty penguins back to her house. She cleaned them three at a time in her bathtub with dish soap and at first filled a trailer with water from a hose for them to swim. As some of the penguins improved, she drove them a couple of times a week in her station wagon to the ocean to practice swimming again. She fed them strips of hake. Westphal was able to rehabilitate a few of them back into the wild. Most died.[45]

Westphal was moved to action by the experience. She got ornithologists and professors at the university involved, and they helped organize a group of private patrons. Soon Westphal and her new colleagues established what would become SANCCOB. The center is now one of the oldest and best-known seabird rehabilitation facilities. It has an educational program and runs a boat trip for school groups. The small, dedicated staff relies on local volunteers and men and women, often right out of university, who come to work for as long as a year, from places as distant as Brazil and California and China.

Althea Westphal stayed involved with SANCCOB until her death at age seventy-one in 2002. Between 1948 and the grounding of the *Treasure* in 2000, there have been at least fourteen known spills in the waters of this part of Africa that oiled at the very least 500 seabirds.[46] Westphal lived to see the rehabilitation of the 19,000 penguins after the *Treasure* disaster. But the event must have been bittersweet for her. Today Westphal would not be pleased to see that SANCCOB remains so necessarily busy still. Next to International Seabird Rescue based in California, SANCCOB is probably the busiest seabird rehabilitation center in the world.

It was not just 19,000 penguins that suffered from the *Treasure* spill. Almost a quarter of the bank cormorants at Robben Island died from the oil that spilled out of the ship after it ran aground. Conservation workers were able to recover only fifty-four adult cormorants, of any species, after the spill. SANCCOB was able to wash and return twenty of these back into the wild. When gathering penguins at Robben Island, conservation staff collected over thirty abandoned bank cormorant chicks and brought them in to be hand-reared at SANCCOB. Of these, they were able to return about half to the island. Hundreds more oiled, dying bank cormorants and abandoned chicks remained on other oiled islands; there simply was not enough manpower. Parsons says more cormorants were not gathered because the penguins had a greater chance of survival. But, if we are honest, most of the energies were also devoted to the penguins because they are simply cuter.[47]

Oil binds to the barbs of the feathers of any bird, disabling its ability to fly and to thermoregulate. Oiled birds starve. They freeze to death. They go blind if the oil gets in their eyes. If they ingest the oil it is toxic to their system.[48]

The rehabilitation center staff do not see too many oiled cormorants of any species on a regular basis, as they do somewhat consistently the penguins — even when there has been no publicized spill. Parsons thinks they do not get many phone calls about oiled cormorants because of their black plumage. Residents, even wildlife biologists, cannot recognize from afar when the birds are suffering. And also, because cormorants can fly if not completely coated, they are harder to find until they are in really rough shape.

"Often by the time the cormorant is weak enough to be captured," Parsons says, "it does not stand much chance of surviving in our care."

Because cormorants of one species or another live along almost every coastline around the earth and certainly along the major shipping routes — which take advantage of ocean currents — these particular seabirds are consistently in harm's way. The 1989 oil spill from the *Exxon Valdez* in Alaska killed directly anywhere from 2,900 to 8,800 cormorants of three different species.[49] As another example, the Socotra cormorants live exclusively on a dozen or so colonies in the Persian Gulf and the Gulf of Aden. They were dangerously depleted by the Gulf Wars. The now officially "vulnerable" Socotra cormorant is still regularly at risk because of — among other factors — the amount of oil drilling, processing, and ship traffic in the region.[50] The Dubai Zoo with local managers has recently been raising and releasing Socotra cormorants in order to bolster the population.[51]

Because of their migratory nature, however, some cormorant populations have escaped reasonably unharmed immediately after a few of the most massive spills around the world. Relatively few cormorants died from the oil that flooded out of the ship *Prestige* off the Iberian Peninsula in 2003. Most cormorants seemed to have survived the BP *Deepwater Horizon* rig explosion in the Gulf of Mexico in 2010. The cormorants were largely elsewhere during these summer events, while together the slicks of petroleum from these two catastrophes killed tens of thousands, if not hundreds of thousands, of birds belonging to other species.[52]

Scientists meanwhile are just beginning to discover how long the damage remains in a given area. Oil spills disturb ecosystems for decades, harming the cormorants and other birds who fish in the area when they return to the region, notably through the diminished health of the water and the web of life living within and around it. Studies in the years after the *Prestige* spill, for example, show that the European shags sharply reduced their reproduction.[53] A recent study at a rookery in Minnesota found in the eggs of returning white pelicans traces of oil and the same chemical dispersant used at the BP spill some eighteen months earlier.[54] Thus perhaps the bank cormorants, as well as the gobies, klipfish, and lobster upon which they rely, might still be suffering from the long-term damage of the *Treasure* and the succession of previous spills.

The staff at SANCCOB see the more direct result of human activities that are lethal to cormorants and other seabirds. These include not only oil spills, but ocean dumping of plastics and other garbage, lost or abandoned fishing gear — both recreational and commercial — and all manner of runoff pollutants, which occur all over the world and reduce fish stocks, prompt algal blooms, and create aquatic dead zones. Cormorants can be especially vulnerable because their range is often close to shore and near urban areas — near cities that may have originally been settled because of easy access to good fishing.

In the region of the Benguela Current, there was an odd case in the 1970s when at least 4,500 cormorants died near the guano platforms off Namibia. It seemed that a dense slick of fish oil had been created when fishermen squeezed an especially large commercial catch of anchovies into their nets and onto the decks of their boats. This oil combined with what spilled into the sea when they offloaded the tons of fish at the dock. Hundreds of the cormorants, rendered flightless by the slick, were run down when they wandered in front of a train or in front of the automobile traffic on the coastal road.[55]

In all our interviews and correspondence, including several direct requests, I could never get Nola Parsons to tell me about a particular cormorant that she cared for at the rehabilitation center. I wanted to tell you about a single day, about one memorable experience of hers with one bird. But she would never share a story like this with me.

"Working with cormorants in rehab is normally heartbreaking," she says. "But I have to try to remain objective."

Parsons works hard to not get emotionally connected to each and every bird — to not pile up the memories and attachments. The job at the rehabilitation center is emotionally exhausting enough as it is. Veteran staff members discourage the vollies from naming the birds unless it is a situation like Bobby and Evil Bobby, animals who will be around in the "home pen" for a while. Some of the named birds they take out to schools and other outreach events in order to teach biology, conservation, and empathy for seabirds.

I think Parsons also submerges most if not all her memories of individual birds because she works with a lot of them after they are dead.

"I try to do a postmortem evaluation on all birds that die at SANCCOB,"

she says. "But this isn't always possible because of time, depending on the number and species of birds. I definitely postmortem any bank cormorants that come in, though. I collect basic morphometric measurements. I've used bank cormorant samples in my Health Survey project, so we do try to contribute to research even with dead bodies if we can."

I ask Parsons if her choice to eat as a vegan isn't entirely for ethical, health, or environmental reasons, but also might be because of her close work with the animals at SANCCOB.

"Well, yes," she laughs. "I find the thought of eating any bird disgusting after doing so many postmortems, and I especially ask people not to eat duck if I'm eating out at a restaurant with them! South Africa is not the most vegan-friendly country around."

All that said, Parsons does permit herself an emotional, more significant personal place for bank cormorants as a *species*, because their population is at such alarmingly low numbers. Another major accident could render the animal functionally extinct in her lifetime.

Fortunately, the plight of this endangered bank cormorant has not been entirely *neglectus*. Both Namibia and South Africa, spurred on by organizations such as SANCCOB, have tried to place restrictions to protect the birds from human disturbance, but still only a quarter of their rookeries have any sort of nature reserve status, and even here enforcement is difficult on small budgets.[56] The cormorants often ride the coattails of the penguins, since the birds usually nest on the same islands. All the Namibian islands were declared a Marine Protected Area in 2009, but illegal fishing remains, and now there seems to be some backroom dealing to allow guano harvesting again on Ichaboe and perhaps other islands. Jessica Kemper of the African Penguins Conservation Project told me recently: "I fear the worst and am somewhat despondent about what might happen next."[57]

The cause of trying to keep the bank cormorant from going extinct has been championed by a conservation charity in southwestern England called Living Coasts. They have been working to start a bank cormorant captive breeding program. It is a notable endeavor on a few different levels, not the least being that rallying conservation energy to educate about cormorants can be difficult. It can be hard to excite a member of the public about a similar-looking relative of a bird that a visitor might see near his or her home. Not to mention all the other negative cultural and literary cormorant baggage I've been telling you about, especially in England.

Undaunted, the keepers at Living Coasts decided to begin keeping bank cormorants in 2004. Nola Parsons had raised some from chicks rescued by conservation staff at Robben Island. Bank cormorant chicks are often brought in to SANCCOB because heavy seas wash their nests off the breakwater at Robben. Imprinting on humans can be a real problem with raising young birds at the rehabilitation center, but if a lot of chicks remain together, this effect is lessened. In this case Parsons knew they were going to a captive facility, so it was not a concern that these birds liked people and wanted to follow them around. The full transfer up to the UK wasn't entirely successful, though. Because of concerns about avian influenza, the second group of cormorants were placed in a quarantine facility, where they all died.

Eventually visitors to Living Coasts in Torquay, England, were able to see, photograph, and learn about two bank cormorants, named Mnandi and Amanzi, who waddled around the little beach in Torquay with the African penguins. The cormorants flew around the open-air aviary. Visitors watched the cormorants diving and capturing fish in the underground tanks. Sometimes they could watch bank cormorants nab fish from the hand of a keeper in scuba gear. The director of Living Coasts, Elaine Hayes, told me that to her a bank cormorant's eye seems "knowing."[58]

After learning how to care for the adult bank cormorants, Living Coasts experimented with establishing an in-house breeding program, which they had done successfully with terns, penguins, and other species. In July 2009, Tony Durkin, the head keeper at the time, flew down to South Africa.

"It's a long way to go for two dozen eggs, but it will be worth it," Durkin explained in a press release. "It's an investment in the future of the species."[59]

With the help of Nola Parsons and several other individuals and groups, Durkin was able to collect ten eggs at Robben Island.[60] He kept them in a portable incubator that South African Airways let him bring aboard and plug in (after a few other airlines turned the concept down because of how bomb-like the incubator appeared).

Once back at the zoo in England, Durkin and the rest of the staff kept careful watch. Nine cormorants hatched, but only one of these survived. As the chick grew and they were able to test for gender — by sending out a tail feather to a lab — it turned out, fortunately, that the fledgling was a female. They named her Mpumi, which is Zulu for "survivor." The keepers hovered over this cormorant as she grew and learned to fly. Living Coasts managed to get visitors and local bird enthusiasts excited about Mpumi's significance. Presumably the two male bank cormorants also watched her progress with interest.

Yet one Saturday morning, a little over a year after she hatched, the staff found Mpumi dead. There were no wounds or outward signs of disease. A careful tissue analysis and autopsy also revealed nothing. Elaine Hayes told me that the bird's death remains a mystery.

Hayes said officially in a press release: "This is truly heartbreaking. The keeping staff—and the rest of the team—are devastated. She is the first bird I have cried over."[61]

Mpumi the bank cormorant represented over two and a half years of preparation, research, and hand rearing. Hayes continued: "We have learned a lot from this. We aim to go back next year, collect more eggs and start again. When you are talking about extinction, you can't admit defeat."[62]

Living Coasts did not start right back up again, though. They decided to step back and conduct more research before collecting eggs again. This included another visit by one of the keepers to SANCCOB in order to learn more about their chick-rearing program and to observe the wild colony on Robben Island.

Not everyone is applauding the efforts of Living Coasts. Some ornithologists and conservationists in South Africa are actually pretty skeptical of the program and would much rather any organizational funding be directed toward saving the bank cormorants' wild habitats, rather than toward what one skeptical scientist based in South Africa calls the "Noah's Ark solution."

Elaine Hayes told me recently that she remains committed to making certain that the bank cormorant will not go extinct anytime soon. "All species are worth saving," she says. "If we have the skills, we should use them."[63]

As befits her profession, Nola Parsons says clinically what the poet John Blight implies wickedly: "We need to save the bank cormorant because of its individual place in the ecosystem as well as to help with ensuring that the ocean as a whole remains healthy."

February

Warm sea breeze, cloudless afternoon. The two surviving juveniles from the first clutch by the two boulders on Gates Island now join dozens of other cormorants and several other species of birds at the mouth of a freshwater rivulet in the Everglades. The cormorant siblings have been here for almost two weeks. The fish have been plentiful. None of the birds is fishing right now, except for an ibis that searches for minnows in the shallows with its beak. The cormorants loaf in a tall shady pine grove with a clear view out toward the Gulf of Mexico. Anhingas, pelicans, egrets, herons, spoonbills, and cormorants all stand in the branches and tend to their feathers. Several birds doze. Butterflies alight on the outermost pine needles.

The male juvenile from Gates Island opens his mouth. His gular pouch flutters for a while. He is still too hot. He stretches out his wings and faces toward the breeze. He decides to drop down into the water by the sawgrass. An alligator snaps him up in one lightning chomp.

Gates Island

UNITED STATES

An attempt is made to point out that the [cormorant]
is fully deserving of our protection and that, except in
scattered, local instances, it is largely neutral if not
actually beneficial in its relationship to man.

HOWARD L. MENDALL,
*The Home-Life and Economic Status of
the Double-Crested Cormorant,* 1936

*I*t's March again. I can see how cold and windy it is out the window,
but I'm comfortable inside sipping coffee. My office is warmly lit by a
large paper lantern hand-painted with a Nagara River ukai scene, and
I'm surrounded by all of my cultural artifacts about the relationship between humans and cormorants.

I have a sort of Victorian curiosities cabinet. It has various skulls and
feathers and eggs and bags of guano collected from my travels. On one
of the shelves of cormorantabilia I display a preserved skin of a juvenile
pelagic cormorant. I had hoped to donate this to the Bird Group at Tring.
Students found the cormorant dead on a beach in Northern California.
We brought it home on the plane in a cooler. It turns out that there is a
taxidermist named Johnny who lives not far from me. In between his work
with deer and ducks and fish, he skinned and preserved the dead bird with
borax. I was ready to ship the skin to Tring, but it turned out that not
only was it illegal to pick up that dead cormorant on the beach without
a permit, but it was even more so to take it out of the state of California.
Despite the kindness and interest of Robert Prŷs-Jones — and my own
eagerness to have something we collected in the historic holdings of the
Natural History Museum — I eventually lost motivation with the amount

243

of paperwork and favors that would have been necessary to ship the specimen. To be honest, the pelagic pelt does not look attractive on the shelf. It can be morbid for the visitor. I do love the feet, though.

I look out the window again. I should really go out to Gates Island. When I walked past the river yesterday I thought I saw in the distance a double-crested cormorant (*Phalacrocorax auritus*). They should just be arriving now. But the brown, bare branches are shivering out there, so I'm dragging my feet. And even as I sit here in the office debating how lazy and wimpy I am for not wanting to go out on the boat this afternoon, the stories I have shared in this cormorant study continue to evolve.

The Army Corps of Engineers just commissioned Bird Research Northwest to "dissuade" over 60 percent of the previous nesting area for the cormorants on East Sand Island this spring. Not without media attention and some minor public protest, they are moving the "Great Wall" dissuasion fence to about the middle of the birds' former colony.[1] Megan Gensler is working at a different Bird Research Northwest site this season, but new researchers like her will be out there on East Sand Island capturing cormorants for tagging and scaring the rest of the birds off, forcing them to disperse and hopefully nest someplace else, away from the large Columbia River runs of salmon smolts. Oregon Fish and Wildlife has asked for permits to go further, to begin culling cormorants on the Columbia River.[2] Meanwhile, hazing programs have already been reinstated to the south along the Oregon coast so that volunteers and wildlife managers will try to scare the birds away from these other salmon runs.[3]

At Louisiana State University's AgCenter, engineers have been developing a "Scarebot." This is an autonomous little catamaran the size of a coffee table that paddles around the surface of a pond and not only frightens away the cormorants but also monitors the water conditions. "Scarebot" has a solar-powered battery, "feelers" to sense the shore, and a GPS in which the farmer can set the pond boundaries. "Scarebot" runs quietly until it senses birds, and then paddles unpredictably toward the cormorants to get them up.[4]

U.S. Representative John Klein, from a southeastern district in Minnesota, just presented to a subcommittee in Congress a new piece of legislation. Klein, working with another Minnesota representative, calls this the "Cormorant Management and Natural Resources Protection Act." It would allow states more latitude in killing cormorants, so state wildlife

THE DEVIL'S CORMORANT

agencies would no longer need to work through the federal application process currently in place with the U.S. Fish and Wildlife Service's management plan.

Representative Klein says officially: "Thousands of constituents have expressed concern over the noise and odor emanating from the excessive cormorant population on an island on Lake Waconia. Residents, marina owners, and local officials believe the cormorants have consumed an entire generation of fish — leaving only fry and trophy fish in the lake." Klein continues: "Carver County residents are frustrated by the damage cormorants inflict on their land and lakes. They are equally frustrated by the merry-go-round of red tape that hinders the ability of states and communities to control these pests. This bipartisan legislation empowers states to better control the overpopulation of cormorants, mitigating their destruction on property and fish populations. As an avid fisherman and outdoorsman, this legislation simply makes sense."[5]

It actually makes no sense at all. It is worth spending some time dissecting each of Klein's statements to the word, but at least remember that currently under the federal usfws plan, Minnesota organizations, businesses, and residents may already file for permits to kill cormorants that destroy property, endanger diversity, or unfairly reduce recreational fisheries. Minnesota is also already the only northern state with open permission for private fish farmers to kill cormorants. In 2011 state wildlife biologists had permission to kill over 1,100 cormorants on Lake Waconia. According to the local television station that ran the story (using YouTube video clips of swarming cormorants from a different part of the continent), there weren't enough cormorants that year to kill more than 309 birds. The state had already been killing hundreds of cormorants under permit in previous years.[6]

So what more exactly does Representative Klein want?

Linda Wires, a leading expert in both double-crested cormorant biology and management, is based at the University of Minnesota. She has worked closely with the usfws as they have assessed the best management plan. Wires and I have kept in touch over the years. She has been deeply disappointed with the federal plan, local wildlife managers, and proposals like that from Representative Klein.

"At the heart of this whole issue," she told me a few years ago, "is nearly a zero tolerance for cormorants. It's a witch hunt."[7]

The federal cormorant plan has been in place almost ten years now, renewed once, and is up for revision in 2014. As I sit here at my desk, the USFWS has been fielding public comments from all manner of private and nonprofit individuals and interest groups representing every facet of the issue.

From 2004 through 2011, under the "public resource depredation order," state, federal, and tribal wildlife biologists in thirteen states killed a reported total of 146,100 of these animals. They oiled hundreds of thousands of eggs in more than 117,842 nests. They destroyed 21,388 other nests. From 2004 through 2010, under the "aquaculture depredation order" and other permits, private fish farmers and federal managers shot 164,417 cormorants.[8] This aquaculture total is based on what was reported by the private farmers, but is certainly an underestimate. In Canada well over 10,000 more cormorants have been killed during this time by park managers and private residents, but this is difficult to total, because Canada has fewer restrictions on killing birds on private property, much less federal oversight, and no requirements for reporting.[9]

So if we use the very rough estimates of one to two million of these birds out there—including all of the Interior, Atlantic, Southern, and Pacific Coast–Alaska populations—then this has been an elimination, not counting any of those shot in Canada, of 15 to 31 percent of all the adult and juvenile double-crested cormorants on the continent over the last seven years.[10]

On the other side of the Atlantic in Britain, the "Ghosts of Gone Birds" exhibition has been popular and is now traveling to a museum in Oxford. Anna Kirk-Smith's *The Unfortunate Repercussions of Discovery and Survival (Spectacled Cormorant)* has not yet found a permanent exhibit space. It could just fit in my office. I've measured. Anna has been slow in accepting my bid.

Also in England, a book of recipes by William M. W. Fowler was recently rediscovered, collected, and published. Fowler was a former World War II bomber pilot and apparent soul mate of the Antarctic explorer-ornithologist Niall Rankin. *Countryman's Cooking* was originally self-published in 1965 for men out camping or for reluctant bachelors. Fowler's advice on the preparation and eating of cormorant is this:

> Having shot your cormorant, hold it well away from you as you carry
> it home; these birds are exceedingly verminous and the lice are said

to be not entirely host-specific. Hang up by the feet with a piece of wire, soak in petrol and set on fire. This treatment both removes most of the feathers and kills the lice. When the smoke has cleared away, take the cormorant down and cut off the beak. Send this to the local Conservancy Board who, if you are in the right area, will give you 3/6d. or sometimes 5/- for it. Bury the carcase, preferably in a light sandy soil, and leave it there for a fortnight. This is said to improve the flavour by removing, in part at least, the taste of rotting fish. Dig up and skin and draw the bird. Place in a strong salt and water solution and soak for forty-eight hours. Remove, dry, stuff with whole, unpeeled onions — the onion skins are supposed to bleach the meat to a small extent, so that it is very dark brown instead of being entirely black. Simmer gently in sea-water, to which two tablespoons of chloride of lime have been added, for six hours. This has a further tenderising effect. Take out of the water and allow to dry, meanwhile mixing up a stiff paste of methylated spirit and curry powder. Spread this mixture liberally over the breast of the bird. Finally roast in a very hot oven for three hours. The result is unbelievable. Throw it away. Not even a starving vulture would eat it.[11]

I keep up my correspondence with Elisa Goya. She and her colleagues say it is still possible to buy guanay cormorant on the black market in Callao. It's popular for family festivals and soccer matches. Thus the guardias on the Chinchas must still protect the birds from poachers.[12] (When I was there I tried to buy a cormorant to eat, but as my taxi driver drove me deeper and deeper into the alleys of Lima, I lost my nerve.) Elisa writes that Hermilio and Claudio have moved on. She has been especially busy as I write this, trying to find the cause of a sudden, unexplained die-off of seabirds off the Peruvian coast.[13]

Across the Pacific in New Zealand, residents are reeling from the oil spill of the cargo ship *Rena*, which wrecked in October 2011. Four hundred metric tons of fuel leaked out and killed thousands of seabirds, including cormorants. Some cormorants and other birds have been admitted and successfully released from a local rehabilitation center. The wreck of the *Rena* is widely considered New Zealand's most significant environmental disaster to date.[14]

On a more positive note, one ukai community in Japan made news by accepting a new female apprentice to be an usho.[15] In Cape Town, South Africa, the Percy Fitzpatrick Institute and the Animal Demography Unit just received funding to support new fieldwork and analysis on the depletion of the bank cormorant, in the hopes of reversing the downward trend.[16]

All right. I stall no longer. I bundle up to walk down to the dock. Beside the boat it's even gustier than I thought. A north wind sweeps up white-caps even here upriver, north of the drawbridge. Sometimes a March cold is worse than a January one because there have been a few warm days already. I climb in the boat.

I've been taking this trip out to Gates Island for over a decade — some years as often as twice a week. I've almost always traveled in this same open wood boat powered by this same unreliable outboard engine. I've gripped these same long splintery oars too many times when this engine has failed over and again.

The engine starts right up. I motor out downriver with three jackets on, gloves, and a wool hat over my PROABONOS cap.

I'm rewarded almost immediately. Just north of the drawbridge: a double-crested cormorant is standing on a piling. I put the engine in neutral. The bird is in full breeding plumage. (I did splurge a couple years ago for an upscale pair of waterproof binoculars.) The cormorant is glossy black with a green iridescence. The skin on the face is a bright vermilion. The eye is a wet emerald green. Two comic little black crests sprout ear-like off the head. I can't tell if it's a male or female, but certainly it is an adult all decked out for, as Audubon called it, "the love season."[17]

The cormorant faces into the north wind. It stands still on the piling, occasionally adjusting to the gusts. The bird spreads its wings half open. I putter the boat ever so slowly closer. I look at it again with the binoculars, bracing my knee and elbow against the console.

This is my first perfect, crisp view of a breeding cormorant this year. The bird adjusts its head with that nonchalant look that really means *I'm watching you carefully*. I keep the cormorant in my binoculars' view as I motor still closer, but ever so slowly so as not to scare it.

The cormorant spreads its wings slightly wider. In the clean telescopic circle of view I see the object of my affection, of so much study and searching, of so much travel and reading and thought. The cormorant leans forward. It perks up its sharp black tail. The bird edges forward farther. Then, just before leaping off the piling to fly upriver, it spritzes a long stream of white shit, which splatters into the water.

The double-crested cormorant is native to the islets around the Mystic River, to Fishers Island Sound, and to New England. The cormorant is not invasive. The cormorant is not alien. The cormorant is not exotic. I agree with the position of Linda Wires and her colleague Francesca Cuthbert, who say that the cormorant population is not overabundant in any region of North America, nor is the cormorant population anywhere beyond what managers call "carrying capacity" in any natural biological or historical sense.[18]

Because North America was largely colonized by English speakers sailing to the eastern coast, our best early accounts of double-crested cormorants are of what managers now call the Atlantic population (*P. auritus auritus*). Some of the early reports of explorers and settlers of the Canadian Maritimes and New England, however, might have not specified or known if they were observing another species of cormorant, the great cormorant (*P. carbo*), which is the same species the men would have recognized from Europe. The great cormorant has lived on the Atlantic coast in smaller numbers than the double-crested, and the great seems to have ranged from Labrador to as far south as Long Island Sound. These days they can go as far as South Carolina for the winter.[19] This winter range might be expanding because of climate change.

Dozens of great cormorants show up in the Mystic River and on Gates Island each winter. They are gone by February. They do not nest any farther south than northernmost Maine.[20] Around November or in the winter months I've often seen greats and double-cresteds standing on Gates Island together, although they usually stand with their own kind on respective ends of the little islet, like middle school girls and boys at a dance.

With the potential confusion between double-crested and great cormorants in mind, there are several early seventeenth-century records of large numbers of cormorants all along the Atlantic coast, as well as of Native Americans capturing these birds for food. In 1604 Samuel de Champlain named an island south of Nova Scotia "the Isle of Cormorants," because "of the infinite number of these birds of whose eggs we took a barrel full." As Champlain continued sailing to the west he wrote: "The abundance of birds of different kinds is so great that no one would believe it possible unless he had seen it — such as cormorants."[21] In 1610, an anonymous author describes cormorants in the Virginia rivers "in such abundance as are not in all the world to be equaled."[22]

In 1634 colonist and author William Wood observed Native Americans in Massachusetts gathering "drowsy" cormorants from the rocks of the rookeries "as easily as women take a hen from a roost."[23] This was a few decades after the plays of Shakespeare, who used cormorants to represent gluttony as consistent with the language of his time. Thus, as far as I've been able to find, William Wood left the first North American complaint about the cormorant's harmful greed. Wood wrote that the "common" cormorants "destroy abundance of small fish."[24]

In 1643 Roger Williams described Native Americans catching and eating cormorants around the Massachusetts–Rhode Island region, a short distance from Gates Island and the Mystic River.[25]

Then a few decades later, John Josselyn described native men hunting cormorants in Saco Bay, Maine: "[Cormorants] roost in the night upon some Rock that lyes out in the Sea: thither the Indian goes in his Birch-Canow when the moon shines clear. . . . So soundly do they sleep that they will snore like so many Piggs. . . . He clambreth to the top of the Rock, where walking softly he takes them as he pleaseth, still wringing off their heads; when he hath slain as many as his Canow can carry he gives a shout which awakens the surviving Cormorants, who are gone in an instant."[26]

In the following eighteenth and nineteenth centuries, colonists and settlers up and down the Atlantic coast shot seabirds for food, killed birds for feathers and sport, and gathered their eggs for food or as a hobby. As firearms improved, Americans and Canadians shot birds for sport. People settled on islands and along the coast for land, lumber, stone, and access to fish and lobster.

Meanwhile, the old literary, cultural attributes ascribed to the cormorant had sailed right across the Atlantic to the colonies. An Irish physician and naturalist named John Brickell, for example, who drew the first known European illustration of the double-crested cormorant, wrote in 1737 about how in North Carolina these birds prey on herring, "whereof they are very ravenous and greedy." Brickell continued that these cormorants were "very strengthening to the Stomach, and cure the Bloody Flux. The Flesh is black, and hard of digestion, therefore seldom made use of."[27]

Up north in New England at least, fishermen sometimes used cormorants as bait for catching groundfish, since the flesh held up well underwater. (Cormorant meat was used for fishing bait in Scotland up through at least the 1960s). As the human population expanded along the Atlantic coast, particularly in New England, people gradually displaced cormorants and other seabirds from their traditional rookeries.

In 1834 the famous ornithologist Thomas Nuttall wrote of "numerous dense flocks" of cormorants gathering in Massachusetts Bay at the approach of winter: "Several dozen have been killed at a [single] shot," he wrote.[28] These were migrating birds, because by this time, as observed by his contemporary John James Audubon, cormorants in this region bred only in Labrador.[29]

In his classic study on the double-crested cormorant, Harrison Flint Lewis wrote in 1929 what had already been true for decades: cormorant numbers were greatly reduced in Québec, were practically gone too from Nova Scotia, and there wasn't a single nesting pair in all of New England, on the coast or inland.[30]

Motoring south, I steer under the drawbridge. I see people watching me from windows of the seafood restaurant. They are warm and drinking wine. They eat salmon, lobster, flounder, maybe even some U.S. farm-raised catfish. I continue toward the railroad bridge, past a few of the yacht marinas. Nearly all the boats are still hauled up in parking lots, covered in plastic.

As the river opens up and I put down the throttle, I pass still more marinas and the mansions that are built down to the water's edge. The clouds are low, gray, and knuckled. The sun appears occasionally over the clouds and then tucks back in. The water is gray green. I zoom past several long-tailed ducks. The birds are probably stopping here on their way to their breeding rookeries in the Arctic.

The mouth of the Mystic River opens into a bay. Now I can run the boat at full speed. There's one guy out on a workboat dredging for scallops. This has been an encouraging local renewal, signaling an improvement in water quality. When I turn past the sandy spit I can see Gates Island off in the distance.

Suddenly I have to cut my wake to slow down. There's a man in a shallow boat floating beside a spit of sand. He's dressed in camouflage and holds a shotgun. A duck hunter. We nod at each other.

Once past, I put the throttle back down. From here Gates Island appears as only a pile of rocks. I can really only see on the horizon two boulders and a few smaller ones a short distance to the south. I don't know if you've ever been scuba diving, but for me Gates Island is like this. When you go diving it doesn't look like much on the surface, but then when you get under the water, like the opening of a curtain at a Broadway show, an entire world of magic immediately opens up. Gates Island doesn't look like much from afar. And that's probably a good thing.

I slow down as I approach, when I can see the full outline of the islet. It looks like there are a couple dozen cormorants and a dozen black-backed gulls. I can tell even from here that these are all breeding adults. With the sun behind the clouds, the light on the river is soft, so the contrast of the birds on the island absolutely pops: the sharp black of the cormorants over the gray rocks and the sharp black wings of the gulls against their crisp white heads and chests. I know that with no chicks yet to worry about, the cormorants will be even quicker to depart if they sense I'm getting too close. During the years I used to visit here every couple of weeks, I swear the cormorants and gulls recognized my boat. When I went a couple of times in a different boat, it took longer for them to fly off.

I still have the wind behind me, so I put the engine in neutral and coast toward the island. Using the binoculars makes me a bit seasick out here, so I have to put them down often. The closer I get, though, the more birds I realize are here. I count thirty cormorants. I see one pair leaning toward each other, touching beaks, nibbling each other's neck.

There doesn't look to be anything at all left on Gates from last year's nesting. Presumably it was washed cleaner than normal by last fall's tropical storm. It seems like even the chimney foundation has been moved twenty feet or so. I still find it hard to believe that there was once a house on Gates Island — the space just seems entirely too small — but that's what the locals recount, and there's the chimney foundation and even some steel rebar as witness. The house was supposedly crushed and washed off Gates in the 1938 hurricane.

Drifting in the boat, I think I'm giving the island a wide enough berth as I steer to the south, but the gulls sound the alarm. All the cormorants fly up.

I count well over fifty deeply black cormorants silently rising into the wind. Most veer around, over my head, look down at me, and then curve into a landing on the water on the other side of the island. I slowly motor around the south and toward where the cormorants have gathered on the surface. They cautiously swim away. Then they take flight nearly all at the same time, via some silent intuitive communication. They fly in a big arc and return back onto the island.

I steer around a couple of buoys that are connected to lobster pots on the bottom. They're privately owned, and no one bothered to pick them up for the winter, although they might have been trapping and retrapping fish, invertebrates, or even a cormorant, a merganser, or a loon who ventures in after the fish inside. I look carefully at the birds on the island again. They are all breeding adults: no juveniles have arrived back yet. I motor around to where I can see the depression beside the two boulders. On this frigid windy March day, it looks to be that a pair of cormorants has already begun to build a nest.

Feeling a bit adventurous and having some spare time, I decide to drive the boat south to check in on South Dumpling Island. It's a couple of nautical miles across open water. I know it's pretty stupid. It's going to be a brutal slog back against the wind. But what would Niall Rankin do?

I make it across the sound just fine. No one else is out on the water, and the clouds are clearing away. I have to watch the boat's stern carefully to not do too much surfing on the following swells. Then once I make it around the bend of South Dumpling, I'm in calm water and out of the wind.

The island of South Dumpling is owned by a Connecticut nonprofit called the Avalonia Land Conservancy. This organization owns several properties that are preserved as undeveloped space, most of which are open to the public. They maintain trails for walking. They sponsor public talks. They do all sorts of good things for the community, for my community. At about three acres, including the hill that makes up most of the island, South Dumpling is more than ten times the size of Gates.

I survey South Dumpling from the boat and see only a few gulls. No cormorants this afternoon. In a month or two about one hundred pairs will settle here to breed—about half the number that are on Gates. They nest in the few remaining trees and on the northwesterly hillside. Gulls nest here, too, in a similar ratio to that on Gates Island, about one gull for every twenty or so cormorants.

Cormorants have not been as large a public issue these days in the

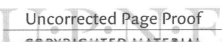

Maritimes or in New England—as compared to on the Columbia, in the Mississippi Delta, and in Henderson Harbor. Yet this little island of South Dumpling provides a nice microcosm of the larger issues that humans have with cormorants all around the world.

The Avalonia Conservancy has been watching the cormorants over the last two decades slowly kill off the trees on South Dumpling with their guano. Some members fear that soil erosion is to follow. For several years, herons, ibis, and egrets nested here on South Dumpling, but when their nesting, along with that of the cormorants, began to deplete the dense foliage, these birds began to leave, while the cormorants could remain because they will nest in bare trees or on the ground. Anne Nalwalk, the former president of the Avalonia Conservancy, explained candidly: "We'd certainly rather see egrets than a bird that's destroying all the vegetation."[31]

The conservancy works with a local marine-science educational program called Project Oceanology, which teaches students of all ages about marine science and Long Island Sound. For several years, Project O, as they're known around here, landed instructors and students on South Dumpling to test various cormorant deterrents. They tried anti-erosion mats, snow fences, and nets draped over the trees.

Recently Project O hosted a television show called *Aqua Kids*. In front of the camera, the Project O director, Thaxter Tewksbury, showed the student-actors around South Dumpling, teaching about the gulls and cormorants.

When they came to a nest of cormorant chicks, Tewksbury told them: "Their chicks are probably the ugliest babies you'll ever see."[32]

The show aired in the fall of 2011. At one point, Tewksbury gives a short history of the island's birdlife.

A girl responds: "So it seems like these cormorants have kind of come in and taken over this entire area. Does that make them an invasive species?"

"It certainly makes them a very successful species," Tewksbury says. "They've moved from an island nearby where they had a small rookery on a beach face [Gates?] and taken over the top of this island. So in the sense that they have invaded it and displaced the other species that were here, that's true. Certainly they are a common species throughout this area and have caused problems in other areas as well, in terms of displacing other birds."

Tewksbury continues: "Whether that's a good thing or not, that's for someone else to say."[33]

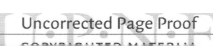

While the group stands around on the colony during the interview, gulls eat cormorant chicks and cormorant eggs. Tewksbury tells the kids this is a normal thing. The gulls are aggressive. The camera focuses in on three gull chicks gobbling up a cormorant chick. Tewksbury does not explain, at least on camera, that this is happening at this moment *because* they, the people, are standing there disturbing the colony.

My point is that South Dumpling Island serves as a small example of a place where people want to displace cormorants solely because of an aesthetic ranking of species. It also serves to show how we gently, subtly pass on these views to our children through our well-intentioned teachers and our stories. There is some local concern that cormorants are affecting fish stocks, such as flounder, but this is largely anecdotal. The small collections of pellets that several student groups and I have collected over a dozen years on Gates Island show that the cormorants eat a large variety of species, a very small percentage of which are flounder or any other fish of commercial interest.[34]

Thaxter Tewksbury did not return my requests for an interview, but Brae Rafferty, a senior instructor at Project Oceanology, and a veteran of over thirty years on the sound, told me a few years ago: "When you're out on the water all the time and seeing the birds, you think 'They've got to be feeding on something.' The winter flounder aren't coming back. I've seen cormorants eating flounder, so between them and the seals, there's got to be some impact."[35]

Anne Nalwalk, the former president of the conservancy, knows that her objection to the South Dumpling cormorants is in "the eye of the beholder." Her concern is not based on the local environment's true carrying capacity, but rather what biologists have termed our "wildlife acceptance capacity."[36] I think Anne derives her opinion primarily from cultural factors: learned aesthetics, negative portrayals in literature and in the media, and a nostalgia for the way things used to look within her memory. Yes, cormorant numbers increased quite dramatically in the Atlantic population in the last third of the twentieth century. But in this part of the continent, without any culling or egg-oiling programs, the cormorant population increase seems to have minimized in New England. Current region-wide counts are hard to come by around Long Island Sound and Southern New England, but in 2001 I counted over two hundred nests on Gates Island. I counted the colony in 2012 and found only seventy-five active nests.[37]

It is hard to prove native nesting in this sliver of Southern New England, but I have found an 1847 chart that names a "Cormorant Reef" less than a nautical mile from Gates. There is also a "Cormorant Rock," which is about four miles to the west of South Dumpling. Archaeologists identified cormorant breeding before European contact as far south as Boston Harbor—about one hundred nautical miles away, as the cormorant flies.[38] So considering early explorers' accounts, these chart names, and this Boston record, it seems more than likely that cormorants once nested in this little part of the coast—or at least spent significant time here each year—whether that was on Gates Island, South Dumpling, or any number of the small islands within the region of Long Island Sound and Block Island Sound.

Yet even though they are clearly native and there is no real evidence they are affecting much of anything in New England, cormorants are still a human target beyond South Dumpling. On Martha's Vineyard, Massachusetts, officials are applying for federal permits to kill cormorants and haze the rest with pyrotechnics in order, they say, to protect their local herring and flounder. Once again, the human-cormorant conflicts are formed and accentuated by subtle, even unconscious word choice. Journalist Remy Tumin opens her article on this situation on the Vineyard by writing that if the group of people gets the permit, "the Island will soon see fewer of the pesky birds, whose voracious appetite threaten fish restoration projects."[39]

In 2003, two brothers on Martha's Vineyard shot a dozen cormorants illegally. Tumin quotes Bret Stearns, the director of the natural resources department for the Wampanoag Tribe of Gay Head, which is one of the groups applying for the permit to kill the birds. Stearns says of the cormorants: "They're unbelievably gluttonous."[40]

An even stronger article came out in the summer of 2012 in a magazine called *On the Water*, New England edition, in which Charley Soares in his final-page column titles a piece on cormorants "Black Death—the Birds from Hell." I'm not making this up. He relays all sorts of biological nonsense, such as that they have "sharp pin-like teeth" on their beaks and that their guano can burn through a rubber boot. Soares tells of himself running on a Rhode Island beach and scaring them off. He calls them "ravenous." The "bandits" come in "black hordes." They "devour." They are "destructive." Soares sees cormorants as decimating local fish stocks. He writes: "The serious threat from cormorants is not exaggerated."[41]

Over the course of my studies to examine the human relationship with cormorants across cultures, across time, and across academic disciplines, I have arrived at six distilled lessons about these birds and what I've learned about our larger connections to the natural world.

First, only a few organisms on earth—a few other seabird species—have anywhere near the capacity to both fly above the clouds and migrate the width of a continent, and also be able to stop, land on the surface, and dive deeply underwater to hunt in the cold and dark.

Not only are cormorants practically unmatched in the animal world, but no human invention has yet been able to successfully replicate this combination, either manned or as a robot. People are trying. One proto-type diving and flying unmanned plane is under development by Lockheed Martin for the American military. It is called the Cormorant, and will be able to be launched from a submarine.[42] (The Canadian helicopter class that is most often used to rescue mariners at sea is also called Cormorant.)

Of those few other seabirds such as mergansers and puffins that have flying and swimming abilities comparable to those of the cormorant, none individually as a species or in their genetic families has the vast global and habitat range of the cormorants. Though they need toxic-free, mammal-free rookery space, the cosmopolitan cormorant as a single species (for example, the great cormorant) and as a family of species (Phalacrocoracidae) can live in fresh water and salt, beside sand, mud, marsh, rocky shore, and urban coastline. Cormorants occupy nearly every shoreline around the world except the islands of the central Pacific. They can live at sea level or at the mind-numbing elevation in Nepal of almost thirteen thousand feet (3,960 m).[43] Cormorants can nest in trees, on dirt, or beside snow. They can perch, waddle, and when darting underwater in the dark can turn on a dime at nearly full speed. They can capture and swallow an eel thicker and wider than their own neck. They can collaboratively herd a school of fish and work together to grab an especially large one.

We are still learning about the intelligence and emotional life of ani-mals. Like cats and dogs, cormorants are curious. They have different in-dividual personality traits. They engage in what can only be termed "play." They recognize group hierarchy. You can train a cormorant to catch fish and cough up the catch into a basket. They can count to at least seven.[44] Recent research shows that pairs of seabirds, such as gannets, spend more time doing mutual preening, neck entwining, and other courtship behaviors in relation to how long the pair has been apart. In other words, a pair of birds seems to reunite after a winter away from the breeding colony or after one has been gone unexpectedly for weeks.[45] Probably very few if any cormorant individuals are monogamous for a lifetime, but after extensive observations of a cormorant colony off the coast of Maine, Howard Mendall wrote in 1936: "Frequently I have seen mated Cormorants continue to spend several minutes at a time doing nothing but showing affection long after their young had left the nest."[46]

For tens of millions of years wild cormorants have been figuring out how to fly long distances and then return to their home several months later. Cormorants can fly in dense fog or in a snowstorm. They can live in freezing temperatures or in equatorial heat.

Just as the human body is a miracle, dung beetles are miracles, kelp forests are miracles, and elephants are miracles, so too cormorants, individually and collectively, are awe-inspiring miracles.

A second lesson I've gathered from this study is how much we still have to learn about the cormorants, or any birds.

This is especially evident with the blue-eyed shags and the Galápagos flightless, but it is true for all the cormorants. Despite this family of birds' global range and relative accessibility, biologists do not know with significant confidence why some cormorant species spread their wings and others do not, how and where exactly they migrate, how long they live on average, why they have the coloration they do, how exactly they stay warm and dry, why one species dives deeper than the other, how exactly they hunt underwater and with which senses, or how, when, and why one cormorant species on an isolated island group lost the ability to fly. New technologies, such as DNA analysis, satellite tagging, and underwater "cormorant cams," continue to open up clues to some of these mysteries.

What is most critical from a cormorant-human conflict perspective, as it has been for centuries, is our imperfect ability to figure out what and how much cormorants are eating or to calculate just how many birds are out there. Population numbers are estimates. They can be misleading or confusing. We need to be discerning when reading consumption estimates, which might overestimate the number of fish taken by seabirds. Some of these estimates might be hasty, back-of-the-envelope calculations using inflated numbers, small sample sizes, or assume a constant rate of eating all season and/or all day long.[47]

Cormorants do not eat garbage or dead stuff. Cormorants eat an extraordinary variety of fish and invertebrates. By some estimates, cormorants, because of their high metabolism—necessary for their ecological niche—can indeed eat slightly more by percentage of their body weight than do other piscivorous birds, such as terns. But to suggest they are gluttonous or greedy or different from any other bird trying to survive is unreasonable. If there is not enough clean, safe food for them to eat, a colony of cormorants, like any animal, will move somewhere else or die. Gull species have been able to thrive on human development and garbage; cormorants cannot.

Nearly every study out there—which all have fundamental methodological limitations—points to the fact that cormorants do not focus on fish that people want to catch in particular. They feed on what is most available. Certainly, if there are a lot of cormorants, in the thousands, they are going to eat a lot of fish of any given species, even if that species is the one the local human population most desires and even if it is a small percentage of the cormorants' total diet. Cormorants eat perhaps about a pound of fish per day, with huge variations by individuals, species, time of year, and within an individual's life cycle.[48] For example, they eat much less when older and when not raising chicks.[49] But, again, cormorants do not aim for any single species, unless the fish are in a great concentration in a single place. Outside of vast upwelling systems such as that of the Humboldt or Benguela Current, these concentrations of single fish species usually happen for heavily anthropogenic reasons, such as at aquaculture sites or in artificially stocked ponds or lakes.

Biologists are continuing to learn how cormorants fit into a given region's ecology and how the birds' population growth or decline impacts other underwater species. Fish-eating birds are a natural if not essential influence on wild fish populations. Many ecologists believe that cormorants and other piscivorous birds help to thin out slower, older fish, and eat a given fish stock's prey. So these birds might increase a wild fish stock's production and vitality in the long run by helping the fish population maintain an optimum density in a dynamic natural system.[50]

In 1929 Harrison Flint Lewis wrote in his seminal cormorant opus: "The principal complaint against the Double-crested Cormorant is that it eats fish. The mere fact that a bird or a mammal eats fish is often considered sufficient to condemn it without more ado, quite regardless of the fact that many fish eat fish, and that some other fish are of no commercial value, so that the habits of a fish-eating creature, when carefully inquired into, may be found to be beneficial, even from the point of view of man's narrow self interest."[51]

Often, in fact, it is other fish species or man that is doing the vast majority of the consuming of a given fish species or their eggs.[52] So, in short, we don't have a clear sense of what's happening under the surface. And whenever we have tried to remove a predator from an ecosystem, it usually backfires.

A third lesson: cormorant taxonomy, like our assessment of almost all life on earth, remains fluid.

Not only is the description and hierarchy of known cormorants continually debated as to the names and classifications of genera, species, or recognition of subspecies, but there have been in recent decades claims of entirely new cormorant species. For example, one potentially new cormorant is the "Kenyon's cormorant" (*P. kenyoni*), identified from an archaeological dig found on Amchitka Island, Alaska, and a few recent skeletons. Not all biologists, however, accept the Kenyon's cormorant. It might be a variation of the pelagic or the red-faced cormorant.[53]

The scientific or common names we give to the cormorants certainly do not matter to the animals. They are, however, relevant to our current navigation of complex management decisions and their urgency. A few cormorant species may be extinct before we have any real opportunity to do something about it. This includes the bank cormorant, the Chatham shag, the Pitt shag, the Galápagos flightless cormorant, the Socotra cormorant, and the cormorants along the Peruvian coast — even the guanay — as well as all some of the other isolated blue-eyed shags fishing in the icy waters throughout the Southern Ocean. It is too easy to forget that less than a century ago populations of double-crested cormorants in North America and great cormorants in Europe were locally eradicated. Managers had declared them endangered and wrote laws to protect them.

We do not know with confidence the historic evolution or distribution of the cormorants. It is interesting to look at where on earth there is a greater diversity of cormorant species living sympatrically — New Zealand, Alaska and the Pacific Northwest, southwestern Africa — and to see that this matches where more diverse groups of fish and invertebrates live: in areas of ocean that are older geologically. Maintaining our global diversity is important, especially given how quickly we have been altering the earth with loss of species, wild habitat, and climate change. As one ecologist who specializes in marine invasions told me: "We just can't be certain which genotypes are better equipped to survive in the future."[54] Liam O'Flaherty and Kurt Vonnegut agreed, emphasizing in their fiction the element of chance in an animal's survival — including ours.

The fourth lesson that comes up again and again with cormorants, across cultures and across time, is that in almost every case our problem with cormorants has been the direct result of our own busy, heavy hand. This is exemplary of so many of our interactions with the natural world.

As evidenced by Russ McCullough's comment that "there is nothing natural in the Great Lakes," and the complexity of restoring salmon runs on the Columbia River, it can be difficult to declare now what, if anything, has *not* been affected by humankind. But the renewal, the recovery of cormorants in what some of us deem commercially or aesthetically inconvenient places, has been regularly assisted, even created by ironic, often direct, human endeavors. We have constructed open-air aquaculture ponds along the migration routes of these birds in North America, Europe, the Middle East, and Australia. We have stocked recreational fishing lakes near protected mammal-free islands. We have built reservoirs and constructed mammal-free rookery space in areas of traditional fish densities. Elsewhere, we have crowded every bit of shoreline and populated every isolated islet, leaving the birds little place to go.

The fifth lesson is that because of how and where they fish, cormorants are scapegoats for much larger, more difficult environmental concerns.

We are shooting the coal mine canary. We are poisoning the messenger. We are taking a pickax to the tip of an iceberg because this is easiest to reach. If there are not wild fish and clean water and safe habitat, cormorants will not be there. Did you notice how few of the stories I tell here have any voice from a coastal or deep-sea fisherman?

Paul Heikkila, a commercial fisherman and policy official in Oregon, told me: "Personally, and I think for most of the fishermen, birds are what we look for. They tell you where the feed is." When I met with him, Heikkila was pleased, for example, to see the murre population rebounding. "Then you know the ocean productivity is coming back," he said. "It's not smart to go shoot all your indicators."[55]

As we well know, our salt and freshwater fisheries all around the world

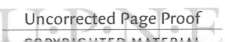

have big, sticky problems: coastal pollution in all its forms, the introduction of species, and a tireless industrial overfishing. The money we spend trying to manage cormorants—which taxpayers will need to cough up indefinitely, unless we wipe the birds out forever or pave over every patch of nesting area—could be much better spent to reduce our runoff, plastics, our destruction of coastal habitat, and to secure more conservation land and marine preserves. To be fair, the Angling Trust in the UK also has many of these sorts of initiatives, but it should take the cormorant bit completely off the table. The organization's campaign against the birds is unethical, illogical, comic, and a distraction from the larger, harder, less visible problems. Their videos, along with others' ill-informed diatribes in popular magazines all over the world, such as Charley Soares's "Black Death" piece, are, at the very least, unhelpful.

I like to fish. I love to eat fish. Over the years, I have found Ron Ditch from Henderson Harbor to be quite an amiable and worthy man, but I obviously do not respect or understand his tirade on cormorants—nor do I understand that perspective from any sport fisherman. Fishing for me has always been about getting outside and tinkering around with the challenge of landing a fish to eat. I would like to see the sport fishermen who rage against birds reconsider what they are really out there for. Recreational fishing is a hobby and a pastime and a First World privilege, especially in stocked systems. Tourists will continue to travel to beautiful, peaceful destinations to fish, and given the right setting and experience, catching two or twenty is less relevant for a clientele that is raised in the sport ethically and progressively.

It is important to note that participation in recreational fishing has been steadily dropping, at least in the United States and the Great Lakes. These sorts of declines do not seem to correlate to towns that have a "cormorant problem."[56]

It seems it has always been the easy visibility of cormorants eating large fish that seizes man's imagination as a simple cause of a fisheries decline. Cormorants catch fish within easy reach of the shoreline. When they have a big one they bring it up to the surface in order to orient their prey to swallow it. Ron Ditch watches cormorants swallow bass in the boat channel just off his front porch.

That said, people do need to eat. I'm inspired by Nola Parsons—and my wife—in their choice to eat vegan or vegetarian. With our wild fisheries in such decline, aquaculture is an important aspect of our modern world. We want it to succeed. Our tax money would do well to work with men such as Jim Steeby to figure out ways to make it work—in the Mississippi Delta, in Romania, and in Israel. If fish farmers could avoid killing birds in the process, all the better. Jim wrote to me recently: "The Chris Nerrin farm has just been reduced to less than one hundred acres. High risk, long hours and low profits have overtaken another long standing catfish operation."[57]

Aquaculture producers are the only interest groups for which I have sympathy when it comes to killing wild birds for any reason besides eating them. I would prefer, frankly, if organizations such as Bird Research Northwest would suspend their killing of cormorants for stomach contents, regardless of the scientific benefits and this being the most accurate way to tell what the birds are eating. I'd like to see the dissuasion programs end on the Columbia River, with the money and energy spent on other, more positive methods to restore salmon. I would prefer also that Russ McCullough and the wildlife biologists on Little Galloo Island suspend their egg oiling and their culling of birds; there seems no convincing evidence at this time that the birds are vastly detrimental to the recreational fishing industry in eastern Lake Ontario.

Hunting seasons are not a management option, because shooting for sport would have little effect on reducing cormorant populations, unless hunters were allowed on the rookeries. Hunting seasons on cormorants, regardless, would set a dangerous precedent.

While I'm on my high horse, I should tell you that I've personally caused the death of scores of cormorant chicks over the years during my trips to Gates Island and South Dumpling and to Stratton Island in Maine. In my own way, I did the same as Thaxter Tewksbury and Project O. I used to rationalize that taking students onto the island during prime nesting time to show them cormorant chicks, or my collecting of pellets or regurgitants, was enough to merit the sacrifice of the chicks to the beaks of the waiting gulls. These days I do not go on Gates until the chicks are too large to be eaten.

I should also state openly that my judgment and opinion here are from a chair, a place, where cormorant densities do not affect me personally. My livelihood is not connected to recreational fishing, to aquaculture, nor do I live near a colony that I believe to be dramatically altering a habitat I care deeply about. Yes, certainly, if my personal property or business was being diminished or even destroyed by a given cormorant population, I might feel differently.

Do I kill the rats that burrow into my basement and into the walls in my house? Yes.

I rank animals, too. We all do.

The sixth and final point from this cormorant study is that literature, story, language, art, and visual imagery have had a significant effect on our collective antipathy or ambivalence toward this bird, as well as our wildlife management policies.

The cormorant's appearance and behavior align with our collective, often subliminal fears: we intuit a dark, even malicious intent from this animal. Most cormorants appear all black from afar. They are usually silent. Their neck is serpentine. Their altricial chicks are lizard-like. And they have the unmistakable notorious behavior of standing still, with their wings spread, on a piling or a rock beside the water. Like other birds, their guano kills the trees in which they roost.

The wise owl, the sly fox, the fun-loving dolphin, the gluttonous cormorant. The traits we attribute culturally to cormorants, as with other animals, have their derivations in witnessed behaviors — we can readily see these birds eating large fish — but also in early writings, which perpetuate themselves into stock characteristics. The stories we read and tell each other, the poems, the novels, the films — they not only reflect our cultural perception of an animal but also help form it. In the Bible, *The Odyssey, Paradise Lost*, in our very earliest texts, cormorants have been depicted in Western, English-language literatures as ravenous and evil, often as an omen for storms and death. Only in recent decades, with modern environmental writing, have cormorants been shown consistently with sympathy, or at least depicted more equitably, with an attempt to consider the perspective of their own world, indifferent to our needs and without our emotions.

In some future study, I or someone else will do a thorough analysis of world literatures to see if this trend is universal. In Japanese, as Junji Yamashita told me, the character for cormorant means "little brother bird." Yet in Hindi, the word for cormorant, *pān-kowwa* or *jăl-kowwa*, means "water crow," and in India the birds are often depicted as gluttons.[58] How has the perception of this animal evolved over time in other parts of the world?

Storytelling in whatever fashion, the building of character, remains the most significant way to shift public, cultural opinion. A work like Anna Kirk-Smith's table can have an impressive reach. Those trying to stop a cormorant cull at Middle Island on the Canadian side of Lake Erie have begun a Twitter account from the voice of a character they call "Cormy the Cormorant." Cormy "tweets" missives like: "OMG. One of my friends was shot and flew out over the water and fell from the sky . . . She must be dead."[59] (OK, yes, even I admit this is silly, but I appreciate the attempt.)

We rarely make decisions on facts alone, of course. We rarely make decisions on facts even primarily. We should not assume, as Russ Mc-Cullough intimated, that our scientists or wildlife managers are any different — on any higher moral ground — than our fishermen or farmers.

Our opinions, our selective morality about the natural world, are usually formed early in life. In 1992 Maine author Jane Weinberger wrote a children's book titled *Cory the Cormorant*. Weinberger told me: "I wrote it to teach children not to accept the status quo. To really understand that most birds are worth preserving. And this one particularly, where a lot of people throw stones at them, shoot at them, and everything else. They're really very good birds. Not pretty to look upon, but I like them."[60]

Celia Thaxter, a nineteenth-century poet who lived much of her life on the Isles of Shoals off the New England coast, wrote a ballad about a cormorant in a collection for children. In this she comments on how a little knowledge, open-mindedness, and experience with this animal, as with anything, can alter our lasting impression. Published in 1878, it is the earliest literary work I've found on either side of the Atlantic that is overtly sympathetic to the cormorant:

THE SHAG

"What is that great bird, sister, tell me,
      Perched high on the top of the crag?"
"'Tis the cormorant, dear little brother;
      The fishermen call it the shag."

"But what does it there, sister, tell me,
      Sitting lonely against the black sky?"
"It has settled to rest, little brother;
      It hears the wild gale wailing high."

"But I am afraid of it, sister,
                    For over the sea and the land
It gazes, so black and so silent!"
                    "Little brother, hold fast to my hand."

"Oh what was that, sister? The thunder?
                    Did the shag bring the storm and the cloud,
The wind and the rain and the lightning?"
                    "Little brother, the thunder roars loud.

"Run fast, for the rain sweeps the ocean;
                    Look! over the light-house it streams;
And the lightning leaps red, and above us
                    The gulls fill the air with their screams."

O'er the beach, o'er the rocks, running swiftly,
                    The little white cottage they gain;
And safely they watch from the window
                    The dance and the rush of the rain.

But the shag kept his place on the headland,
                    And when the brief storm had gone by,
He shook his loose plumes, and they saw him
                    Rise splendid and strong in the sky.

Clinging fast to the gown of his sister,
                    The little boy laughed as he flew:
"He is gone with the wind and the lightning!
                    And — I am not frightened, — are you?"[61]

I finish circling around South Dumpling Island. I turn the boat north and against the winds and the chop. It's going to be a long, cold, wet ride home.

I'm almost across the sound when the engine putters out. I try to start it up again. I check a few things, but still no luck. I'm slowly drifting back toward open water, so I reluctantly grab the oars. The boat is a bit too clunky to row swiftly, but it does make headway, and the time I might waste trying to get the engine going would be better spent at least making slow progress toward the first marina at the head of the Mystic River. It's too deep here to anchor.

I'm out of breath quickly, yet making slow progress. There is nobody else in sight on the water. It's nice to be spared the embarrassment, but I'm feeling a little insecure out here without a working engine and the wind and seas stronger than I'm comfortable with.

Then, I'm rewarded once more.

When I look over to the west, toward Long Island Sound, I see a thin black line wavering across the sky. Perhaps it's a flock of undisciplined geese, but even from here, from a distance, I can tell they might be cormorants. They are smaller, more slender, the wing-beats quicker. The string of birds seems to almost pulse within its fluid formation. As they gradually descend from their high flight, I can see they are definitely cormorants. Several of the birds in the line pause from flapping for several seconds. They soar, then resume beating their wings.

The cormorants now fly close overhead. I can see by their light-brown chests and plumage that several of these birds are juveniles. I can just hear the shushing of their wings as they descend toward Gates Island.

March

The one remaining female from the first clutch, after migrating nearly the length of the continent, veers with the flock around and down toward Gates Island. As have her relatives for tens of millions of years, eons before *Homo sapiens* walked the earth, the cormorant pulls herself up vertically. She opens her wings and splays her feet wide. She lands softly on one of the two boulders, beneath which a pair of adults has already begun crafting a new nest.

What does a cormorant feel then? Is there any measure at all of relief, of satisfaction, of a warmth toward a remembered home?

Acknowledgments

This project would never have taken flight if not for my tolerant master's thesis adviser, Jim McKenna. As I studied all things cormorant and traveled around, I dragged in many friends and colleagues as members of the Brave Research Team. These kind figures provided curiosity, patience, and good humor while helping with fieldwork, logistics, and expertise in a wide range of disciplines: Munro Johnson, Jon "Hoss" Mitchell, Jenny Doak, Ret Talbot, Sean Donaghy, Nicole Dobroski, and Rush Hambleton.

A number of people over the years have given generously with their time and assisted with logistics and local contacts, helped locate images, and read drafts of previous cormorant articles and the chapters of this book. A few of them I have bothered further still to include directly in this text, for which required still more of their patience and expertise. Thank you especially to Kazuto Hino, Mari Takada, Junji Yamashita, Ron Ditch, Russell McCullough, Irene Mazzocchi, Linda Wires, Chip Weseloh, Elisa Goya, Jose Carlos Marques, Adam Peck-Richardson, Megan Gensler, Don Lyons, Daniel Roby, Jim Steeby, Chris Nerrin, Larry Brown, Neil Bernstein, Bob Madison, Hugh Powell, Robert Prŷs-Jones, Jennifer Doucette, Anna Kirk-Smith, Carlos Valle, Elaine Hayes, Kevin De Ornellas, Nola Parsons, Terry Doyle, Brian Dorr, Rónadh Cox, David Oehler, Tony Amos, Gail Kern, and Shauna Hanisch. Thank you to the readers of the full manuscript whose expert feedback at different stages informed and enormously improved this book: Becky Kessler, Jan Hodder, Svati Narula, Timothée Cook, and editor Stephen Hull at University Press of New England, who so kindly supported and developed the project. I'm grateful for the great care, eye for detail, and far-reaching knowledge of production editor Peter Fong and copy editor Glenn Novak.

I received financial and institutional support in various stages from the Maritime Studies Program of Williams College and Mystic Seaport, the American Sail Training Association, Wesleyan University, the Overseas Research Scholarship, the University of St Andrews, and the Text and Academic Authors Association. Thank you to organizations that hosted me on various expeditions, notably IMARPE, PROABONOS, Bird Research Northwest, the National Audubon Society / Stratton Island, Sea Education Association, the *RV Professor Multanovskiy*, Ocean Classroom Foundation, and the film crew of *Catch It!*

I am indebted to various libraries and their staff, especially Shain Library at Connecticut College, Sawyer Library at Williams College, the University Library at St Andrews, the Blunt White Library at Mystic Seaport, and the Kemble Special Collections Library in Mystic. I owe a special and continual thank you to Alison O'Grady, interlibrary loan supervisor at Williams.

None of this could have been accomplished without the institutional, academic, logistic, and social support of colleagues and students at the Maritime Studies Program of Williams College and Mystic Seaport. For reasons unknown, the director, Jim Carlton, has remained a supporter of my cormorant studies in dozens of ways. He and mentor Mary K. Bercaw Edwards allowed me a year's research leave.

The largest thanks goes to my wife, Lisa, of course. I'm looking forward to many new adventures with Alice Day, where you look at rocks and I look at birds.

Appendix

CORMORANT SPECIES OF THE WORLD AND IUCN RED LIST STATUS

See chapter 6, among others, for a discussion of this
family's much-debated taxonomy and common names. See also
the map on page 000 for more specific ranges of North American
cormorant species. This list is derived from Bryan Nelson's *Pelicans,
Cormorants and Their Relatives* (Oxford University Press, 2005).
The Red List status is as of 2012.

EXTINCT

Spectacled cormorant (*Phalacrocorax perspicillatus*): Commander Islands,
    North Pacific

CRITICALLY ENDANGERED

Chatham shag (*P. onslowi*): Chatham Islands, New Zealand

ENDANGERED

Bank cormorant (*P. neglectus*): coast of Namibia and South Africa
Pitt shag (*P. featherstoni*): Chatham Islands, New Zealand

VULNERABLE

Auckland shag (*P. colensoi*): Auckland Islands, Southern Ocean
Bounty shag (*P. ranfurlyi*): Bounty Islands, Southern Ocean
Campbell shag (*P. campbelli*): Campbell Island, Southern Ocean
Crozet shag (*P. melanogenis*): Crozet, Marion, and Prince Edward Islands, Southern
    Ocean (not recognized/evaluated by IUCN, author's placement under "vulnerable")
Flightless cormorant (*P. harrisi*): Fernandina and Isabela, Galápagos Islands
Heard shag (*P. nivalis*): Heard Island, Southern Ocean (not recognized/evaluated by
    IUCN, author's placement under "vulnerable")
King shag (*P. carunculatus*): coast of northern South Island, New Zealand
Macquarie shag (*P. purpurascens*): Macquarie, Bishop, and Clerk Islands, Southern
    Ocean (not recognized/evaluated by IUCN, author's placement under "vulnerable")

Socotra cormorant (*P. nigrogularis*): coast of northwestern Arabian Sea, Persian Gulf

Stewart shag (*P. chalconatus*): coast of southern South Island, New Zealand

Cape cormorant (*P. capensis*): coast of Angola, Namibia, South Africa

Crowned cormorant (*P. coronatus*): coast of Namibia, South Africa

Guanay cormorant (*P. bougainvillii*): coast of western South America

Red-legged cormorant (*P. gaimardi*): coast of Peru, Chile, southern Argentina

Antarctic shag (*P. bransfieldensis*): Antarctic Peninsula, South Shetlands (included as part of *atriceps* evaluation by IUCN)

Black-faced cormorant (*P. fuscescens*): coast of southern Australia, Tasmania

Brandt's cormorant (*P. penicillatus*): coast of western North America, Gulf of California, as far north as southern Alaska

Double-crested cormorant (*P. auritus*): throughout North America, coast and inland

European shag (*P. aristotelis*): coast of Europe into Arctic Circle and as far south as Mediterranean

Great cormorant (*Phalacrocorax carbo*): throughout Asia, Africa, Australia, northeast coast of North America, coast and inland

Imperial shag (*P. atriceps*): Patagonia, Falklands, South Georgia, Antarctic Peninsula

Indian cormorant (*P. fuscicollis*): coast and inland India, Pakistan, Southeast Asia

Japanese cormorant (*P. capillatus*): coast of Japan, northeast Asia

Javanese cormorant (*P. niger*): coast and inland Pakistan, India, and throughout southeast Asia

Kerguelen shag (*P. verrucosus*): Kerguelen Islands, Southern Ocean (included as part of *atriceps* evaluation by IUCN)

Little black cormorant (*P. sulcirostris*): coast and inland Australia, New Zealand, Papua New Guinea, Indonesia

Little pied cormorant (*P. melanoleucos*): coast and inland Australia, New Zealand, Papua New Guinea and nearby islands

Long-tailed cormorant (*P. africanus*): coast and inland throughout Madagascar and most of the southern half of Africa

Neotropic cormorant (*P. brasilianus*): Southern United States to Tierra del Fuego, throughout South America, Caribbean, coast and inland

Pelagic cormorant (*P. pelagicus*): northwest coast of North America and across islands of North Pacific, Kamchatka

Pied cormorant (*P. varius*): coast and inland Australia, New Zealand

Pygmy cormorant (*P. pygmaeus*): coast and inland eastern Europe and Middle East, including Italy, Uzbekistan, and Iraq

Red-faced cormorant (*P. urile*): far northwest coast of North America and across islands of North Pacific, Kamchatka

Rock shag (*P. magellanicus*): coast of southern Chile, Argentina, Falklands

South Georgia(n) shag (*P. georgianus*): South Georgia, South Orkney, South Sandwich Islands (included as part of *atriceps* evaluation by IUCN)

Spotted shag (*P. punctatus*): coast of New Zealand

Notes

1. The epigraph is from a personal communication with Junji Yamashita, translated by Munro Johnson, August 7, 2001. All of Yamashita's quotations in this chapter are from this interview.

2. Henry Spencer Palmer, "Cormorant Fishing in Japan," in *Letters from the Land of the Rising Sun* (Yokohama: Japan Mail, 1894), 172; Kiyoshi Otsuka, personal communication, August 9, 2001.

3. David Landis Barnhill, *Bashō's Journey: The Literary Prose of Matsuo Bashō* (Albany: SUNY Press, 2005), 110.

4. Zempei Yamashita, "Cormorant Fishing on the Nagara River," *Oceanus* 30, no. 1 (Spring 1987): 84. In Japanese it is *omoshirōte / yagate / kanashiki / ubune / kana*. Barnhill translates it this way:

> fascinating,
> and then sorrowful:
> cormorant boat.

5. "Ukai: Cormorant Fishing on the Nagara River," 2012, Gifu Convention and Visitors Bureau, Gifu, Japan, www.gifucvb.or.jp/en/01_sightseeing/01_01.html.

6. India and Korea: Andres von Brandt, *Fish Catching Methods of the World* (Surrey, England: Fishing News Books Ltd., 1984), 27; Egypt: Darryl Wheye and Donald Kennedy, *Humans, Nature, and Birds: Science Art from Cave Walls to Computer Screens* (New Haven, CT: Yale University Press, 2008), 21.

7. Hermann Leight, *Pre-Inca Art and Culture*, trans. Mervyn Savill (New York: Orion Press, 1960), 49–50.

8. Christine E. Jackson, "Fishing with Cormorants," *Archives of Natural History* 24, no. 2 (1997): 189; Pamela Egremont and Miriam Rothschild, "The Calculating Cormorants," *Biological Journal of the Linnean Society* 12 (September 1979): 181.

9. Maya Manzi and Oliver T. Coomes, "Cormorant Fishing in Southwestern China: A Traditional Fishery under Siege," *American Geographical Society* 92, no. 4 (October 2002): 602; Erling Ho, "Flying Fishes of Wucheng," *Natural History* 107, no. 8 (October 1988): 66.

10. Berthold Laufer, "The Domestication of the Cormorant in China and Japan," *Field Museum of Natural History No. 300: Anthropological Series* 18, no. 3 (1931): 209.

11. The *Sui shu* was completed in 636 CE. Laufer, "Domestication of the Cormorant," 212, 233. Cormorants figured in some of Japan's earliest mythology and writing. Ukai is

275

mentioned in the *Manyoshu*, the earliest book of Japanese poems, compiled in the eighth century. There is a reference to ukai in the *Tale of Genji*, normally credited to Lady Murasaki Shikibu in the early eleventh century.

12. In 702 CE, census records in a town near Gifu City show a person with the name Ukaibe or Uyobe, suggesting this would have been an usho. Census records during the Engi period (901–22 CE) reference seven houses of cormorant fishermen on the Nagara River. See Laufer, "Domestication of the Cormorant," 213; and E. W. Gudger, "Fishing with the Cormorant in Japan," *Scientific Monthly* 29, no. 1 (July 1929): 10.

13. S. Ikenoya, "Cormorant Fishing," *Japan Magazine*, May 1917, 31–32, as cited in Gudger, "Fishing with the Cormorant," 10.

14. Yamashita, "Cormorant Fishing," 83.

15. Gudger, "Fishing with the Cormorant," 11; Ukai Exhibit, Gifu History Museum, Gifu, Japan, 2001.

16. Yamashita, "Cormorant Fishing," 83–84.

17. In the eighteenth century, the scholar Norinaga Moroori wrote of ukai in Gifu: "Nowhere but in the Nagara can we see / That antique sight of cormorant fishing, / So picturesque and impressive, / Bonfires reflected in the water rushing." See ibid., 83.

18. Laufer, "Domestication of the Cormorant," 213, 255; see also a record of cormorant feathers used to roof birthing huts in a Japanese folktale titled "The Lost Fish-Hook," as told in C. Pfoundes, "The Lost Fish-Hook," *Folk-Lore Record* 1 (1878): 128.

19. Gudger, "Fishing with the Cormorant," 11–12.

20. *The Diary of Richard Cocks*, ed. Edward Maunde Thompson, vol. 1 (London: Hakluyt Society, 1883), 285.

21. Palmer, "Cormorant Fishing in Japan," 171.

22. Ibid., 172–73.

23. Jackson, "Fishing with Cormorants," 199.

24. See Elfriede R. Knauer, "Fishing with Cormorants: A Note on Vittore Carpaccio's *Hunting on the Lagoon*," *Apollo* 158, no. 499 (September 2003): 32–39; and Jackson, "Fishing with Cormorants," 199–201. Both Knauer and Jackson point out that the practice was also recorded in Pietro Longhi's eighteenth-century painting. The two scholars differ on whether the clay balls are for the fish or to make the cormorants cough up the fish. The former makes more sense to me. See Von Brandt, *Fish Catching Methods*, 49–51.

25. Jackson, "Fishing with Cormorants," 202; Laufer, "Domestication of the Cormorant," 206.

26. James Edmund Harting, "Fishing with Cormorants," in *Essays on Sport and Natural History* (London: Horace Cox, 1883), 429.

27. Jackson, "Fishing with Cormorants," 203–6.

28. Ibid., 201.

29. Ibid.

30. See Jean Baptiste Emmanuel Hector Le Couteulx de Canteleu, *La Pêche au Cormoran* (Paris: Bureaux de la Revue Britannique, 1870). For nineteenth-century cor-

morant fishing in France, see also Pierre-Amédée Pichot, *Les Oiseaux de Sport* (Paris: Librairie ad Legoupy, 1903), 27–35.

31. Laufer, "Domestication of the Cormorant," 207.

32. Franck Maubert, *Toulouse-Lautrec in Paris* (New York: Assouline, 2004), 9; Gerstle Mack, *Toulouse-Lautrec* (New York: Alfred A. Knopf, 1938), 226.

33. Laufer, "Domestication of the Cormorant," 236. Translated from Pichot, *Les Oiseaux de Sport* (1903).

34. U.S.: Daniel Mannix, *A Sporting Chance: Unusual Methods of Hunting* (New York: E. P. Dutton & Co., 1967), 143–47; Macedonia (Lake Dojran): Lonely Planet, *The Europe Book: A Journey through Every Country on the Continent* (Victoria, Australia: Lonely Planet, 2010), 148; Von Brandt, *Fish Catching Methods*, 30–31.

35. Von Brandt, *Fish Catching Methods*, 29; Carolyn Thornton, "Peruvian Villagers on Floating Islands Welcome Tour Bus Visitors," *Dallas Morning News*, February 19, 2009, www.newsbank.com. Adolph F. Bandelier in 1910 wrote of the native peoples "robbing the nests," presumably to eat the eggs and/or chicks; in Bandelier, *The Islands of Titicaca and Koati* (New York: Hispanic Society of America, 1910), 34. Decades later, Jacques Cousteau wrote of the Urus eating their domesticated cormorants. In Jacques-Yves Cousteau and Philippe Diolé, *Three Adventures: Galápagos, Titicaca, the Blue Holes*, trans. J. F. Bernard (Garden City, NY: Doubleday, 1973), 133.

36. Tomoya Akimichi, "The Japanese Table: Queen of Freshwater Streams," *Food Forum* (c. 2005) Back Issues, Kikkoman, www.kikkoman.com/foodforum/thejapanese tablebackissues/11.shtml.

37. Also known with the common name of Temmick's cormorants.

38. See, for example, Gal Ribak, Daniel Weihs, and Zeev Arad, "How Do Cormorants Counter Buoyancy during Submerged Swimming?" *Journal of Experimental Biology* 207 (2004): 2102.

39. Egremont and Rothschild, "Calculating Cormorants," 181.

40. See, for example, J. Bryan Nelson, *Pelicans, Cormorants and Their Relatives* (Oxford: Oxford University Press, 2005), 99; and Mannix, *Sporting Chance*, 143–47.

41. Frédéric Fougea (director/cowriter) and Zhu Xiao Ling (cowriter), *He Dances for His Cormorants* (Boréales Production, 1993), 26 mins. See also a children's novel, set on the Li River: Elizabeth Starr Hill, *Bird Boy* (New York: Farrar, Straus and Giroux, 1999).

42. Palmer, "Cormorant Fishing in Japan," 173–74; Gudger, "Fishing with the Cormorant," 28.

43. Laufer, "Domestication of the Cormorant," 238; Fougea and Zhu Xiao Ling, *He Dances for His Cormorants*.

44. Lawrence E. Joseph, "Man and Cormorant," *Audubon* 88 (May 1986): 40.

45. Shiro Kasuya, personal communication, January 3, 2006.

46. Arthur Waley, *The Nō Plays of Japan* (London: George Allen & Unwin, 1921), 163.

47. Ibid.

48. Kiyoshi Otsuka, personal communication, August 9, 2001.

49. Kazuto Hino, personal communication, August 19, 2001, and February 25, 2012;

"Japan: Gifu: Gifu City," 1995–2010, City Population, www.citypopulation.de/Japan
-Gifu.html.

50. Kazuto Hino, personal communication, February 25, 2012.

51. Kazuto Hino, personal communication, January 18, 2006.

52. Shiro Kasuya, Soichi Murase, and Yuichi Miyano, "Destructive Effects of the Estuary Dam on the Nagara River's Environment, and the Program for Its Regeneration," *Bulletin of the Faculty of Regional Studies, Gifu University* 20 (2007): 2.

53. Ibid., 13.

54. Ibid., 15.

2. HENDERSON HARBOR, UNITED STATES

Mark Clymer is quoted in Jodi Wilgoren's "A Bird That's on a Lot of Hit Lists," *New York Times,* January 18, 2002, A12.

1. Ron Ditch, personal communications, July 7, 2001, March 18, 2012, and April 17, 2012. All quotations unless otherwise specified are from one of these three interviews. For other accounts of the evening see Susan McGrath, "Shoot-out at Little Galloo," *Smithsonian,* February 2003, 72–78, and Dennis Wild, *The Double-Crested Cormorant: Symbol of Ecological Conflict* (Ann Arbor: University of Michigan Press, 2012), 5.

2. The others are the pelagic cormorant (*P. pelagicus*), the red-faced (*P. urile*), the Brandt's (*P. penicillatus*), the neotropic (*P. brasilianus*), and the great cormorant (*carbo*).

3. Linda R. Wires and Francesca J. Cuthbert, "Historic Populations of the Double-Crested Cormorant (*Phalacrocorax auritus*): Implications for Conservation and Management in the 21st Century," *Waterbirds* 29, no. 1 (2006): 24–27; Jeremy J. Hatch, "Changing Populations of Double-Crested Cormorants," in *The Double-Crested Cormorant: Biology, Conservation and Management,* ed. David N. Nettleship and David C. Duffy, *Colonial Waterbirds* 18, Special Publication 1 (1995), 10.

4. Hatch, "Changing Populations," 10; Shauna Hanisch, personal communication, March 11, 2008.

5. Paul Johnsgard, *Cormorants, Darters, and Pelicans of the World* (Washington, DC: Smithsonian Institute Press, 1993), 108; Jeremy J. Hatch and D. V. Weseloh, "Double-Crested Cormorant (*Phalacrocorax auritus*)," in *The Birds of North America,* ed. A. Poole and F. Gill, no. 441 (Philadelphia: Birds of North America Inc., 1999), 17.

6. For calculations of double-crested cormorant survivorship of eggs and adults see U.S. Fish and Wildlife Service, "Final Environmental Impact Statement: Double-Crested Cormorant Management in the United States" (Arlington, VA, 2003), 29.

7. Hatch and Weseloh, "Double-Crested Cormorant," 20; Nelson, *Pelicans, Cormorants and Their Relatives,* 181.

8. Hatch and Weseloh, "Double-Crested Cormorant," 20.

9. M. Kathleen Klimkiewicz and Anthony G. Futcher, "Longevity Records of North American Birds, Supplement 1," *Journal of Field Ornithology* 60, no. 4 (Autumn 1989): 473. David Bird gives the oldest recorded double-crested cormorant as twenty-three

years old, but without any reference. See David Bird, *The Bird Almanac* (Buffalo, NY: Firefly Books, 1999), 267.

10. T. Fransson, T. Kolehmainen, C. Kroon, L. Jansson, and T. Wenninger, "EURING List of Longevity Records for European Birds," November 20, 2010, EURING Bird Ringing Databank, www.euring.org/data_and_codes/longevity-voous.htm.

11. U.S. Fish and Wildlife Service, "Final Environmental Assessment: Extended Management of Double-Crested Cormorants" (Arlington, VA, 2009), 6. Lewis did the same. See Harrison Flint Lewis, *The Natural History of the Double-Crested Cormorant* (Ottawa: H. C. Miller, 1929), 8.

12. USFWS, "Final Environmental Impact Statement: Double-Crested Cormorant Management in the United States," 23; Laura A. Tyson, Jerrold L. Belant, Francesca J. Cuthbert, and D. V. (Chip) Weseloh, "Nesting Populations of Double-Crested Cormorants in the United States and Canada," *Symposium on Double-Crested Cormorants: Population Status and Management Issues in the Midwest,* Technical Bulletin 1879, ed. Mark E. Tobin (USDA-APHIS, 1999), 20; USFWS, "Final Environmental Assessment: Extended Management of Double-Crested Cormorants," 5.

13. R. J. Pierotti and T. P. Good, "Herring Gull (*Larus argentatus*)," 1994, in *The Birds of North America Online,* ed. A. Poole, Cornell Lab of Ornithology, http://bna.birds .cornell.edu/bna/species/124; Thomas B. Mowbray, Craig R. Ely, James S. Sedinger, and Robert E. Trost, "Canada Goose (*Branta canadensis*)," 2002, in *Birds of North America Online,* http://bna.birds.cornell.edu/bna/species/682.

14. *Phalacrocorax auritus auritus* [Interior and Atlantic]; *P.a. cincinatus* [Alaska/BC]; *P.a. albociliatus* [BC to Baja, inland to Rockies]; *P.a. floridanus* [Florida, Caribbean]; *P.a. heuretus* [Bahamas, Cuba]. In Hatch and Weseloh, "Double-Crested Cormorant," 5. New research suggests this might soon be changing. See Dacey Mercer, "Phylogeography and Population Genetic Structure of Double-Crested Cormorants (*Phalacrocorax auritus*)" (PhD diss., Oregon State University, 2008).

15. USFWS, "Final Environmental Assessment: Extended Management of Double-Crested Cormorants," 5. The Pacific and Alaska are sometimes split into two population management groups, as they are separate subspecies. See the discussion about this in my later chapter 5, "East Sand Island."

16. Andrew C. Revkin, "A Slaughter of Cormorants in Angler Country," *New York Times,* August 1, 1998, B5.

17. Tommy L. Brown and Nancy A. Connelly, "Lake Ontario Sportfishing: Trends, Analysis, and Outlook," Human Dimensions Research Unit, Cornell University, Series No. 09–3 (June 2009), 7.

18. Ibid., 2.

19. Ron Ditch, personal communication, April 17, 2012.

20. Robert M. Ross and James H. Johnson, "Fish Losses to Double-Crested Cormorant Predation in Eastern Lake Ontario, 1992–97," Symposium on Double-Crested Cormorants: Population Status and Management Issues in the Midwest, Technical Bulletin 1879, ed. Mark E. Tobin (USDA-APHIS, 1999), 61.

21. Russell D. McCullough, James F. Farquhar, and Irene M. Mazzocchi, "Cormorant Management Activities in Lake Ontario's Eastern Basin," NYSDEC Lake Ontario Annual Report, Section 13 (2010), 2; Irene Mazzocchi, personal communication, April 18, 2012.

22. Ross and Johnson, "Fish Losses," 61; Irene Mazzocchi, personal communication, April 18, 2012.

23. Chip Weseloh, personal communication, April 20, 2012; Jennifer Doucette, personal communication, April 10, 2012; C. M. Somers, V. A. Kjoss, F. A. Leighton, and D. Fransden, "American White Pelicans and Double-Crested Cormorants in Saskatchewan: Population Trends over Five Decades," *Blue Jay* 68, no. 2 (June 2010): 85; Keith A. Hobson, Richard W. Knapton, and Walter Lysack, "Population, Diet and Reproductive Success of Double-Crested Cormorants Breeding on Lake Winnipegosis, Manitoba, in 1987," *Colonial Waterbirds* 12, no. 2 (1989): 193; Scott Wilson, personal communication, September 10, 2012.

24. D. V. Weseloh and B. Collier, "The Rise of the Double-Crested Cormorant on the Great Lakes: Winning the War against Contaminants," Great Lakes Fact Sheet, Environment Canada, Ontario Region, Catalogue No. EN 40–222/2–1995E (1995), 7. This source reports 10,000 nests in Canadian Lake Ontario in 1993. Irene Mazzocchi of NYSDEC records 2,477 pairs at five Canadian sites in the Eastern Basin in 1996 (personal communication, February 20, 2013). In the 1970s there were very few nests in the region of the Great Lakes. By 2000, cormorant populations had skyrocketed to about 115,000 pairs throughout the Canadian and U.S. Great Lakes. See Shauna L. Hanisch and Paul R. Schmidt, "Resolving Double-Crested Cormorant *Phalacrocorax auritus* Conflicts in the United States: Past, Present, and Future," in *Waterbirds around the World*, ed. G. C. Boere, C. A. Galbraith, and D. A. Stroud (Edinburgh: Stationery Office, 2006), 826.

25. B. F. Lantry, T. H. Eckert, and C. P. Schneider, "The Relationship between the Abundance of Smallmouth Bass and Double-Crested Cormorants in the Eastern Basin of Lake Ontario," in New York State Department of Environmental Conservation Bureau of Fisheries and USGS, Biological Resources Division, "Final Report: To Assess the Impact of Double-Crested Cormorant Predation on the Smallmouth Bass and Other Fishes of the Eastern Basin of Lake Ontario" (February 1, 1999), Section 12, 1–4.

26. James H. Johnson, Russell D. McCullough, and James Farquhar, "Double-Crested Cormorant Studies at Little Galloo Island, Lake Ontario in 2010: Diet Composition, Fish Consumption and the Efficacy of Management Activities in Reducing Fish Predation," NYSDEC Lake Ontario Annual Report (2010), section 14, 1.

27. Ron Ditch, personal communication, July 7, 2001.

28. Russell McCullough, personal communication, April 18, 2012.

29. Johnson, McCullough, and Farquhar, "Double-Crested Cormorant Studies," section 14, 2.

30. For a thorough review of all methods see David N. Carss and the Diet Assessment and Food Intake Working Group, "Techniques for Assessing Cormorant Diet and

Food Intake: Towards a Consensus View," February 21, 2001, Wetlands International Cormorant Research Group, http://web.tiscalinet.it/sv2001/WI%20-%20CRSG/diet_foodintake.htm. A new, fourth method is emerging, using stable isotopes from the muscle tissue of cormorants, but this also requires killing the birds. See Jennifer L. Doucette, Björn Wissel, and Christopher M. Somers, "Cormorant-Fisheries Conflicts: Stable Isotopes Reveal a Consistent Niche for Avian Piscivores in Diverse Food Webs," *Ecological Applications* 21, no. 8 (2011): 2987–3001.

31. Hatch and Weseloh, "Double-Crested Cormorant," 7.

32. Howard L. Mendall, "The Home-Life and Economic Status of the Double-Crested Cormorant, *Phalacrocorax auritus auritus* (Orono: University [of Maine] Press, 1936), 113.

33. Zzrmags, "Cormorant vs Rat," January 6, 2009, YouTube, www.youtube.com; Johnsgard, *Cormorants, Darters, and Pelicans*, 149, 162.

34. Hatch and Weseloh, "Double-Crested Cormorant," 10.

35. D. V. Weseloh and J. Casselman, "Calculated Fish Consumption by Double-Crested Cormorants in Eastern Lake Ontario" (abstract), *Colonial Waterbird Society Bulletin* 16, no. 2 (1992): 64. A 1927 account of double-crested cormorants, of the slightly smaller Florida subspecies, said the cormorants in captivity ate between 0.75 to 1.0 pounds of fish per day, with none on Sunday. See A. Wetmore, "The Amount of Food Consumed by Cormorants," *Condor* 29 (November 1927): 274. For an introduction in studying cormorant diet based on caloric intake rather than weight of food, see Erica H. Dunn, "Caloric Intake of Nestling Double-Crested Cormorants," *Auk* 92, no. 3 (1975): 553–65.

36. Hatch and Weseloh, "Double-Crested Cormorant," 2; Nelson, *Pelicans, Cormorants and Their Relatives*, 589.

37. Thomas McIlwraith, *The Birds of Ontario* (Hamilton, ON: A. Lawson, 1886), 47.

38. T. S. Roberts, *Birds of Minnesota*, vol. 1 (Minneapolis: University of Minnesota Press, 1932), 171. See also David R. Cline and Eric Dornfeld, "The Agassiz Refuge Cormorant Colony," *Loon* 40, no. 3 (September 1978): 68.

39. Jennifer Doucette, personal communication, July 12, 2012.

40. Howard White and Karen Allanach (contacts), "Press Release: HSUS Offers Reward for Information on Cormorant Mass Slaying in Michigan," Humane Society of the United States (June 15, 2000), www.hsus.org/news/pr/061500_michigan.html; Les Line, "A Conflict of Cormorants," *Wildlife Conservation*, February 2002, 47.

41. Wilgoren, "Bird That's on a Lot of Hit Lists," A12.

42. Jim Moodie, "Cormorant Nests Dwindle on Huron Due to Natural Ebb and Vigilante Culls: MNR Survey Shows Nine Colonies 'Shot Up' on Georgian Bay, N. Channel," *Manitoulin Expositor*, March 19, 2008, www.manitoulin.ca/Expositor/oldfiles/mar19_2008.htm.

43. Judith Minty, "Destroying the Cormorant Eggs," *Walking with the Bear* (East Lansing: Michigan State University Press, 2000), 162. See the Associated Press and Special to the Sentinel, "Untouched Cormorant Eggs Found," *Milwaukee Sentinel*, June 19, 1987, pt. 2, 10.

44. Wires and Cuthbert, "Historic Populations," 29.

45. Barry Kent MacKay and Liz White, "A Critical Analysis of Point Pelee National Park's Rationale for Killing the Middle Island Cormorants," Cormorant Defenders International, February 2008, 13, 31–32; Wires and Cuthbert, "Historic Populations," 15.

46. MacKay and White, "Critical Analysis," 30.

47. Chas. S. Whitehead, "Loon Lake," *Ornithologist and Oölogist* 12, no. 2 (February 1887): 21. This source, among several others I cite, identified earlier by Wires and Cuthbert, "Historic Populations."

48. Frank W. Langdon, "Observations of Cincinnati Birds," *Journal of the Cincinnati Society of Natural History* 1 (October 1878): 117–18.

49. Mendall, "Home-Life," 4–6; Wires and Cuthbert, "Historic Populations," 9–10, 15–16; Jerome A. Jackson and Bette J. S. Jackson, "The Double-Crested Cormorant in the South-Central United States: Habitat and Population Changes of a Feathered Pariah," in Nettleship and Duffy, *Double-Crested Cormorant*, 119–21.

50. Lewis, *Natural History*, 88.

51. Ibid., 67; see also P. A. Taverner, "The Double-Crested Cormorant (*Phalacrocorax auritus*) and Its Relation to the Salmon Industries on the Gulf of St. Lawrence," *Museum Bulletin* no. 13 / Biological Series, no. 5 (April 30, 1915): 1–24.

52. The Interior population has regularly been by far the largest in North America, more so than the other three combined.

53. Lewis, *Natural History*, 7–8.

54. Ibid., 5.

55. See, for example, Hatch, "Changing Populations," 13.

56. D. V. Weseloh, P. J. Ewins, J. Struger, P. Mineau, C. A. Bishop, S. Postupalsky, and J. P. Ludwig, "Double-Crested Cormorants of the Great Lakes: Changes in Population Size, Breeding Distribution and Reproductive Output between 1913 and 1991," in Nettleship and Duffy, *Double-Crested Cormorant*, 50. At this time there were fewer than one thousand nests across all of the Great Lakes.

57. Hatch, "Changing Populations," 13.

58. Weseloh and Collier, "Rise of the Double-Crested Cormorant," 4–5; Hatch and Weseloh, "Double-Crested Cormorant," 23.

59. It used to be taught that cormorants incubated eggs *under* their webbed feet — which made their eggs especially vulnerable to these shell-thinning pesticides, especially when they took off out of the nest in a hurry — but, according to Nelson, this brooding style in cormorants has been discounted recently. Cormorants warm their eggs with the heat of their bodies over the skin of their feet, but they do not have as cozy a featherless brood patch as do other birds. See Nelson, *Pelicans, Cormorants and Their Relatives*, 176.

60. Weseloh and Collier, "Rise of the Double-Crested Cormorant," 8.

61. Weseloh, Ewins, Struger, et al., "Double-Crested Cormorants," 50.

62. Hatch, "Changing Populations," 12–13; "Double-Crested Cormorant: Conservation Status," National Audubon Society, http://birds.audubon.org/species/doucor (accessed August 29, 2012).

63. John L. Trapp, Thomas J. Dwyer, John J. Doggett, and John G. Nickum, "Management Responsibilities and Policies for Cormorants: United States Fish and Wildlife Service," in Nettleship and Duffy, *Double-Crested Cormorant*, 226.

64. For details on pre-2003 activities see USFWS, "Final Environmental Impact Statement: Double-Crested Cormorant Management in the United States," 11.

65. USFWS, "Final Environmental Assessment: Extended Management of Double-Crested Cormorants," 1.

66. Ontario Parks, "Resource Management Implementation Plan High Bluff and Gull Islands" (Queen's Printer for Ontario, 2011), 19; Ontario Parks, "Annual Report on the Management of Double-Crested Cormorants on Presqu'ile Provincial Park" (Queens Printer for Ontario, 2006), 18.

67. For public response see, for example, Thomas Walkolm, "Government Takes Aim at Cormorants: Ministry Showing a Crazy Kind of George W. Bush Logic," *Toronto Star*, May 28, 2005, www.zoocheck.com/cdiwebsite.

68. See, for example, Cormorant Defenders International, "Bloody Carnage in Provincial Park," media release, May 29, 2006, www.zoocheck.com/CDIwebsite/CorMedia releaseCarnageMay2906.shtml.

69. See, for example, CBC News / Prince Edward Island, "Fishermen Want Cormorant Cull," CBC News, January 30, 2012, www.cbc.ca.

70. Chip Weseloh, personal communication, April 20, 2012; McGrath, "Shoot-out at Little Galloo," 78.

71. Russell McCullough, personal communications, July 9, 2001, April 18, 2012. All the following quotations are from these two interviews unless otherwise specified.

72. Evidence is emerging that lampreys might be native. See, for example, P. Fuller, L. Nico, E. Maynard, J. Larson, and A. Fusaro, "*Petromyzon marinus*," USGS Nonindigenous Aquatic Species Database (Gainesville, FL, 2012), http://nas.er.usgs.gov.

73. Russell McCullough, "Lake Ontario Milestones," personal notes, 2012; Weseloh, Ewins, Struger, et al., "Double-Crested Cormorants," 55–56.

74. See Alban Guillaumet, Brian Dorr, Guiming Wang, Jimmy D. Taylor 2nd, Richard B. Chipman, Heidi Scherr, Jeff Bowman, Kenneth F. Abraham, Terry J. Doyle, and Elizabeth Cranker, "Determinants of Local and Migratory Movements of Great Lakes Double-Crested Cormorants," *Behavioral Ecology* 22 (2011): 1096–1103.

75. McCullough, Farquhar, and Mazzocchi, "Cormorant Management Activities," section 13, 1; Russ McCullough and Irene Mazzocchi, personal communication, September 13, 2012. McCullough and Mazzocchi explained to me that "the cormorant predation impact target (feeding days) is based on an unmanaged population associated with 1,500 nesting pairs (3,000 nesting adults, 300 subadults, 3,000 fledged chicks) or approximately 780,000 feeding days."

76. New York State scientists report: "Smallmouth bass abundance in the Eastern Basin as measured in index gill nets (CPUE = 6.1) decreased 30.0% compared to the 2006–2010 average, but was 45.9% higher than the record-low 2000–2004 period. Improved growth and a reduction in cormorant predation pressure likely contributed to

the increased CPUEs observed from 2005–2011." (CPUE is catch per unit of effort.) See NYSDEC, "Executive Summary," 2011 Annual Report, Bureau of Fisheries, Lake Ontario Unit and St. Lawrence River Unit (March 2012), 6.

77. Johnson, McCullough, and Farquhar, "Double-Crested Cormorant Studies," section 14, 3; Russ McCullough explained to me: "Gobies were first confirmed from Lake Ontario in 1999. They were first found in cormorant diets in Canadian colonies in 2002 and at Little Galloo Island in 2004. They were the primary prey in all three monitored Lake Ontario colonies by 2005." See NYSDEC annual reports. Russ McCullough, personal communication, May 1, 2012.

78. Jim Farquhar, personal communication, April 18, 2012; Irene Mazzocchi, personal communication, April 18, 2012.

3. ARAN ISLANDS, IRELAND

Synge's epigraph is from *The Aran Islands* (London: George Allen, 1934), 48; Nicholson's is from *Sea Room: An Island in the Hebrides* (New York: Harper Perennial, 2007), 184.

1. Liam O'Flaherty's "The Wounded Cormorant" first appeared in the *Nation [and the Athenaeum]*, November 28, 1925, 317–18. O'Flaherty then published it in his collection *The Tent* (London: Cape, 1926). See Angeline A. Kelly, *Liam O'Flaherty the Storyteller* (New York: Harper & Row, 1976), 149.

2. Rónadh Cox, personal communication, August 28, 2012. See also Rónadh Cox, Danielle B. Zentner, Brian J. Kirchner, and Mea S. Cook, "Boulder Ridges on the Aran Islands (Ireland): Recent Movements Caused by Storm Waves, Not Tsunami," *Journal of Geology* 120, no. 3 (May 2012): 249–72.

3. Liam O'Flaherty, *Liam O'Flaherty's Short Stories* (London: New English Library, 1981), 2:30–32. All quotations from this story are from this edition.

4. Hatch, "Changing Populations," 8.

5. *Oxford English Dictionary*, 2nd ed., online version, s.v. "cormorant, *n.*," www.oed .com/viewdictionaryentry/Entry/41582.

6. See, for example, John M. Marzluff and Tony Angell, *In the Company of Crows and Ravens* (New Haven, CT: Yale University Press, 2005), 1–10.

7. Edward A. Armstrong, *The Folklore of Birds* (London: Collins, 1958), 83.

8. Kevin De Ornellas, "'Fowle Fowles?': The Sacred Pelican and the Profane Cormorant in Early Modern Culture," in *A Cultural History of Animals*, vol. 3, *In the Renaissance*, ed. Bruce Boehrer (New York: Berg, 2007), 36; Kevin De Ornellas, personal communication, October 16, 2004.

9. See King James Bible: Deuteronomy 14.17, Leviticus 11.17, Isaiah 34.11, and Zephaniah 2.14.

10. James Spedding, *The Letters and the Life of Francis Bacon* (London: Longmans, Green, and Co., 1890), 1:245. The context of this has "cormorant" here as not just connected with evil, but too with greed: she is complaining about the servants of her other,

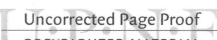

more famous son Francis, believing that they were after his money. See also Walter Begley, *Is It Shakespeare?* (London: John Murray, 1903), 49.

11. Professor Gordon McMullan, King's College, London, has recently lectured expertly on the subject.

12. De Ornellas, "'Fowle Fowles?'" 45.

13. John Milton, *Paradise Lost*, ed. Alastair Fowler, 2nd ed. (London: Longman, 1998), 227–28.

14. Ibid., 228 n. Consider also Andrea Mantegna's fifteenth-century painting *Agony in the Garden*, in which a vulture-like bird, but potentially a cormorant, looms on a bare tree limb, symbolizing death, watching Christ pray to angels as Judas arrives with Roman soldiers. Is this where Milton got the idea?

15. Homer, *The Odyssey of Homer*, trans. Richmond Lattimore (New York: Harper-Perennial, 1991), 196. This is toward the end of book 12, lines 415–19. See also J. MacLair Boraston, "The Birds of Homer," *Journal of Hellenic Studies* 31 (1991): 216–50.

16. Scott, *The Antiquary* (1816); Shelley, "The Witch of Atlas" (1820); Melville, *Moby-Dick* (1851), discussed further in chapter 11; Browning, "Balaustian's Adventure" (1871); Verne, *Off on a Comet* or *Hector Servadac* (1877); Yeats, "The Madness of King Goll" (1887). See also Andrew Marvell, "The Unfortunate Lover" (1648–49).

17. See Susan B. Taylor, "Brontë's *Jane Eyre*," *Explicator* 59, no. 4 (2001): 182–85.

18. Henry Wadsworth Longfellow, "The Skeleton in Armor," in *The Complete Poetical Works of Longfellow* (Boston: Houghton Mifflin, 1922), 13.

19. Charles Swainson, *The Folklore and Provincial Names of British Birds* (London: Elliot Stock, 1886), 143; see Peter Coleman, "History and Legends of St. Botolph's Church," *Lincolnshire Poacher* 11, no. 2 (Summer 2010): 12. Cormorants on top of towers are part of the mythology and culture of Liverpool, England, where their "Liver Bird" is often shown as a cormorant-like bird. This dates back to at least the fourteenth century, is depicted on the tower of the city's Royal Liver Building, and is on the seal of the Liverpool Football Club.

20. Gordon D'Arcy, "The Wildlife of the Aran Islands," in *The Book of Aran*, ed. John Feehan et al. (Tír Eolas: Newtownlynch, 1994), 63; A. C. Haddon and C. R. Browne, "The Ethnography of the Aran Islands, County Galway," *Proceedings of the Royal Irish Academy* 2 (1891–93): 810–11. Tom O'Flaherty, Liam's brother, wrote in 1934 that the last man who used to lower for eggs was still alive but no longer did so. He wrote that the islanders really only ate, because of access and taste, the guillemots. "The cormorant's flesh was tough, and he was an ill-smelling bird anyhow." In Tom O'Flaherty, "Coming Home," in *An Aran Reader* (Dublin: Lilliput Press, 1991), 178, 183.

21. Synge, *Aran Islands*, 250–51.

22. J. M. Synge, *"Riders to the Sea" and "In the Shadow of the Glen"* (London: Methuen & Co., 1961), 94. Apparently Liam O'Flaherty when he got older did not approve of Synge's work. See Peter Costello, *Liam O'Flaherty's Ireland* (Dublin: Wolfhound Press, 1996), 15.

23. Ted Hughes, *The Mermaid's Purse* (New York: Alfred A. Knopf, 1999), 21.

24. Emily Lawless, "The Cormorant," *All Ireland Review* 3, no. 25 (August 23, 1902): 389. This poem was originally published in her collection *With the Wild Geese* in 1902.

25. Editorial comments on Lawless's "The Cormorant," in *All Ireland Review*, 389.

26. Geoffrey Chaucer, "The Parlement of Foules," in *The English Poets*, vol. 1, *Chaucer to Donne*, ed. Thomas H. Ward (New York: Macmillan, 1880), 37.

27. William Turner, *Turner on Birds: A Short and Succinct History of the Principal Birds Noticed by Pliny and Aristotle*, ed. A. H. Evans (Cambridge: Cambridge University Press, 1903), 207.

28. Olaus Magnus, *Description of the Northern Peoples* [*Historia de Gentibus Septentrionalibus*]: *Rome 1555*, ed. Peter Foote, trans. Peter Fisher and Humphrey Higgens (London: Hakluyt Society, 1998), 3:969.

29. Ibid., 970.

30. Ibid., 1000.

31. Ibid., 969–70.

32. De Ornellas, "'Fowle Fowles?'" 37–38; *OED*, 2nd ed., online version, s.v. "cormorant."

33. Milton, *Paradise Lost*, 228.

34. *OED*, 2nd ed., s.v. "cormorous, *adj.*," www.oed.com/viewdictionaryentry/Entry/41583.

35. *OED*, 2nd ed., s.v. "cormorancy, *n.*," www.oed.com/viewdictionaryentry/Entry/41581.

36. Sidney, "The Last Eclogues" (1590); Sylvester, "The Fifth Day of the First Week" (late 16th to early 17th century); Drayton, "The Twenty-Fifth Song" (1622), Congreve, *The Old Batchelour* (1693); and Alexander Pope, "The Odyssey" (1725).

37. William Shakespeare, *Love's Labour's Lost*, in *The Complete Works*, ed. Stanley Wells and Gary Taylor (Oxford: Clarendon Press, 1998), 1.1.4, 281.

38. William Shakespeare, *The Tragedy of Coriolanus*, in *The Complete Works*, 1.1.119, 1068. See also *King Richard II* (2.2) and *Troilus and Cressida* (2.2).

39. See, for example, Barbara K. Lewalski, "Biblical Allusion and Allegory in *The Merchant of Venice*," *Shakespeare Quarterly* 13, no. 3 (Summer 1962): 333.

40. Samuel Taylor Coleridge, "The Devil's Thoughts," in *The Complete Poetical Works of Samuel Taylor Coleridge*, ed. Ernest Hartley Coleridge (Oxford: Clarendon Press, 1975), 1:321. In case you miss the pun here, "sate" serves to mean both satiate and sat down.

41. Bram Stoker, *Dracula* (New York: Bantam, 1989), 113.

42. According to local ornithologist Oscar Merne (personal communication, May 10, 12, 2005), formal seabird nest censuses of the Aran Islands didn't begin until 1969–70, and he has found only European shags breeding there for his several visits over thirty years. Cramp, Bourne, and Saunders (1974) found that only shags nested on the Aran Islands, but Sharrock (1977) recorded great cormorants there, as does D'Arcy in Feehan et al., *Book of Aran* (1994). A list of birds of the Aran Islands appears in the *Gaelic Journal* in 1899, listing both cormorant and shag, in both English and Irish.

43. Nelson, *Pelicans, Cormorants and Their Relatives*, 446–47.

44. For one clear, taxonomical differentiation between cormorants and shags based on morphology and behavior see G. F. van Tets, "Australia and the Origin of Shags and Cormorants, Phalacrocoracidae," *Proceedings of the 16th International Ornithological Congress* (1974): 121.

45. *OED*, 2nd ed., s.v. "shag, *n.*," www.oed.com/view/Entry/177257. There are over ten definitions for "shag" as noun and verb in the *OED*, and "shagadelic" was added in 1997. "Shag" to mean sex goes back to at least the 1930s. Phrases such as "wet as a shag" or as "miserable as a shag on a rock" appeared as early as the nineteenth century. Another theory about the derivation of shag for the bird suggests that the name comes from the Norse word *skegg*, meaning a beard, as put forth in Francesca Greenoak, *British Birds: Their Folklore, Names and Literature* (London: Christopher Helm, 1997), 24. See also Hugh MacDairmid, "Shags' Nests," in *Stony Limits and Other Poems* (London: V. Gollancz, 1934), 88.

46. Meg Bateman, personal communication, January 13, 2005; Andrew Murphy, personal communication, March 25, 2005; Robin Hull, *Scottish Birds: Culture and Tradition* (Edinburgh: Mercat Press, 2001), 96; Rónadh Cox, personal communication, August 28, 2012; Tomás De Bhaldraithe, ed., *English-Irish Dictionary* (Dublin: Oifig an tSoláthair, 1959), 151, 301. The cormorant and shag in Irish Gaelic and Scots Gaelic are also both called *sgarbh*, with some people distinguishing the European shag as *sgarbh-an-sgumain*. In Scots Gaelic the cormorant can be *geòcaire*, meaning "glutton." De Bhaldraithe found the ornithological word for shag in particular to be *seaga*. See also *An Irish-English Dictionary*, ed. Edward O'Reilly (Dublin: James Duffy, 1864).

47. Jeremy Bentham, *An Introduction to the Principles and Morals of Legislation* (Oxford: Clarendon Press, 1879), 311.

48. See Judith Washburn, "Objective Narration in Liam O'Flaherty's Short Stories," *Éire-Ireland* 24 (Fall 1989): 120–25.

49. Helene O'Connor, "Liam O'Flaherty: Literary Ecologist," *Éire-Ireland* 7, no. 2 (Summer 1972): 47.

50. James M. Cahalan, *Liam O'Flaherty: A Study of Short Fiction* (Boston: Twayne, 1991), 58. Cahalan was referring to O'Flaherty's "An Charraid Dhubh" ("The Black Rock").

51. Nelson, *Pelicans, Cormorants and Their Relatives*, 443.

52. Scottish ornithologist Jeffrey Graves (personal communications, February 10, March 10, 2005) and Irish ornithologist Oscar Merne (personal communications, May 10, 12, 2005) have doubts and have not witnessed anything like this in cormorants, but neither said it would be impossible. In 1940 F. Goethe records the behavior of birds such as storks, geese, cranes, gulls, and crows attacking an odd, sick, or injured bird of their own species ("Ueber 'Anstoss-Nehmen' bei Vögeln," *Zeitshrift fur Tierpsychologie* 3, 371–74). Crows and ravens are well known for their mobbing behavior, which has been anecdotally recorded around a sick bird of their own kind. See, for example, "Territory and Mobbing," *CorvidAid*, www.corvidaid.org/about-corvids/territory-and-mobbing.

53. Helen O'Connor writes that O'Flaherty is "absorbed by the inexorable ecology of the universe which changes everything in a matter of moments," in "Liam O'Flaherty: Literary Ecologist," 53.

54. Liam O'Flaherty, *The Black Soul* (London: Jonathan Cape, 1936), 89–93.

55. Nelson, *Pelicans, Cormorants and Their Relatives*, 445.

56. Amy Clampitt, *The Kingfisher* (New York: Alfred A. Knopf, 1983), 46. First printed in *Atlantic Monthly*.

57. Donald J. Borror, *Dictionary of Word Roots and Combining Forms* (Palo Alto, CA: National Press Books, 1971), 43.

58. Animated version: Oliver Postgate, Peter Firmin, and Smallfilms, *Noggin and the Ice Dragon*, BBC, c. 1982, www.youtube.com. Book version: Oliver Postgate and Peter Firmin, *The Saga of Noggin the Nog, 2: The Ice Dragon* (New York: Holiday House, 1968). Quotation is from the animated version. Postgate and Firmin are certainly not out to save the reputations of all black birds. In their story of *The Pie*, the evil character Nogbad the Bad is surrounded by a flock of his dark crows as he plots dastardly deeds.

59. As cited in Stephen Gregory, *The Cormorant* (New York: St. Martin's Press, 1986).

60. Ibid.

61. Stephen Gregory, *The Cormorant* (Clarkston, GA: White Wolf, 1986), 23.

62. Ibid., 26.

63. Ibid., 27.

64. Ibid., 99–100.

65. Peter Markham (director) and Peter Ransley (screenplay), *The Cormorant*, BBC television, 1993.

66. O'Connor, "Liam O'Flaherty: Literary Ecologist," 49.

67. Kelly, *Liam O'Flaherty the Storyteller*, 6.

68. Cahalan, *Liam O'Flaherty*, 54. I believe "great" is just an adjective here, not a declaration of the species. O'Flaherty wrote in 1930 about showing a filmmaker around: "I shall never forget looking down over a precipice near my native village with [Heinrich] Hauser. It was the scene of one of my short stories, the one I love best." As cited in Costello, 28.

69. O'Flaherty, *Black Soul*, 85–86.

4. SOUTH GEORGIA, ANTARCTICA

The epigraph is from Cook's *The Voyages of Captain James Cook* (London: William Smith, 1846), 1:570.

1. "Lady Jean Rankin: Obituary," October 5, 2001, *Telegraph*, www.telegraph.co.uk.

2. Niall Rankin, *Antarctic Isle: Wild Life in South Georgia* (London: Collins, 1951), 29.

3. Ibid., 17–25.

4. Robert Cushman Murphy, *Oceanic Birds of South America* (New York: Macmillan, 1936), 2:871; Johnsgard, *Cormorants, Darters, and Pelicans*, 261.

5. See, for example, Murphy, *Oceanic Birds of South America*, 2:878–99; and F. Behn,

J. D. Goodall, A. W. Johnson, and R. A. Phillippi, "The Geographic Distribution of the Blue-Eyed Shags, *Phalacrocorax albiventer* and *Phalacrocorax atriceps*," *Auk* 72 (January 1955): 6.

6. Nelson, *Pelicans, Cormorants and Their Relatives*, 476.

7. Rankin, *Antarctic Isle*, 304.

8. S. Wanless, M. P. Harris, and J. A. Morris, "Diving Behaviour and Diet of the Blue-Eyed Shag at South Georgia," *Polar Biology* 12 (1992): 713.

9. Nelson, *Pelicans, Cormorants and Their Relatives*, 498.

10. Rankin, *Antarctic Isle*, 304.

11. Otto Nordenskjöld and Johan Gunnar Andersson, *Antarctica or Two Years amongst the Ice of the South Pole* (New York: Macmillan, 1905), 21, 202.

12. W. E. Clarke (1913) as cited in Murphy, *Oceanic Birds of South America*, 2:889.

13. Rankin, *Antarctic Isle*, 197–98.

14. These could arguably have been any one of half a dozen or so species.

15. J. C. Beaglehole, ed., *The Endeavour Journal of Joseph Banks, 1768–1771* (London: Angus and Robertson, 1863), 1:430.

16. Robert F. Jones, ed., *Astorian Adventure: The Journal of Alfred Seton, 1811–1815* (New York: Fordham University Press, 1993), 50–51.

17. E. Lucas Bridges, *Uttermost Part of the Earth* (London: Hodder and Stoughton, 1963), 87. Yaghan Indians are also known as the Yamana, among other names.

18. Ibid., 98. For recent distribution of the cormorants of Tierra del Fuego, see A. C. M. Schiavini, P. Yorio, and E. Frere, "Distribución reproductiva y abundancia de las aves marinas de la Isla Grande de Tierra del Fuego, Isla de los Estados e Islas de Año Nuevo (Provincia de Tierra del Fuego, Antártida e Islas del Atlántico Sur)," in *Atlas de la distribucion reproductiva de aves marinas en el litoral Patagonico Argentino*, ed. Pablo Yorio et al. (Puerto Madryn: Fundación Patagonia Natural, 1998): 179–213.

19. Bridges, *Uttermost Part of the Earth*, 98.

20. See "The Selfish Cormorant" and "The Revenge of the Tufted Cormorants," in *Folk Literature of the Yamana Indians*, ed. Johannes Wilbert (Berkeley: University of California Press, 1977).

21. Luis Abel Orquera and Ernesto Luis Piana, *La vida material y social de los Yamana* (Buenos Aires: Universidad de Buenos Aires, 1999), 145.

22. Exhibit, December 30, 2001, Museo del Fin del Mundo, Ushuaia, Argentina; cormorant bones regularly appear in the shell middens of the Beagle Channel dating back to 4500 BCE. See Jordi Estevez, Ernesto Piana, Adrian Schiavini, and Nuria Juan-Muns, "Archaeological Analysis of Shell Middens in the Beagle Channel, Tierra del Fuego Island," *International Journal of Osteoarchaeology* 11 (2001): 28. See also Colin McEwan, Luis A. Borrero, and Alfredo Prieto, eds. *Patagonia: Natural History, Prehistory and Ethnography at the Uttermost End of the Earth* (Princeton, NJ: Princeton University Press, 1977).

23. R. N. Rudmose Brown, *A Naturalist at the Poles: The Life, Work and Voyages of Dr. W. S. Bruce, the Polar Explorer* (Philadelphia: J. B. Lippincott, 1924), 159.

24. Robert Cushman Murphy, *Logbook for Grace* (Chicago: Time-Life, 1982), 235. This

is not to suggest that all blue-eyed shags are placid. Neil Bernstein told me: "I only made friends with one bird, and she would let me handle her chicks and eggs. I never tried to lift her. The rest of them tore at me as I went to weigh their chicks and eggs."

25. Robert Cushman Murphy, "Notes on American Subantarctic Cormorants," *Bulletin of the American Museum of Natural History* 35 (1916): 47. Biologists sometimes refer to this "throat trembling" as "gular flutter."

26. L. Harrison Matthews, "The Birds of South Georgia," in *The Discovery Reports*, vol. 1 (Cambridge: Cambridge University Press, 1929), 584. It seems almost certain that both Rankin and Murphy's anecdotes about this behavior are directly from Matthews.

27. A. G. Bennett, *Whaling in the Antarctic* (New York: Henry Holt, 1932), 208. Bennett, I should add, was not one to use exclamation points liberally.

28. Alister Doyle, "Seal Brain and Penguin Breasts off Antarctic Menus," January 26, 2009, Reuters UK, www.reuters.com.

29. Sanford Moss, *Natural History of the Antarctic Peninsula* (New York: Columbia University Press, 1988), 109.

30. Nelson, *Pelicans, Cormorants and Their Relatives*, 481.

31. Ibid., 499.

32. Ibid.; Murphy, *Oceanic Birds of South America*, 2:886, 890.

33. Rankin, *Antarctic Isle*, 304.

34. P. Shaw, "Factors Affecting the Breeding Performance of Antarctic Blue-Eyed Shags *Phalacrocorax atriceps*," *Ornis Scandinavica* 17, no. 2 (May 1986): 141; Johnsgard, *Cormorants, Darters, and Pelicans*, 123. Penguins tend to lay two eggs, one of them a sort of "insurance" egg, as one of the chicks inevitably dies soon after hatching. See Frank B. Gill, *Ornithology* (New York: W. H. Freeman, 2007), 429.

35. Claire Léger and Raymond McNeil, "Nest Attendance and Care of Young in Double-Crested Cormorants," *Colonial Waterbirds* 8, no. 2 (1985): 99; see also Mendall "Home-Life," 52, on his observing a similar colony in Maine.

36. Neil P. Bernstein and Stephen J. Maxson, "Sexually Distinct Daily Activity Patterns of Blue-Eyed Shags in Antarctica," *Condor* 86 (1984): 151.

37. Timothée Cook, personal communication, July 21, 2012. Cook, by contrast, has observed shifts at a variety of different times and duration in blue-eyed cormorants.

38. For cormorant phylogeny see Douglas Siegel-Causey, "Phylogeny of the Phalacrocoracidae," *Condor* 90 (1988): 887; and Martyn Kennedy, Carlos Valle, and Hamish G. Spencer, "The Phylogenetic Position of the Galápagos Cormorant," *Molecular Phylogenetics and Evolution* 53 (2009): 97.

39. Pamela C. Rasmussen and Philip S. Humphrey, "Wing-Spreading in Chilean Blue-Eyed Shags (*Phalacrocorax atriceps*)," *Wilson Bulletin* 100, no. 1 (March 1988): 140. Matthews describes cormorants spreading their wings at South Georgia, but this is a rare observation. See Matthews, "Birds of South Georgia," 1:584.

40. Neil P. Bernstein and Stephen J. Maxson, "Absence of Wing-Spreading Behavior in the Antarctic Blue-Eyed Shag (*Phalacrocorax atriceps brandsfieldensis*)," *Auk* 99, no. 3 (July 1982): 588.

41. Ibid.; see also Thor Hanson, *Feathers: The Evolution of a Natural Miracle* (New York: Basic Books, 2011), 218–21.

42. Neil Bernstein, personal communication, December 15, 2011. See also Timothée R. Cook and Guillaume LeBlanc, "Why Is Wing-Spreading Behaviour Absent in Blue-Eyed Shags?" *Animal Behaviour* 74 (2007): 650.

43. Y. Tremblay, T. R. Cook, and Y. Cherel, "Time Budget and Diving Behaviour of Chick-Rearing Crozet Shags," *Canadian Journal of Zoology* 83 (2005): 971.

44. R. M. Bevan, I. I. Boyd, P. Butler, K. Reid, A. Woakes, and J. Croxall, "Heart Rates and Abdominal Temperatures of Free-Ranging South Georgian Shags, *Phalacrocorax georgianus*," *Journal of Experimental Biology* 200 (1997): 661. Another study of South Georgian shags had one male diving, as a mean, 275 feet (83.9 m). See A. Kato, J. P. Croxall, Y. Watanuki, and Y. Naito, "Diving Patterns and Performance in Male and Female Blue-Eyed Cormorants *Phalacrocorax atriceps* at South Georgia," *Marine Ornithology* 19 (1992): 117–29. See also R. Casaux, M. Favero, P. Silva, A. Baroni, "Sex Differences in Diving Depths and Diet of Antarctic Shags at the South Shetland Islands," *Journal of Field Ornithology* 72, no. 1 (Winter 2001): 22–29; and Timothée R. Cook, Amelie Lescroël, Yann Tremblay, and Charles-Andre Bost, "To Breathe or Not to Breathe? Optimal Breathing, Aerobic Dive Limit and Oxygen Stores in Deep-Diving Blue-Eyed Shags," *Animal Behavior* 76 (2008): 565–76.

45. Flavio Quintana, Rory P. Wilson, and Pablo Yorio, "Dive Depth and Plumage Air in Wettable Birds: The Extraordinary Case of the Imperial Cormorant," *Marine Ecology Progress Series* 334 (2007): 304; S. Wanless, T. Corfield, M. P. Harris, S. T. Buckland, and J. A. Morris, "Diving Behaviour of the Shag *Phalacrocorax aristotelis* (Aves: Pelecaniformes) in Relation to Water Depth and Prey Size," *Journal of Zoology* (London) 231 (1993): 15.

46. Timothée Cook, personal communication, July 21, 2012. The biological term for this is "regional hypothermia." See also Bevan, Boyd, Butler, et. al., "Heart Rates," 661–74.

47. See, for example, Quintana, Wilson, and Yorio, "Dive Depth," 304.

48. Timothée Cook, personal communication, July 21, 2012.

49. David Oehler, personal communications, December 9, 2011, and March 30, 2012.

50. See, for example, Elizabeth R. Thomas, Gareth J. Marshall, and Joseph R. Mc-Connell, "A Doubling in Snow Accumulation in the Western Antarctic Peninsula since 1850," *Geophysical Research Letters* 35, no. L01076 (2008): 1–5.

51. David Burkitt, personal communication, December 24, 2001.

52. Jo Hardy, personal communication, December 25, 2001.

53. Johnsgard, *Cormorants, Darters, and Pelicans*, 257. Timothée Cook explains: "The only known case is *P. bransfieldensis* which is known to do partial latitudinal migrations along the Antarctic peninsula, presumably to stay near waters that are ice-free, thus preventing starvation." Personal communication, July 21, 2012.

54. Cook, *Voyages*, 1:570.

55. Ibid., 2:362. (November 7, 1778, south of Unalaska).

56. Ibid., 1:570.

57. John Barrow, *Captain Cook: Voyages of Discovery* (Chicago: Academy Chicago, 1993), 215. His sailors went ashore at Staten Island and loaded "the Boat with young Shags." First Lieutenant R. P. Cooper wrote in his own journal: "Today boil'd Shags & Penguins in the Coppers for the Ships Company's Dinner." But then two days later in the ship's log: "The people tired of eating Penguins and Young Shags, they prefer Salt Beef and Pork to either." See James Cook, *The Journals of Captain James Cook on His Voyages of Discovery*, vol. 2 (Cambridge: Cambridge University Press, 1969), 606, 615–17.

58. S. Wanless and M. P. Harris, "Use of Mutually Exclusive Foraging Areas by Adjacent Colonies of Blue-Eyed Shags (*Phalacrocorax atriceps*) at South Georgia," *Colonial Waterbirds* 16, no. 2 (1993): 176; A similar range has been observed with Crozet Island blue-eyed shags. See Timothée R. Cook, Yves Cherel, and Yann Tremblay, "Foraging Tactics of Chick-Rearing Crozet Shags: Individuals Display Repetitive Activity and Diving Patterns over Time," *Polar Biology* 29, no. 7 (2006): 563. Harris reports that the blue-eyed shags of Patagonia (*P. atriceps*) can range as far as 18 miles (30 km). See Graham Harris, *A Guide to the Birds and Mammals of Coastal Patagonia* (Princeton, NJ: Princeton University Press, 1998), 50.

59. Fred G. Alberts, ed., *Geographic Names of the Antarctic* (Washington, DC: National Science Foundation, 1981), 764. These are believed to be the infamous Aurora Islands.

60. Ibid., 175, 764.

61. F. A. Worsley, *Shackleton's Boat Journey* (New York: W. W. Norton, 1998), 140.

62. Ernest Shackleton, *South* (New York: Signet, 1999), 195.

63. Rankin, *Antarctic Isle*, 204.

64. Ibid.

65. Ibid., 210.

5. EAST SAND ISLAND, UNITED STATES

The epigraph is from a personal communication with a man who would not give his name, March 13, 2002.

1. East Sand Island currently seems to be just barely the second-largest colony in North America, behind Balabas Island in Manitoba. These two seem far and away the largest across the continent. In the summer of 2012, the Canadian Wildlife Service conducted a count by aerial photograph of Balabas Island, Lake Winnipegosis, counting 13,864 active double-crested cormorant nests. Provided by Scott Wilson, personal communication, September 10, 2012. In the summer of 2011 Bird Research Northwest counted by aerial photography 13,045 active nests. Management on East Sand Island will cause the breeding population to reduce in the following years. See Daniel D. Roby and Ken Collis, principal investigators, "Research, Monitoring, and Evaluation of Avian Predation on Salmonid Smolts in the Lower and Mid-Columbia River: Final 2011 Report," U.S. Geological Survey, Oregon Cooperative Fish and Wildlife Research Unit and Real Time Research Inc. (August 2012), 43.

Notes to Chapter 00

2. Megan Gensler, personal communications, August 8, 9, 2011; September 10, 23, 2011.

3. William Clark's 1805 map of Baker's Bay does not show East Sand Island.

4. See U.S. Coast Survey, "Mouth of the Columbia River," 1851 nautical chart, http://historicalcharts.noaa.gov/historicals/preview/image/H00273-00-0000. By 1887 it is a sizable elbow and labeled Sand Island. In 1853 James G. Swan describes a shoal in the same location: James G. Swan, *The Northwest Coast; or, Three Years' Residence in Washington Territory* (New York: Harper & Brothers, 1857), 100. See also "Columbia River: A Photographic Journey" http://columbiariverimages.com/Regions/Places/sand_island.html (accessed September 3, 2012).

5. See U.S. Coast and Geodetic Survey, "Columbia River Entrance to Harrington Point," 1913 nautical chart, http://historicalcharts.noaa.gov/historicals/preview/image/6151-00-1913.

6. Excavations of shell middens and remains at sites on the coast just south of the Columbia suggest cormorants were a part of the Native American diet and/or culture as early as 3,500 BCE through until c. 1800 CE.

7. George A. Jobaneck and David B. Marshall, "John K. Townsend's 1836 Report of the Birds of the Lower Columbia River Region, Oregon and Washington," *Northwestern Naturalist* 73, no. 1 (Spring 1992): 6, 10. Townsend calls the birds "Black Cormorant (*Phalacrocorax carbo*)," but this is almost certainly what we know today as the double-crested cormorant.

8. John James Audubon, *Ornithological Biography, or an Account of the Habits of the Birds of the United States of America*, vol. 5 (Edinburgh: Adam & Charles Black, 1849), 148.

9. Nelson, *Pelicans, Cormorants and Their Relatives*, 459.

10. See David G. Ainley, Daniel W. Anderson, and Paul R. Kelly, "Feeding Ecology of Marine Cormorants in Southwestern North America," *Condor* 83, no. 2 (1981): 120–31.

11. Again, some of the double-crested cormorant classification might be changing, owing in part to genetic analysis. See Mercer, "Phylogeography."

12. Jessica Y. Adkins and Daniel D. Roby, "A Status Assessment of the Double-Crested Cormorant (*Phalacrocorax auritus*) in Western North America: 1998–2009," final report submitted to U.S. Army Corps of Engineers, USGS-Oregon Cooperative Fish and Wildlife Research Unit (March 2010), 35, 44.

13. See Terry McEneaney, "Piscivorous Birds of Yellowstone Lake: Their History, Ecology, and Status," in *Yellowstone Lake: Hotbed of Chaos or Reservoir of Resilience?* Conference Proceedings, ed. Roger J. Anderson and David Harmon (Yellowstone, WY: Yellowstone Center for Resources and the George Wright Society, 2002), 121–34.

14. *David Thompson's Narrative of His Explorations in Western America, 1784–1812*, ed. J. B. Tyrell (Toronto: Champlain Society, 1916), 459. Thompson writes "fishy tasted" and "it's."

15. Adkins and Roby, "Status Assessment," 18; Adam Peck-Richardson, personal communication, February 19, 2013. See also Karen N. Courtot, Daniel D. Roby, Jessica Y. Adkins, Donald E. Lyons, D. Tommy King, and R. Scott Larsen, "Colony Connectivity

of Pacific Coast Double-Crested Cormorants Based on Post-Breeding Dispersal from the Region's Largest Colony," *Journal of Wildlife Management* 76, no. 7 (September 2012): 1462–71.

16. Pelagics in Nelson, *Pelicans, Cormorants and Their Relatives* (p. 454) summarizing Johnsgard, *Cormorants, Darters, and Pelicans of the World* (published in 1993): "c. 130,00 adults in N American population," including all Aleutians; *Birds of North America* writes data "only crudely known": Keith A. Hobson, "Pelagic Cormorant (*Phalacrocorax pelagicus*)," *The Birds of North America Online*, ed. A. Poole (Ithaca, NY: Cornell Lab of Ornithology, 1997), http://bna.birds.cornell.edu/bna/species/282; red-faced "> c.200,000 individuals," (2006): BirdLife International, "*Phalacrocorax urile*," 2012, IUCN Red List of Threatened Species, version 2012.2, www.iucnredlist.org.

17. Brandt's roughly worldwide at 75,000 pairs: Elizabeth A. Wallace and George E. Wallace, "Brandt's Cormorant (*Phalacrocorax penicillatus*)," *Birds of North America Online*, http://bna.birds.cornell.edu/bna/species/362. Double-crested: Daniel Roby, personal communication, May 10, 2012.

18. The number was probably closer to 213,500 pairs on San Martín in 1913. See Wires and Cuthbert, "Historic Populations," 14. For further information see Harry R. Carter, Arthur L. Sowls, Michael S. Rodway, et al., "Population Size, Trends, and Conservation Problems of the Double-Crested Cormorant on the Pacific Coast of North America," in Nettleship and Duffy, *Double-Crested Cormorant*, 202.

19. Eduardo Palacios and Eric Mellink, "Nesting Waterbirds on Islas San Martín and Todos Santos," *Western Birds* 31 (2000): 185.

20. For the last fifteen years East Sand Island has hosted an average of 10,000 nesting pairs of cormorants at the peak of summer each year: Bird Research Northwest, "Annual Colony Size (1997–2009)." www.birdresearchnw.org. See earlier note about Balabas Island, Manitoba, which is now likely the largest in all of North America.

21. Wires and Cuthbert, "Historic Populations," 14.

22. U.S. Army Corps of Engineers, Portland District, "Draft Environmental Assessment: Double-Crested Cormorant Dissuasion Research on East Sand Island in the Columbia River Estuary" (February 2, 2012), 28–29.

23. See, for example, "Grant McOmie: Salmon Killers Flock to Bird Island," KGW NewsChannel 8 (May 19, 2009), www.kgw.com/archive/59516752.html; sometimes sea lions, harbor seals, and terns are also called "salmon killers."

24. Megan Gensler, personal communications, August 8, 2011, September 10, 2011, September 23, 2011.

25. Robinson Jeffers, "Birds and Fishes," in *The Wild God of the World: An Anthology of Robinson Jeffers*, ed. Albert Gelpi (Stanford, CA: Stanford University Press, 2003), 182. First published in *The Beginning and the End and Other Poems* (1963).

26. U.S. Fish and Wildlife Service, "Caspian Tern Management to Reduce Predation of Juvenile Salmonids in the Columbia River, Final Environmental Impact Statement" (Portland, OR, 2005), 3–3.

27. William G. Robbins, "The World of Columbia River Salmon: Nature, Culture,

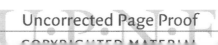

and the Great River of the West," in *The Northwest Salmon Crisis: A Documentary History*, ed. Joseph Cone and Sandy Ridlington (Corvallis: Oregon State University Press, 1996), 2.

28. Courtland L. Smith, *Salmon Fishers on the Columbia* (Corvalis: Oregon State University Press, 1979), 21, 25, 27.

29. Shown on an 1887 map from W. A. Jones, *The Salmon Fisheries of the Columbia River*, Senate Document 123, Serial Set No. 2510, in Smith, *Salmon Fishers*, 32. This map does not show East Sand Island.

30. Robbins, "World of Columbia River Salmon," 11–14.

31. Roy Lowe, personal communication, March 14, 2002; "Chinook Salmon Catch Best in Years," *Curry Coastal Pilot* (Brookings, OR), March 13, 2002, 3A.

32. See Matthew Elliot, "Seafood Watch: Pacific Salmon, U.S. West Coast Region," Monterey Bay Aquarium (Monterey, May 10, 2011), 1–128.

33. See, for example, William L. Lang and Robert C. Carriker, "A Resurgent Columbia River: An Introduction," in *Great River of the West: Essays on the Columbia River*, ed. Williams L. Lang and Robert C. Carriker (Seattle: Washington University Press, 1999), 3–17.

34. As of 2012, thirteen of the twenty salmonid ESUs that spend some time in the Columbia River Basin are listed by the National Marine Fisheries Service as either threatened or endangered. See Northwest Regional Office, "ESA Salmon Listing," NOAA, National Marine Fisheries Service, www.nwr.noaa.gov/ESA-Salmon-Listings/Index .cfm (accessed August 25, 2012); USFWS, "Caspian Tern Management," 3–14; Bird Research Northwest, "Overview," Real Time Research Inc., www.birdresearchnw.org (accessed August 25, 2012).

35. Intermountain Communications, "Mitchell Act Economic Impact: 17 Hatcheries, 70 Million Juveniles, Almost Half the Basin Harvest," *Columbia Basin Bulletin*, January 21, 2011, www.cbbulletin.com/404190.aspx.

36. "2013 Pikeminnow Sport Reward Program," Bonneville Power Administration, www.pikeminnow.org/info.html (accessed February 17, 2013).

37. Bird Research Northwest is a collaboration between Oregon State University, the U.S. Geological Survey–Oregon Cooperative Fish and Wildlife Research Unit out of Corvalis, and Real Time Research out of Bend.

38. Donald Lyons, Daniel Roby, and Ken Collis, "Foraging Patterns of Caspian Terns and Double-Crested Cormorants in the Columbia River Estuary," *Northwest Science* 81, no. 2 (2007): 100.

39. USFWS, "Caspian Tern Management," 1–3; Bird Research Northwest, "East Sand Island," Real Time Research Inc., www.birdresearchnw.org (accessed August 25, 2012).

40. Daniel Roby, personal communication, May 10, 2012; USFWS, "Caspian Tern Management," 1–3, 3–5, 4–5; BirdLife International, "*Sterna caspia*," 2012, IUCN Red List of Threatened Species, version 2012.1, www.iucnredlist.org.

41. Bird Research Northwest, "Background on the Research and Monitoring of Caspian Terns Nesting in the Columbia River Estuary," December 30, 2011, Real Time Research Inc.,

www.birdresearchnw.org/Project-Info/Project-Background/Caspian-Terns/Columbia-River-Estuary-Terns/default.aspx; Don Lyons, personal communication, April 1, 2012.

42. Bird Research Northwest was first called Columbia Bird Research.

43. Thank you to Hugh Powell for putting it this way.

44. Citing Doug Bell: Dan Roby, personal communication, May 10, 2012; Carter, Sowls, Rodway, et al., "Population Size, Trends," 198–99.

45. Donald Lyons, "Bioenergetics-Based Predator-Prey Relationships between Piscivorous Birds and Juvenile Salmonids in the Columbia River Estuary" (PhD diss., Oregon State University, 2010), 266. See also Lyons, Roby, and Collis, "Foraging Patterns," 91–103.

46. Cynthia D. Anderson, Daniel D. Roby, and Ken Collis, "Foraging Patterns of Male and Female Double-Crested Cormorants Nesting in the Columbia River Estuary," *Canadian Journal of Zoology* 82 (2004): 541.

47. Lyons, Roby, and Collis, "Foraging Patterns," 100.

48. Daniel Roby, personal communications, May 10, 2012, October 3, 8, 2012; Don Lyons, personal communication, October 4, 2012.

49. Daniel Roby, personal communication, May 10, 2012; Roby and Collis, "Avian Predation on Salmonid Smolts," 7.

50. Daniel Roby, personal communication, May 10, 2012. Yet Roby adds: "In cases where both wild and hatchery-reared smolts from the same stock are PIT-tagged, hatchery-raised fish are not significantly more susceptible to cormorant predation than their wild counterparts, a surprising result."

51. For example, from 2000 to 2009, the federal government allocated $804.9 million to the tribes and states of the Pacific Coast, as well as Idaho, through the Pacific Coast Salmon Recovery Fund. See Pacific Coastal Salmon Recovery Fund, "2010 Report to Congress," NOAA Fisheries Service (2010), 1. For the size of the hatchery operation specifically, see Intermountain Communications, "Mitchell Act Economic Impact."

52. Range D. Bayer, personal communication, March 14, 2002. All subsequent quotations are from this interview.

53. Range D. Bayer, "Cormorant Harassment to Protect Juvenile Salmonids in Tillamook County, Oregon: Studies in Oregon Ornithology, No. 9" (Gahmken Press, 2000), 2.

54. Anonymous Portland man, personal communication, March 13, 2002.

55. The net pen facilities experimented with feeding the smolt under the surface with underwater pipes — to teach the fish to be less vulnerable — but they saw no significant survival gains. Alan Dietrichs, personal communication, October 11, 2012.

56. Anonymous CCEDC Fisheries Project employee, personal communication, March 13, 2002. He is probably thinking of Arctic terns here, who migrate much farther than the Caspian terns.

57. Adam Peck-Richardson, personal communication, September 27, 2011; Army Corps of Engineers, Portland District, "Draft Environmental Assessment," 21.

58. Anonymous CCEDC Fisheries Project Employee, personal communication, March 13, 2002.

59. Mike Stahlberg, "Three Little Pigs Could Save the Salmon? No Fooling," *Eugene (OR) Register-Guard*, August 1, 2008, www.highbeam.com/doc/1G1-178644926.html.

60. Adkins and Roby, "Status Assessment," 8.

61. Ibid.

62. Don Lyons, personal communication, April 1, 2012.

63. Adkins and Roby, "Status Assessment," 10–11; Carter, Sowls, Rodway, et al., "Population Size, Trends," 189–215.

64. Bird Research Northwest, "East Sand Island," www.birdresearchnw.org/Project -Info/Study-Area/Columbia-Basin/East-Sand-Island/default.aspx; Army Corps of Engineers, Portland District, "Draft Environmental Assessment," 30.

65. Army Corps of Engineers, Portland District, "Draft Environmental Assessment," 33; American white pelicans are also in the estuary, and they eat salmon upriver, but the birds do not occur in large numbers.

66. Steve Grinley, "Rare Birds Are Objects of Pursuit," *NewburyportNews.com* (July 16, 2011), www.newburyportnews.com/local/x967735479/Rare-birds-are-objects -of-pursuit; Lawrence Pyne, "Outdoors: 'Bully' Eagles Keep Cormorant Population in Check," August 7, 2011, *Burlington Free Press.com*, www.mychamplain.net/forum/bully -eagles-keep-cormorant-population-check; Jack Knox, "Seagulls on a Bombing Mission over Victoria," *Victoria (BC) Times Colonist*, August 14, 2011, www.timescolonist .com/technology/Seagulls+bombing+mission+over+Victoria/5253351/story.html; Tom Banse, "Bald Eagle Comeback Pressures Coastal Seabirds," *Oregon Public Broadcasting News*, July 28, 2010, http://news.opb.org/article/bald-eagle-comeback-pressures-coastal -seabirds.

67. For simplicity throughout this book, I do not use the terms "sub-adult" or "immature" but use "juvenile" to mean any bird that is independent of parental care but not yet old enough to breed.

68. Paul Wolf and LeAnn White, "Newcastle Disease Surveillance in Minnesota," *Carrier* 4, no. 1 (February 2012): 3.

6. TRING, ENGLAND

The epigraph is from *Dawson's Avian Kingdom*, ed. Anna Neher (Berkeley, CA: Heyday Books, 2007), 11.

1. For a more detailed and accurate description of the traditional skinning process, see C. J. O. Harrison and G. S. Cowles, *Instructions for Collectors No. 2A: Birds* (London: Trustees of the British Museum [Natural History], 1970).

2. C. S. Roselaar, "An Inventory of Major European Bird Collections," *Bulletin of the British Ornithologists' Club* 123A (2003): 291–92; Robert Prŷs-Jones, personal communication, February 16, 2012.

3. Roselaar, "Inventory," 291; "Bird Group," February 20, 2012, *Natural History Museum at Tring*, www.nhm.ac.uk/research-curation/departments/zoology/bird-group/index .html; Robert Prŷs-Jones, personal communication, April 13, 2012.

4. Robert Prŷs-Jones, personal communications, March 22, 2002, and April 13, 15, 2012. All subsequent quotations are from these interviews unless otherwise noted.

5. R. Bowdler Sharpe and W. R. Ogilvie-Grant, *Catalogue of the Birds of the British Museum*, vol. 26 (London: Order of the Trustees of the British Museum [Natural History], 1898), 405.

6. Nelson, *Pelicans, Cormorants and Their Relatives*, 595.

7. Ibid., 530; BirdLife International, "*Phalacrocorax pygmeus*," 2012, IUCN Red List of Threatened Species, version 2012.1, www.iucnredlist.org.

8. Nelson, *Pelicans, Cormorants and Their Relatives*, 529.

9. Ibid., 528; Sharpe and Ogilvie-Grant, *Catalogue of the Birds*, 402–3.

10. Johnsgard, *Cormorants, Darters, and Pelicans*, vi–vii; Peter Harrison, *Seabirds: An Identification Guide*, rev. ed. (Boston: Houghton Mifflin, 1985), 293.

11. Aristotle, Book 8, *The History of Animals*, trans. D'Arcy Wentworth Thompson (Cambridge, MA: Internet Classics Archive and Daniel C. Stevenson, 1994–2000), http://classics.mit.edu/Aristotle/history_anim.8.viii.html. See a slightly different translation in Aristotle, *History of Animals, Books VII–X*, ed. and trans. D. M. Balme (Cambridge, MA: Harvard University Press, 1991), 105–9.

12. Johnsgard, *Cormorants, Darters, and Pelicans*, 10–11.

13. Ibid., 6.

14. Van Tets, "Australia and the Origin of Shags and Cormorants," 121.

15. See, for example, Kennedy, Valle, and Spencer, "Phylogenetic Position of the Galápagos Cormorant," 94–98; and Charles G. Sibley and Jon E. Ahlquist, *Phylogeny and Classification of Birds: A Study in Molecular Evolution* (New Haven, CT: Yale University Press, 1990), 491–503.

16. Jan Hodder, personal communication, March 16, 2002.

17. Nelson, *Pelicans, Cormorants and Their Relatives*, 4; Sir A. Landsborough Thomson, *A New Dictionary of Birds* (London: Thomas Nelson, 1964), 535.

18. See a helpful summary of this argument in Hanson, *Feathers*, or more briefly in Chris Elphick, John B. Dunning Jr., and David Allen Sibley, eds., *The Sibley Guide to Bird Life and Behavior* (New York: Alfred A. Knopf, 2009), 40–42.

19. Van Tets, "Australia," 124. As to the age of cormorants, Nelson writes: "The oldest possible 'cormorant' was the related plotopterid [early Miocene, 23 to 15 million years ago]. . . . Undisputed cormorants have been found in the Miocene [23 to 5.3 million years ago]." See Nelson, *Pelicans, Cormorants and Their Relatives*, 9–10, 17.

20. An exception is that tropicbirds have no true gular pouch.

21. Nelson, *Pelicans, Cormorants and Their Relatives*, 7.

22. See a helpful, updated summary of cormorant phylogeny in Les Christidis and Walter E. Boles, *Systematics and Taxonomy of Australian Birds* (Collingwood, Victoria: CSIRO, 2008), 49–50. For many cormologists, this is the new go-to taxonomy.

23. Registration Number: 1893.12.31.2, H. O. Forbes, 1893. See Sharpe and Ogilvie-Grant, *Catalogue of the Birds*, 385.

24. Miriam Rothschild, *Dear Lord Rothschild: Bird, Butterfly, and History* (Philadelphia: Balaban Publishers, 1983), 108–9.

25. Several of the most expert and recent sources cite a further specimen in Dresden, but Steven van der Mije, senior collections manager, the Netherlands Centre for Biodiversity Naturalis, wrote to me on January 2, 2012: "The museum in Dresden can't find any specimen [of the spectacled cormorant]. Where this information came from is still under investigation. For the moment it is best to assume there are six confirmed specimens." See "Spectacled Cormorant *Phalacrocorax perspicillatus* Pallas, 1811," 2005, National Museum of Natural History Naturalis, Leiden, http://nlbif.eti.uva.nl/naturalis. This website has a three-dimensional photograph of their mounted specimen. For mention of the seventh specimen in Dresden, see James C. Greenway Jr., *Extinct and Vanishing Birds of the World*, 2nd ed. (New York: Dover, 1967), 160; and Julian P. Hume and Michael Walters, *Extinct Birds* (London: T & A. D. Poyser, 2012), 76.

26. Graham Satchell (reporter), "Thousands Sign Petition Calling for Cormorant Cull," February 22, 2012, BBC News, www.bbc.co.uk.

27. The Canal and River Trust (formerly British Waterways) is the organization in charge of the licensing for recreational fishing in the reservoirs. John Ellis, National Fisheries and Angling manager, Canal and River Trust, personal communication, July 30, 2012.

28. The Tring Anglers, *NewsLine*, Winter 2010–11, 1. After several queries, I was unable to get a reply from the club.

29. Mark Cocker and Richard Mabey, *Birds Britannica* (London: Chatto & Windus, 2005), 36.

30. Robert A. Lambert, "Seabird Control and Fishery Protection in Cornwall, 1900–1950," *British Birds* 96, no. 1 (January 2003): 30, 32.

31. Matthew Heydon, "Licensing the Shooting of Cormorants to Prevent Serious Damage to Fisheries and Inland Waters," Natural England–Wildlife Management and Licensing (January 14, 2008), 5. For some of the controversy regarding this see, for example, Paul Brown, "Cull 'will wipe out cormorants': RSPB Will Go to Court against Minister's Move to Help Anglers," *Guardian*, February 17, 2004, www.guardian.co.uk.

32. "Cormorant Busters," July 24, 2012, www.cormorantbusters.co.uk.

33. Richard Lee, "These Birds Must Be Killed," *Angling Times*, no. 2266, December 4, 1996, 1, 8–9. See also Cocker and Mabey, *Birds Britannica*, 36.

34. See, for example, Kathy Marks and Mathew Brace, "Editor Urged Cormorant Cull," June 5, 1997, *Independent*, www.independent.co.uk/news/editor-urged-cormorant-cull-1254212.html.

35. Nelson, *Pelicans, Cormorants and Their Relatives*, 411–12.

36. Stuart E. Newson, Baz Hughes, Ian C. Russell, Graham R. Ekins, and Robin M. Sellers, "Sub-specific Differentiation and Distribution of Great Cormorants *Phalacrocorax carbo* in Europe," *Ardea* 92, no. 1 (2005): 3; Moran Joint Bird Group, "Cormorants: The Facts," Environment Agency (undated), 1.

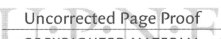

37. Angling Trust, "Cormorant Watch," video, 2011, www.cormorantwatch.org/video .html.

38. Trevor Harrop (contact), "Press Release: Cormorant Campaign Goes to Parliament," Angling Trust, February 22, 2012, www.anglingtrust.net.

39. James Fisher, *The Shell Bird Book* (London: Ebury Press, 1966), 24, 28, 30–31, 38; D. W. Yalden and U. Albarella, *The History of British Birds* (Oxford: Oxford University Press, 2009) 34–36, 74, 87, 210. The Norfolk coastal deposits are perhaps the oldest record, at approximately 350,000 years old.

40. Matthew Heydon, "Licensing the Shooting of Cormorants to Prevent Serious Damage to Fisheries and Inland Waters," Natural England–Wildlife Management and Licensing (January 14, 2008), 5. See discussion of population trends and breeding versus wintering in Rhys E. Green, Will J. Peach, and Norman Ratcliffe, "RSPB Comments on CSL's Revised Cormorant Cull Analysis," Royal Society for the Protection of Birds (March 31, 2005), 3–4.

41. Sarah Dove, personal communication, September 21, 2012.

42. Vivien Behrens, Felix Rauschmayer, and Heidi Wittmer, "Managing International 'Problem' Species: Why Pan-European Cormorant Management Is So Difficult," *Environmental Conservation* 35, no. 1 (2008): 59.

43. Environment Directorate General and CorMan, "The EU Cormorant Platform," February 22, 2012, http://ec.europa.eu/environment/nature/cormorants/home_en.htm; Stefano Volponi (webmaster), "Cormorants," updated July 7, 2012, IUCN Wetlands International Cormorant Research Group, http://cormorants.freehostia.com/index.htm.

44. Wetlands International Cormorant Research Group, "Cormorants in the Western Palearctic" (2008 leaflet), 1–2; Thomas Bregenballe, Stephano Volponi, Mennobart R. van Eerden, et al., "Status of the Breeding Population of Great Cormorants *Phalacrocorax carbo* in the Western Palearctic in 2006," *Proceedings 7th International Conference on Cormorants 2005*, ed. M. R. van Eerden, Stef van Rihn, and V. Keller (2011), 8–20; Environment Directorate General and CorMan, "Distribution of the Breeding Population around 2006," February 22, 2012, http://ec.europa.eu/environment/nature/ cormorants/breeding-distribution-2006.htm. "The size of the cormorant population in Europe varies between half a million and one and a half million birds, depending on who provides the data," in Tilo Arnhold (contact), "Press Release: The Cormorant — the 'Black Plague' or an Example of Successful Species Conservation," *Helmholtz Association of German Research Centres,* June 4, 2008, 1.

45. Environment Directorate General and CorMan, http://ec.europa.eu/environment/ nature/cormorants/breeding-distribution-2006.htm; Bregenballe, Volponi, van Eerden, et al., "Status of the Breeding Population," 14; Stuart E. Newson, Graham R. Ekins, John H. Marchant, Mark M. Rehfisch, and Robin M. Sellers, "The Status of Inland and Coastal Breeding Great Cormorants *Phalacrocorax carbo* in England," BTO Research Report, no. 443 (April 2006), 5, 13, 29, 31, 36.

46. Newson, Hughes, Russell, et al., "Sub-specific Differentiation," 3.

47. Newson, Ekins, Marchant, et al., "Status of Inland and Coastal Breeding Great

Cormorants," 9; Bregenballe, Volponi, van Eerden, et al., "Status of the Breeding Population," 9.

48. Behrens, Rauschmayer, and Wittmer, "Managing International 'Problem' Species," 56, 58.

7. BERING ISLAND, RUSSIA

The epigraph is from Leonhard Stejneger and Frederic A. Lucas, "Contributions to the Natural History of the Commander Islands," *Proceedings of the United States National Museum* 12 (Washington, DC: Government Printing Office, 1890), 83.

1. Andrew Wyeth, *Northern Coast* (1942), Portland Museum of Art, Maine.

2. Terry Thornsley, *Laguna Locals* (2008), Crescent Bay Point Park, Laguna Beach, CA; Duncan McKiernan, *Cormorants* (1980), City Pier at the Feiro Marine Life Center, Port Angeles, WA.

3. Tom Harvey, *Cormorant* (2008), Cotswold Country Park, Cirencester; Kevin Herlihy, *Cormorant* (2008), Chiswick Park, London; Elizabeth Cook, *Cormorant*, Thorpe Meadows, Peterborough. See also *Taking Flight*, by Craig Knowles (1997), steel girders and a steel cormorant in Sunderland, also vertical with wing-spread; and *Cormorant* by Christian Funnell (2000, 2012), seven feet (2 m) tall, standing with wings spread in Newhaven Harbor (see Chris Conil, "Christian Funnell: Cormorant," August 16, 2012, www.youtube.com/watch?v=D80LEZMUIhg). And this is not an exhaustive list.

4. Anna Kirk-Smith, personal communications, January 6, April 13, 2012. All quotations in this chapter are from this correspondence, or from Anna Kirk-Smith, "Easel Words: 'The Unfortunate Repercussions of Discovery and Survival,'" *Jackdaw* 101 (January/February 2012): 21. For images of the sculpture, see www.anna.kirk-smith.com.

5. Victor Zviagin, "The Unknown Vitus Bering," 2010 International Conference on Russian America (Sitka, AK: August 18–22, 2010), 1–3. See http://2010rac.com/papers .shtml.

6. Georg Wilhelm Steller, *Journal of a Voyage with Bering, 1741–1742*, trans. and ed. O. W. Frost, trans. Margritt A. Engel (Stanford, CA: Stanford University Press, 1988), 113.

7. Ibid., 114.

8. Ibid., 120.

9. Ibid.

10. Sven Waxell, *The American Expedition* (London: William Hodge, 1952), 123.

11. Steller, *Journal*, 122.

12. Anonymous Englishman, personal communication, March 20, 2002.

13. Steller, *Journal*, 130.

14. Ibid., 141, 215–16; Zviagin, "Unknown Vitus Bering," 1–3. See http://2010rac.com/ papers.shtml.

15. Steller, *Journal*, 129. For an image and description of this type of knife see Frank A. Golder, *Bering's Voyages*, trans. Leonhard Stejneger, vol. 2 (New York: American Geographical Society, 1925), 47.

16. Leonhard Stejneger, *Georg Wilhelm Steller: The Pioneer of Alaskan Natural History* (Cambridge, MA: Harvard University Press, 1936), 369.

17. BirdLife International, "*Haliaeetus pelagicus*," 2012, IUCN Red List of Threatened Species, version 2012.1, www.iucnredlist.org.

18. Steller, *Journal*, 162. See also Georg Wilhelm Steller, "De Bestis Marinis, or, the Beasts of the Sea (1751)," trans. Walter Miller and Jennie Emerson Miller, ed. and transcriber Paul Royster, *Faculty Publications, UNL Libraries* 17 (2011).

19. Errol Fuller, *Extinct Birds* (Ithaca, NY: Comstock Books, 2001), 70; translated as "so that one single bird was sufficient for three starving men" in Stejneger, *Georg Wilhelm Steller*, 351.

20. Golder, *Bering's Voyages*, 237.

21. Fuller, *Extinct Birds*, 70.

22. This was written by Pallas in Latin and translated in Walter Rothschild, *Extinct Birds* (London: Hutchinson & Co., 1907), 87. This citation includes: "Female smaller, without crest and spectacles." This would be unique among cormorants if indeed the female had different plumage and different skin around the eye.

23. Stejneger and Lucas, "Contributions to the Natural History of the Commander Islands," 84.

24. Nelson, *Pelicans, Cormorants and Their Relatives*, 590; Stejneger and Lucas, "Contributions to the Natural History of the Commander Islands," 84.

25. Greenway, *Extinct and Vanishing Birds*, 159.

26. Stejneger, *Georg Wilhelm Steller*, 351.

27. Steller, *Journal*, 6, 27–31. This text is probably as faithful an edition of Steller's journal as possible, working not from Pallas, but from a different copy of the presumed manuscript found by Frank Golder.

28. Stejneger, *Georg Wilhelm Steller*, 370.

29. BirdLife International, "Table 4a: Red List Category Summary for All Animal Classes and Orders," 2012, IUCN Red List, version 2012.1, www.iucnredlist.org; Stuart Pimm, Peter Raven, Alan Peterson, and Paul R. Ehrlich, "Human Impacts on the Rates of Recent, Present, and Future Bird Extinctions," *National Academy of Sciences of the USA* 103 (July 18, 2006): 10941.

30. BirdLife International, "*Corvus hawaiiensis*," "*Gallirallus owstoni*," "*Mitu mitu*," and "*Zenaida graysoni*," 2012, IUCN Red List, version 2012.1, www.iucnredlist.org.

31. BirdLife International, "Table 4a."

32. See, for example, Stuart H. M. Butchart, Alison J. Stattersfield, Leon A. Bennun, et al., "Measuring Global Trends in the Status of Biodiversity: Red List Indices for Birds," *PLoS Biology* 2, no. 12 (2004): 2299; and Carl Safina, *The Eye of the Albatross* (New York: Henry Holt, 2002).

33. BirdLife International, "*Phalacrocorax featherstoni*," 2012, IUCN Red List of Threatened Species, version 2012.2, www.iucnredlist.org.

34. Pimm, Raven, Peterson, and Ehrlich, "Human Impacts," 10941–43.

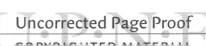

35. Duke University, "Birds Going Extinct Faster Due to Human Activities," *Science Daily*, July 5, 2006, www.sciencedaily.com.

36. The red-faced cormorant (*P. urile*) and the pelagic cormorant (*P. pelagicus*) breed on the Komandorski Islands. See Nelson, *Pelicans, Cormorants and Their Relatives*, 454, 459.

37. Steller wrote, via Pallas: "Of rarer birds not seen on the Siberian coast I have met with a special sea eagle with a white head and tail [and] the white sea raven (*Pelec. Bassanus*). It is impossible to reach the latter because it only alights singly on the cliffs facing the sea." In Golder, *Bering's Voyages*, 237.

38. Elphick, Dunning, and Sibley, *Sibley Guide*, 33–34; Gill, *Ornithology*, 188–90.

39. Elphick, Dunning, and Sibley, *Sibley Guide*, 34.

40. Ibid.

41. Gill, *Ornithology*, 185.

42. Nelson, *Pelicans, Cormorants and Their Relatives*, 160; Craig R. White, Norman Day, Patrick J. Butler, and Graham R. Martin, "Vision and Foraging in Cormorants: More Like Herons Than Hawks," *PloS ONE* 7 (July 2007): 1; Gill, *Ornithology*, 186; Johnsgard, *Cormorants, Darters, and Pelicans*, 23.

43. For altitude of cormorant flight from 1,000 to 1,100 meters: Nelson, *Pelicans, Cormorants and Their Relatives*, 166; Brian Dorr, personal communication, July 17, 2012.

44. White, Day, Butler, and Martin, "Vision and Foraging in Cormorants," 1.

45. See, for example, Yutaka Watanuki, Francis Daunt, Akinori Takhashi, et al., "Microhabitat Use and Prey Capture of a Bottom-Feeding Top Predator, the European Shag, Shown by Camera Loggers," *Marine Ecology Progress Series* 356 (2008): 283–93. For more discussion of cormorant vision underwater, see, for example, Tamir Strod, Ido Izhaki, Zeev Arad, and Gadi Katzir, "Prey Detection by Great Cormorant (*Phalacrocorax carbo sinensis*) in Clear and in Turbid Water," *Journal of Experimental Biology* 211 (2008): 866–72.

46. White, Day, Butler, and Martin, "Vision and Foraging in Cormorants," 3.

47. Stejneger, *Georg Wilhelm Steller*, 373–74.

48. Steller, *Journal*, 148.

49. Ibid.

50. See, for example, W. Michael Mathes, *The Russian-Mexican Frontier: Mexican Documents regarding the Russian Establishments in California, 1808–1842* (Berkeley, CA: Fort Ross Interpretive Association, 2008), 11.

51. See, for example, William Clark, November 20, 1805, in Frank Bergon, ed., *The Journals of Lewis and Clark* (New York: Penguin, 1995), 323.

52. Scott O'Dell, *Island of the Blue Dolphins* (New York: Dell, 1971), 136; this would likely be Brandt's cormorants (*P. penicillatus*), based on location and description.

53. The Vatican Museum does not hold this dress now, confirmed by Jan Timbrook, curator of ethnography, Santa Barbara Museum of Natural History, in Jan Timbrook, "The Lone Woman of San Nicolas Island," www.sbnature.org/research/anthro/chumash/lowom.htm (accessed July 12, 2012).

54. O'Dell, *Island of the Blue Dolphins*, 156.

55. Alexander Wetmore, "Leonhard Hess Stejneger, 1851–1943: Presented to the Academy at the Autumn Meeting, 1945," *Biographical Memoirs* 24 (Washington, DC: National Academy of Sciences): 149; Waldo L. Schmitt, "Leonhard Stejneger," *Society of Systematic Biologists* 13, no. 4 (December 1964): 246.

56. Schmitt, "Leonhard Stejneger," 246.

57. Ibid.

58. Leonhard Stejneger, "Contributions to the History of the Commander Islands: No.1, Notes on the Natural History, Including Descriptions of New Cetaceans," *Proceedings of the United States National Museum* 6, no. 5 (1883): 65.

59. Leonhard Stejneger, "Results of Ornithological Explorations in the Commander Islands and in Kamtschatka" *Bulletin No. 29 of the United States National Museum* 39 (Washington, DC: Government Printing Office, 1885), 180.

60. Douglas Siegel-Causey, Christine Lefevre, and Arkadii B. Savinetskii, "Historical Diversity of Cormorants and Shags from Amchitka Island, Alaska," *Condor* 93 (1991): 850. See also an eighteenth-century Russian account by S. P. Krashnaminnikov about the native use of what are believed to be cormorants. This includes an account of catching cormorants with a hook in a fish, and the use of cormorant bladders for net buoys and the use of bones for needle cases. Theed Pearse, *Birds of the Early Explorers in the Northern Pacific* (Comox, BC: Theed Pearse / Centennial of Canadian Confederation Project, 1968), 55–56.

61. Stejneger and Lucas, "Contributions to the Natural History of the Commander Islands," 84.

62. Surprisingly, Leonhard Stejneger wrote little about the flightlessness of this cormorant, but a close reading of his work shows that he came to conclude that the spectacled cormorant was "practically flightless." See Stejneger, *Georg Wilhelm Steller*, 351, pl. 20.

63. Siegel-Causey, Lefevre, and Savinetskii, "Historical Diversity of Cormorants," 849.

64. Lucien M. Turner, "Contributions to the Natural History of Alaska" (Washington DC: Government Printing Office, 1886), 7, 130.

65. Siegel-Causey, Lefevre, and Savinetskii, "Historical Diversity of Cormorants," 843, 846.

66. Richard Brinsley Hinds, ed., *Zoology of the Voyage of the H.M.S. Sulphur*, vol. 1 (London: Smith and Elder, 1843), 49.

67. See Fuller, *Extinct Birds*.

68. Stejneger, *Georg Wilhelm Steller*, 511.

69. Hinds, *Zoology of the Voyage of the H.M.S. Sulphur*, pl. 32.

70. First printed in Daniel Giraud Eliot, *The New and Heretofore Unfigured Species of Birds of North America*, vol. 2 (New York: published by the author, 1869), pl. 50.

71. The Keulemans painting is in Walter Rothschild, *Extinct Birds* (London: Hutchinson & Co., 1907), pl. 39. The silhouette of this painting is the logo for the "Ghosts of Gone Birds" project.

The epigraph is from William Rodney Allen, "An Interview with Kurt Vonnegut," in *Conversations with Kurt Vonnegut*, ed. William Rodney Allen (Jackson: University Press of Mississippi, 1999), 292.

1. This entry was on June 26, 1826, in Charles Darwin, "Edinburgh Diary," ed. and transcribed John van Whye, rev. 2009, *Darwin Online*, http://darwin-online.org.uk.

2. Richard Keynes, ed., *Charles Darwin's Zoology Notes and Specimen Lists from H.M.S. Beagle* (Cambridge: Cambridge University Press, 2000), xxii, 211.

3. Charles Darwin, *Voyage of the* Beagle (New York: Penguin, 1989), 272–73.

4. See, for example, Charles Haskins Townsend, "The Galapagos Tortoises in Their Relation to the Whaling Industry: A Study of Old Logbooks," in *Logbook Tales: Mostly about Galapagos Tortoises* . . . (New Bedford, MA: Reynolds, 1936).

5. Wilson Heflin, *Herman Melville's Whaling Years*, ed. Mary K. Bercaw Edwards and Thomas Farel Heffernan (Nashville, TN: Vanderbilt University Press, 2004), 90–91, 98–99.

6. *Great Short Works of Herman Melville*, ed Warner Berthoff (New York: Harper & Row, 1969), 106.

7. See, for example, Charles Haskins Townsend, "The Flightless Cormorant in Captivity," *Auk* 46, no. 2 (April 1929): 211.

8. James T. Carlton and Mary K. Bercaw Edwards found this reference of the cormorant in Porter's account.

9. David Porter, *Journal of a Cruise Made to the Pacific Ocean* (New York: Wiley & Halsted, 1822), 1:136. He catches the cormorants a second time along with penguins and other birds and animals near a northern cavern of Isabela.

10. J.A.A., "Rothschild and Hartert's 'Review of the Ornithology of the Galapagos Islands,'" *Auk* 17, no. 3 (July 1900): 301.

11. R. Bowdler Sharpe, ed., "The Hon. Walter Rothschild Sent . . ." *Bulletin of the British Ornithologists' Club* 7, no. 54 (May 25, 1898): 52.

12. Ibid.

13. Joseph R. Slevin, "Log of the Schooner 'Academy' on a Voyage of Scientific Research to the Galapagos Islands 1905–1906," *Occasional Papers of the California Academy of Sciences* 17 (San Francisco: California Academy of Sciences, 1931), 102.

14. William Beebe, *Galápagos: World's End* (New York, Dover: 1951), 170. This is on Isabela.

15. Ibid., 178.

16. Ibid., 176.

17. "Flightless Birds in Cargo for Zoo," *New York Times*, May 17, 1923, 14.

18. Townsend, "Flightless Cormorant in Captivity," 211.

19. "Vanderbilt Back with Sea Trophies," *New York Times*, May 17, 1928, 9.

20. "Again the Galapagos," *New York Times*, April 25, 1930, 17.

21. M. H. Jackson, *Galápagos: A Natural History Guide* (Calgary: Calgary University

Press, 1985), 6, 230–31. See also "Who We Are," Charles Darwin Foundation, www.darwinfoundation.org (accessed July 31, 2012).

22. "Statistics of Visitors to the Galapagos," 2012, Parque Nacional Galápagos, www.galapagospark.org; "Ingreso de visitantes," *Al día con la Dirección del Parque Nacional Galápagos*, no. 12 (November/December 2011): 6.

23. Kurt Vonnegut, *Galápagos* (New York: Laurel, 1988), 90–91.

24. Ibid., 34. In this full quotation, the narrator doesn't recognize, or isn't aware, that all other cormorant species swim under the surface to catch fish. I forgive Vonnegut.

25. Hank Nuwer, "A Skull Session with Kurt Vonnegut," in *Conversations with Kurt Vonnegut*, ed. William Rodney Allen (Jackson: University Press of Mississippi, 1999), 252.

26. Stephen Jay Gould, *Wonderful Life: The Burgess Shale and the Nature of History* (New York, W. W. Norton, 1990), 286.

27. Kurt Vonnegut, *Fates Worse Than Death: An Autobiographical Collage of the 1980s* (New York: G. P. Putnam's Sons, 1991), 15.

28. Gould, *Wonderful Life*, 286. See also Daniel Cordle, "Changing of the Old Guard: Time Travel and Literary Technique in the Work of Kurt Vonnegut," *Yearbook of English Studies* 30 (2000): 166–76.

29. Murphy, *Oceanic Birds of South America*, 2:870. Douglas Siegel-Causey crafted a taxonomy of cormorants based on skeletal features. He traced the relationship of the flightless cormorant as most closely in the Pacific to that of the extinct spectacled as well as the Brandt's cormorant (*P. penicillatus*). See Siegel-Causey, "Phylogeny of the Phalacrocoracidae," 887.

30. Ibid., 96.

31. Ibid., 97.

32 Richard Dawkins, *The Greatest Show on Earth: The Evidence for Evolution* (New York: Free Press, 2009), 345; Richard Dawkins, "Vestigial Organs: Wings of the Flightless Cormorant," May 5, 2010, Richard Dawkins Foundation, www.youtube.com/user/richarddawkinsdotnet.

33. Brian K. McNab, "Energy Conservation and the Evolution of Flightlessness in Birds," *American Naturalist* 144, no. 4 (October 1994): 629.

34. Ibid.

35. BirdLife International, "*Phalacrocorax harrisi*," 2012, IUCN Red List of Threatened Species, version 2012.1, www.iucnredlist.org. This species was listed as endangered from 2000 through 2010.

36. "Census of Native and Endemic Bird Populations of the Galápagos," February 23, 2010, Parque Nacional Galápagos, www.galapagospark.org. Last census was 2008, recording a total of 1,321 individual flightless cormorants.

37. Carlos Valle, personal communication, April 24, 2012; Daniel K. Rosenberg, Carlos A. Valle, Malcolm C. Coulter, and Sylvia A. Harcourt, "Monitoring Galapagos Penguins and Flightless Cormorants in the Galapagos Islands," *Wilson Bulletin* 102, no. 3 (September 1990): 530.

38. David Day, *The Doomsday Book of Animals* (New York: John Wiley & Sons Canada, 1981), 13.

39. Olson, however, has argued convincingly: "The span of time needed to evolve flightlessness in rails can probably be measured in generations rather than in millennia." See Storrs L. Olson, "Evolution of the Rails of the South Atlantic Islands," *Smithsonian Contributions to Zoology* 152 (1973): 34.

40. See some assessments of current threats to flightless cormorants in Matthias Wolff and Mark Gardener, eds., *Proceedings of the Galápagos Science Symposium 2009* (Galápagos: Charles Darwin Foundation, 2009), 87, 106, etc. See also Carlos A. Valle, "The Flightless Cormorant: The Evolution of Female Rule," in *Galápagos: Preserving Darwin's Legacy*, ed. Tui De Roy (Buffalo, NY: Firefly Books, 2009), 169.

41. Peter Weir (director and screenwriter), *Master and Commander: The Far Side of the World*, Twentieth Century Fox (2003). Quotations transcribed from the film. See also Patrick O'Brian, *The Far Side of the World* (New York: W. W. Norton, 1984).

42. Heather Carr, personal communication, December 5, 2012.

43. Rory P. Wilson, F. Hernán Vargas, Antje Steinfurth, Philip Riordan, Yan Ropert-Coudert, and David W. Macdonald, "What Grounds Some Birds for Life? Movement and Diving in the Sexually Dimorphic Galápagos Cormorant," *Ecological Monographs* 78, no. 4 (2008): 633.

44. Ibid.

45. Carlos Valle, personal communication, April 24, 2012; Nelson, *Pelicans, Cormorants and Their Relatives*, 533.

46. Flightless cormorant weight: Nelson, *Pelicans, Cormorants and Their Relatives*, 595. Spectacled cormorant weight: Johnsgard, *Cormorants, Darters, and Pelicans*, 171.

47. Wilson, Vargas, Steinfurth, et al., "What Grounds Some Birds for Life?" 633; Timothée Cook, personal communication, October 30, 2012; see Bernhard Rensch, *Evolution above the Species Level* (New York: Cambridge University Press, 1960).

48. M. P. Harris, "Population Dynamics of the Flightless Cormorant *Nannopterum harrisi*," *Ibis* 121, no. 2 (1979): 145. Carlos Valle sees this as a form of "group selection," but he tells me that not too many other biologists support this. Murphy in *Oceanic Birds of South America*, 2:871–72, wrote that a study of great cormorants in Amsterdam in the 1920s saw females "take the initiative in expression of courtship mannerisms about the time that both sexes assume the details of pre-nuptial plumage."

49. Valle, "Flightless Cormorant," 168.

50. Ibid., 167.

51. Ibid., 164.

52. Ibid., 166. It is rare, according to Valle, that there is a brood of two chicks. If so, the female will postpone desertion (Carlos Valle, personal communication, April 24, 2012).

53. Nelson, *Pelicans, Cormorants and Their Relatives*, 174–75.

54. Snow believed that the courtship dance began in the water, but Valle has observed activities beforehand on shore. Mendall reported this behavior in double-crested

cormorants, as did Audubon of double-crested in Florida, but no one since has corroborated this, as far as I am aware.

55. Johnsgard, *Cormorants, Darters, and Pelicans*, 406.

56. Charles J. Shields, *And So It Goes: Kurt Vonnegut, A Life* (New York: Henry Holt, 2011), 369.

57. Lorrie Moore, "How Humans Got Flippers and Beaks," *New York Times*, October 6, 1985, www.nytimes.com/1985/10/06/books/how-humans-got-flippers-and-beaks.html.

58. Ibid.

59. Barbara K. Snow, "Observations on the Behaviour and Ecology of the Flightless Cormorant *Nannopterum harrisi*," *Ibis* 108, no. 2 (April 1966): 267.

9. BELZONI, UNITED STATES

The epigraph is from David Bird's *The Bird Almanac* (Buffalo, NY: Firefly Books, 1999), 280.

1. Chris Nerrin, personal communication, February 27, 2012.

2. Research has substantiated this. Glahn and King write: "Almost 70% of all catfish consumed were stocker-size catfish ranging from 10 to 20 cm (ca. 4 to 8 inches)." See James F. Glahn and D. Tommy King, "Bird Depredation," in *Biology and Culture of Channel Catfish*, ed. C. S. Tucker and J. A. Hargreaves (Amsterdam: Elsevier B.V., 2004), 506.

3. "Sight and Motion," 2008, Reed-Joseph International Co., www.reedjoseph.com; unnamed receptionist at Reed-Joseph, personal communication, April 5, 2012; For a study of this effigy's effectiveness see Allen R. Stickley Jr. and Junior O. King, "Long-Term Trial of an Inflatable Effigy Scare Device for Repelling Cormorants from Catfish Ponds," *Proceedings of the Eastern Wildlife Damage Control Conference* 6 (1995): 89–92.

4. Jim Steeby, personal communications, February 27–29, 2012. All following quotations are from these days of interviews unless otherwise noted.

5. David J. Harvey (contact), "Aquaculture Background," June 2, 2009, USDA Economic Research Service, www.ers.usda.gov/Briefing/Aquaculture/background.htm; "Aquaculture: 1998 statistics," 2009, Catfish Institute, www.uscatfish.com/aquaculture .html; 2007 US Aquaculture Statistics: USDA, "Quick Stats," 2007, USDA Census of Agriculture, quickstats.nass.usda.gov; Gail Keirn, USDA-APHIS, personal communication, September 10, 2012; Jimmy Avery, personal communication, September 28, 2012.

6. Terrill R. Hanson, "Catfish Farming in Mississippi," *Mississippi History Now*, Mississippi Historical Society (April 2006), mshistory.k12.ms.us.

7. Monterey Bay Aquarium Foundation, "Catfish, U.S. Farmed," 2012, Monterey Bay Aquarium Seafood Watch, www.montereybayaquarium.org.

8. "Commercial Catfish Production," February 15, 2012, Mississippi Agricultural and Forestry Experiment Station and Mississippi State University Extension Service, http://msucares.com/aquaculture/catfish/index.html; National Agriculture and Statistics Service, "Catfish Production" (Washington, DC: U.S. Department of Agriculture, 2004), 11, 16.

9. "Commercial Catfish Production."

10. Monterey Bay Aquarium Foundation, www.montereybayaquarium.org; Hanson, "Catfish Farming in Mississippi," mshistory.k12.ms.us.

11. Jim Steeby, personal communication, April 5, 2012. William Tackett died in August 2012 at the age of eighty-seven.

12. Audubon, *Ornithological Biography*, vol. 3 (1835), 387.

13. "Mississippi River Journal," in *John James Audubon: Writings and Drawings*, ed. Christoph Irmscher (New York: Library of America, 1999), 6, 65.

14. Ibid., 58; Italics here represent Audubon's underlining.

15. Audubon, *Ornithological Biography*, vol. 4 (1838), 145.

16. Ibid., 139.

17. Audubon, *Ornithological Biography*, 3:389.

18. Malcolm Margolin, *The Ohlone Way: Indian Life in the San Francisco–Monterey Bay Area* (Berkeley, CA: Heyday Books, 1978), 37, 54, 95.

19. Audubon, *Ornithological Biography*, 3:393.

20. Ibid., 423.

21. Oliver Luther Austin Jr., *The Birds of Newfoundland Labrador*, Memoirs of the Nuttall Ornithological Club 7 (Cambridge, MA: Nuttall Ornithological Club, 1932), 33.

22. See, for example, Kathryn Blaze Carlson, "New York Solves Its Canada Goose Problem by Serving Them to Pennsylvania's Poor," *National Post*, August 25, 2011, http://news.nationalpost.com; U.S. Fish and Wildlife Service, "Geese Management," April 11, 2012, Migratory Bird Program, www.fws.gov/migratorybirds/RegulationsPolicies/geese.html.

23. Larry Brown, personal communication, February 28, 2012.

24. This cormorant first landed aboard the SSV *Westward* on January 31, 2002, 1730 EST, 24°11.7" N x 83°17.9" W.

25. Biologists differentiate a fifth double-crested cormorant subspecies (*P. a. heuretus*) living in the Bahamas and Cuba all year round.

26. Francesco Tarducci, *The Life of Christopher Columbus*, trans. Henry F. Brownson, vol. 1 (Detroit: H. F. Brownson, 1891), 327; Samuel Eliot Morison, *Admiral of the Ocean Sea: A Life of Christopher Columbus*, vol. 2 (Boston: Little, Brown, 1942), 150.

27. Audubon, *Ornithological Biography*, 3:387.

28. Guillaumet, Dorr, Wang, et al., "Determinants of Local and Migratory Movements," 1097; Hatch and Weseloh, "Double-Crested Cormorant," 5.

29. Elphick, Dunning, and Sibley, *Sibley Guide*, 64–65. See also Paul Kerlinger, *How Birds Migrate*, 2nd ed. (Mechanicsburg, PA: Stackpole Books, 2009), 148–64.

30. See, for example, USFWS, "Final Environmental Impact Statement: Double-Crested Cormorant Management in the United States," 121–22; and Guillaumet, Dorr, Wang, et al., "Determinants of Local and Migratory Movements," 1097.

31. D. Tommy King, Bradley F. Blackwell, Brian S. Dorr, and Jerrold L. Belant, "Effects of Aquaculture on Migration and Movement Patterns of Double-Crested Cormorants," *Human-Wildlife Conflicts* 4, no. 1 (Spring 2010): 77; J. F. Glahn, D. S. Reinhold, and

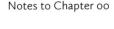

C. A. Sloan, "Recent Population Trends of Double-Crested Cormorants Wintering in the Delta Region of Mississippi: Responses to Roost Dispersal and Removal under a Recent Depredation Order," *Waterbirds* 23, no. 1 (2000): 38.

32. Glahn and King, "Bird Depredation," 505. USDA-APHIS did not have more recent estimates at time of publication.

33. Ibid. This 9.5 percent of all North American double-crested cormorants comes from the 1–2 million estimate often used and Glahn and King's total of 127,000 in the late 1990s. Thus between 6.35 percent and 12.7 percent is 9.5 percent. Gail Keirn of USDA-APHIS pointed out to me that, by the nature of the "snapshot" wintering counts, this estimate is probably less than how many are actually there. Personal communication, September 10, 2012. In other words, she thinks more than one of ten of all double-crested cormorants came through each winter.

34. Jackson and Jackson, "Double-Crested Cormorant," 125–26. It is not a universally accepted conclusion, however, that the catfish farms directly increased the Interior population of cormorants. Brian Dorr explained to me: "I think cormorants do benefit from passing through aquaculture areas, such as by building higher fat reserves, but I'm not sure (and likely no one is) that it's a major driver of population increase." Personal communication, July 5, 2012.

35. Gail Keirn, USDA-APHIS, personal communication, September 10, 2012.

36. *Living with Wildlife: Cormorants*, USDA Wildlife Services (pamphlet, n.d.), 3.

37. You can download the activity sheet online. See www.aphis.usda.gov/wildlife_damage/nwrc/publications/living/cormorants.pdf.

38. King, Blackwell, Dorr, and Belant, "Effects of Aquaculture," 80.

39. Gail Keirn, USDA-APHIS, personal communication, September 10, 2012.

40. King, Blackwell, Dorr, and Belant, "Effects of Aquaculture," 84.

41. Guillaumet, Dorr, Wang, et al., "Determinants of Local and Migratory Movements, 1097–1098.

42. Ibid., 1102.

43. Glahn and King, "Bird Depredation," 505.

44. Wires and Cuthbert, "Historic Populations," 26.

45. Brian Dorr, personal communication, February 19, 2013. Dorr points out that management might be part of why nesting has not increased.

46. Jackson and Jackson, "Double-Crested Cormorant," 118; *Living with Wildlife: Cormorants*, USDA Wildlife Services (pamphlet, n.d.), 3. See also Mark Twain, *Life on the Mississippi*, ed. James R. Osgood (New York: Penguin, 1986), 417.

47. Berlin Heck, personal communication, c. June 2009.

48. Louisiana license for resident hunter, issued August 28, 1958, Louisiana Wildlife and Fisheries Commission.

49. D. N. Carss, ed. "Reducing the Conflict between Cormorants and Fisheries on a Pan-European Scale, Final Report: Summary," REDCAFE (Arberdeenshire: Natural Environment Research Council Centre for Ecology and Hydrology, 2003), 3.

50. See Takaharu Natsumeda, Tetsuya Tsuruta, Kayoko Kameda, and Kei'ichiro Igu-

chi, "Winter Feeding of the Common Cormorant (*Phalacrocorax carbo hanedae*) in a Temperate River System in Japan," *Journal of Freshwater Ecology* 25, no. 1 (2010): 41–48; Tetsumi Takahashi, Kayoko Kameda, Megumi Kawamura, and Tsuneo Nakajima, "Food Habits of Great Cormorant *Phalacrocorax carbo hanedae* at Lake Biwa, Japan, with Special Reference to Ayu *Plecoglossus altivelis altivelis*," *Fisheries Science* 72, no. 3 (2006): 477–84.

51. See, for example, "NSW Land Based Sustainable Aquaculture Strategy," Industry and Investment New South Wales (December 2009), 1–113.

52. Nelson, *Pelicans, Cormorants and Their Relatives*, 530.

53. S. C. Nemtzov, "Relocation of Pygmy Cormorants *Phalacrocorax pygmeus* Using Scare Tactics to Reduce Conflict with Fish Farmers in the Bet She-an Valley, Israel," *Conservation Evidence* 2 (2005): 3.

54. Meredith Price, "Big Fish in a Small Pond," *Jerusalem Post*, May 20, 2005, www .jpost.com/Cooperations/Archives; Nemtzov, "Relocation of Pygmy Cormorants," 3.

55. Nemtzov, "Relocation of Pygmy Cormorants," 4.

56. Price, "Big Fish in a Small Pond."

57. Larry Brown, personal communication, February 28, 2012.

58. Glahn and King, "Bird Depredation," 506.

59. Ibid., 508.

60. Brian S. Dorr, Loren W. Burger, Scott C. Barras, and Kristina C. Godwin, "Economic Impact of Double-Crested Cormorant, *Phalacrocorax auritus*, Depredation on Channel Catfish, *Ictalurus punctatus*, Aquaculture in Mississippi, USA," *Journal of the World Aquaculture Society* 43, no. 4 (2012): 502.

61. Glahn and King, "Bird Depredation," 506.

62. Ibid., 507.

63. See Cynthia M. Doffitt, Linda M. Pote, and D. Tommy King, "Experimental *Bolbophorus damnificus* (Digenea: Bolbophoridae) Infections in Piscivorous Birds," *Journal of Wildlife Diseases* 45, no. 3 (2009): 681–91.

64. The Wildlife Service roost dispersal program is based on this concept, hoping to push cormorants closer to the river, where they feed less on farm-raised fish. Glahn and King, "Bird Depredation," 506. See also Dorr, Burger, Barras, and Godwin, "Economic Impact of Double-Crested Cormorant," 502–13; and Brian S. Dorr, Loren W. Burger, Scott C. Barras, and Kristina C. Godwin, "Double-Crested Cormorant Distribution on Catfish Aquaculture in the Yazoo River Basin of Mississippi," *Wildlife Society Bulletin* 36 (2012): 70–77.

65. Glahn, Reinhold, and Sloan, "Recent Population Trends," 38, 40.

66. Brady Thompson, personal communication, February 28, 2012.

10. ISLA CHINCHA CENTRO, PERU

The epigraph is from Murphy's *Bird Islands of Peru* (New York: G. P. Putnam, 1925), 72–73.

1. In 2008 PROABONOS was absorbed by a larger organization of all national agricultural programs, titled AGRORURAL, which is derived from "Programa de Desarrollo Productivo Agrario Rural," which translates roughly to the Rural Agricultural Productivity Development Program.

2. Elisa Goya, personal communication, September 19, 2012; Simon Romero, "Peru Guards Its Guano as Demand Soars Again," *New York Times*, May 30, 2008, A10.

3. J. Jahncke, D. M. Checkley Jr., and G. L. Hunt Jr., "Trends in Carbon Flux to Seabirds in the Peruvian Upwelling System: Effects of Wind and Fisheries on Population Regulation," *Fisheries Oceanography* 13, no. 3 (2004): 209.

4. The Peruvian booby might have once or twice been more populous on the coast than the cormorant. Today they occur in about equal numbers. See Robert E. Coker, "The Protection of Birds Made Profitable," *Science* 82, no. 2114 (July 5, 1935): 11; Henri Weimerskirch, Sophie Bertrand, Jaime Silva, Jose Carlos Marques, and Elisa Goya, "Use of Social Information in Seabirds: Compass Rafts Indicate the Heading of Food Patches," *PLoS ONE* 5, no. 3 (2010): 2.

5. Nelson, *Pelicans, Cormorants and Their Relatives*, 467.

6. Ibid.

7. As published in Murphy, *Bird Islands of Peru*, 81. Isla Chincha Sur hosted more than one million cormorants, including adults and chicks earlier in the twentieth century: Robert E. Coker, "Regarding the Future of the Guano Industry and the Guano-Producing Birds of Peru," *Science* 28, no. 706 (July 10, 1908): 64.

8. Murphy, *Bird Islands of Peru*, 81. Research suggests that *on average* guanay cormorants consume about 17.5 percent of their body weight in food per day, and they weigh 1.8 kg each, thus .31 kg/day x 5 million = 1,550,000 kg (1,708.6 tons), so Forbes's number of 1,000 tons is possible, even likely, if his number of birds was correct. Consumption data from Jahncke, Checkley, and Hunt, "Trends in Carbon Flux," 215.

9. Murphy, *Bird Islands of Peru*, 74–75.

10. First published in *Arte de pájaros* (1966). Translation by Jack Schmitt in Pablo Neruda, *The Poetry of Pablo Neruda*, ed. Ilan Stavans (New York: Farrar, Straus and Giroux, 2003), 751–52.

11. *OED*, 2nd ed., s.v. "guano, n.," www.oed.com/view/Entry/82113.

12. Garth Bawden, *The Moche* (Cambridge, MA: Blackwell, 1996), 73.

13. Jimmy M. Skaggs, *The Great Guano Rush: Entrepeneurs and American Overseas Expansion* (New York: St. Martin's Griffin, 1995), 4.

14. James W. Reid, *Textile Masterpieces of Ancient Peru* (New York: Dover, 1986), 7.

15. David C. Duffy, "The Guano Islands of Peru: The Once and Future Management of a Renewable Resource," *Birdlife Conservation Series* 1 (1994): 70; Skaggs, *Great Guano Rush*, 4; William M. Denevan, *Cultivated Landscapes of Native Amazonia and the Andes* (Oxford: Oxford University Press, 2001), 35.

16. Robert E. Coker, "Peru's Wealth-Producing Birds," *National Geographic*, June 1920, 541; Victor Wolfgang von Hagen, *South America Called Them: Explorations of the Great Naturalists* (New York: Duell, Sloan, and Pearce, 1955), 154; Denevan, *Cultivated Land-*

scapes, 35. Some of the Incan use and culture surrounding guano has been questioned; see Vogt as cited in Duffy, "Guano Islands of Peru."

17. See, for example, Reid, *Textile Masterpieces of Ancient Peru*; Jürgen Golte, *Los dioses de Sipán* (Lima: IEP Ediciones, 1993); and Hope B. Werness, *The Continuum Encyclopedia of Animal Symbolism in Art* (New York: Continuum International, 2006).

18. Golte, *Los dioses de Sipán*, 72–74.

19. Cuthbert W. Johnson, "On Guano," *Farmer's Magazine* 7, no. 3 (March 1843): 171.

20. George Shelvocke, *A Voyage Round the World by the Way of the Great South Sea* (London: Senex, 1726), 270–71. I have added here the apostrophe in "birds' feathers."

21. Von Hagen, *South America Called Them*, 154; Skaggs, *Great Guano Rush*, 4.

22. Skaggs, *Great Guano Rush*, 11.

23. David Hollett, *More Precious Than Gold: The Story of the Peruvian Guano Trade* (Madison, NJ: Farleigh Dickinson University Press, 2008), 85–86; see also Skaggs, *Great Guano Rush*, 5.

24. For example, Johnson, "On Guano," 171.

25. Gregory T. Cushman, "'The Most Valuable Birds in the Word': International Conservation Science and the Revival of Peru's Guano Industry, 1909–1965," *Environmental History* 10, no. 3 (July 2005): 478.

26. John Peter Olinger, "The Guano Age in Peru," *History Today*, June 1980, 15.

27. Skaggs, *Great Guano Rush*, 14.

28. Ibid., 39–42.

29. G.W.P. [George Washington Peck], "From the Chincha Islands," *New York Times*, January 7, 1854, 2. Correspondence was dated November 10, 1853. Peck wrote of six acres covered densely with guano birds.

30. Nelson Crowell, journal entry January 1853, Log 850, Mystic Seaport Museum, Mystic, CT.

31. Ibid.

32. Ibid.

33. Crowell, journal entry, August 18, 1856.

34. Ibid., September 6, 1856. In this entry Crowell underlined where I have used italics.

35. Simeon G. Fish, journal entries, March 14, June 7, 1857, Log 893, Mystic Seaport Museum, Mystic, CT.

36. Ibid., April 4, 1857.

37. Ibid., May 29, 1857; Sargent also describes a body being interred in guano: see Henry Jackson Sargent Jr., *The Captain of the* Phantom: *The Story of Henry Jackson Sargent, Jr. 1834–1862* (Mystic, CT: Mystic Seaport, 1967), 26.

38. Fish, journal entry, June 6, 1857.

39. Octavius T. Howe and Frederick C. Matthews, *American Clipper Ships, 1833–1858* (New York: Dover, 1986), 1:253–59.

40. Basil Lubbock, *The Down Easters: American Deep-Water Sailing Ships, 1869–1929* (New York: Dover, 1987), 53.

41. Skaggs, *Great Guano Rush*, 155.

42. "Spontaneous Combustion of Guano," *Chemist* 3, no. 26 (February 1845): 78.

43. See, for example, a c. 1860 advertising poster for Soluble Pacific Guano that features a cormorant, although a rather penguin-looking one. Held by Mystic Seaport Museum (1952.1046), Mystic, CT.

44. Nelson writes that the Socotra and Cape cormorants also have "aerial scanners" that enter the water from the air. See Nelson, *Pelicans, Cormorants and Their Relatives*, 430.

45. Murphy, *Bird Islands of Peru*, 78–79.

46. Nelson, *Pelicans, Cormorants and Their Relatives*, 468.

47. Murphy, *Bird Islands of Peru*, 85.

48. Herman Melville, *Moby-Dick*, ed. Harrison Hayford, Hershel Parker, G. Thomas Tanselle (Evanston, IL: Northwestern University Press, 2001), 63.

49. See, among others: Nelson, *Pelicans, Cormorants and Their Relatives*, 466 [330 ft / 100 m]; Murphy, *Bird Islands of Peru*, 104 [up to 150 ft / 46 m]; Skaggs, *Great Guano Rush*, 4, 160 [60–100 ft or more / 18–30 m].

50. Crowell, journal entry, January 1853.

51. Ibid.

52. Ibid.

53. Sargent, *Captain of the* Phantom, 34.

54. This vessel was 210 feet (64 m) long and 43 feet (13 m) at the beam.

55. Lawrence Clayton, "Chinese Indentured Labour in Peru," *History Today*, June 1980, 21.

56. Ibid.

57. Gaddis Smith, "The Agricultural Roots of Maritime History," *American Neptune* 64, no. 1 (Winter 1984): 9.

58. Skaggs, *Great Guano Rush*, 163. Skaggs estimates that by 1875 between eighteen thousand and thirty thousand Chinese men went to the Chincha Islands to mine guano.

59. Ibid., 159; Clayton, "Chinese Indentured Labour in Peru," 22.

60. Robert E. Coker, "Habits and Economic Relations of the Guano Birds of Peru," *Proceedings of the United States National Museum*, 56, no. 2298 (1920): 509; Duffy, "Guano Islands of Peru," 70.

61. Clayton, "Chinese Indentured Labour in Peru," 22; W. M. Mathew, "A Primitive Export Sector: Guano Production in Mid-Nineteenth-Century Peru," *Journal of Latin American Studies* 9, no. 1 (May 1977): 54.

62. Mathew, "Primitive Export Sector," 44.

63. Skaggs, *Great Guano Rush*, 161.

64. See, for example, G.W.P. [George Washington Peck], "From the Chincha Islands." 2. Correspondence was dated November 10, 1853; and Murphy, *Bird Islands of Peru*, 124–25.

65. George W. Peck, *Melbourne, and the Chincha Islands; with Sketches of Lima and a Voyage Round the World* (New York: Scribner, 1854), 206, 212.

66. Ibid., 208.

67. Boston Traveler, "Interesting from the Chincha Islands," *American Farmer* 6, no. 4 (June 1, 1854): 86.

68. Raphael Semmes, *My Adventures Afloat: A Personal Memoir of My Cruises and Services in "The Sumter" and "Alabama"* (London: Richard Bentley, 1869), 579.

69. Henry F. Dobyns and Paul L. Doughty, *Peru: A Cultural History* (New York: Oxford University Press, 1976), 178–79.

70. Olinger, "Guano Age in Peru," 16.

71. Murphy, *Bird Islands of Peru*, 104.

72. Cushman, "'Most Valuable Birds,'" 479.

73. Duffy, "Guano Islands of Peru," 70–71; Cushman, "'Most Valuable Birds,'" 490.

74. Duffy, "Guano Islands of Peru," 71; Cushman, "'Most Valuable Birds,'" 486.

75. Duffy, "Guano Islands of Peru," 70.

76. Cushman, "'Most Valuable Birds,'" 488.

77. Duffy, "Guano Islands of Peru," 71.

78. Weimerskirch, Bertrand, Silva, et al., "Use of Social Information in Seabirds," 2.

79. Duffy, "Guano Islands of Peru," 72; Elisa Goya, "Abundance of Guano Birds and Its Relation with Fishery of Peruvian Anchoveta from 1953 to 1999" (abstract), *IMARPE Boletin* 19, nos. 1, 2 (2000): 3–4.

80. Elisa Goya, personal communication, September 19, 2012.

81. Weimerskirch, Bertrand, Silva, et al., "Use of Social Information in Seabirds," 2.

82. Simon Romero, "Peru Guards Its Guano as Demand Soars Again," *New York Times*, May 30, 2008, A6, 10. See also Dan Collyns, "How a Peruvian Island Is Making Money from Bird Poo," September 2, 2010, BBC News (video), www.bbc.co.uk/news/business.

83. Nelson, *Pelicans, Cormorants and Their Relatives*, 513–14.

84. Cushman, "'Most Valuable Birds,'" 478.

85. Sophie Bertrand, Elisa Goya, and Jaime Silva, "Fishers and Seabirds Competing for the Same Fish: Foraging Strategies, Interactions and Consequences" (abstract), *IMARPE Boletin* 25, nos. 1, 2 (2010): 2. As a past strategy, see Cushman, "'Most Valuable Birds,'" 493.

86. Weimerskirch, Bertrand, Silva, et al., "Use of Social Information in Seabirds," 3–4.

11. CAPE TOWN, SOUTH AFRICA

Elaine Hayes is quoted in a press release, "Heartbreak for Living Coasts," September 28, 2010, Living Coasts, www.livingcoasts.org.uk/news/359.

1. See Dyan deNapoli, *The Great Penguin Rescue* (New York: Free Press, 2011).

2. N. J. Parsons and L. G. Underhill, "Oiled and Injured African Penguins *Spheniscus demersus* and Other Seabirds Admitted for Rehabilitation in the Western Cape, South Africa, 2001 and 2002," *African Journal of Marine Science* 27, no. 1 (2005): 289.

3. Debate remains about splitting the crowned cormorant (*P. coronatus*) from the

long-tailed, known locally as the "reed" cormorant (*P. africanus*), and whether this great cormorant (*P. carbo lucidus*) should be its own species.

4. Nola Parsons, personal communications, a series of interviews from March through June 2012. All quotations are from these interviews.

5. David Cameron Duffy and W. Roy Siegfried, "Historical Variations in Food Consumption by Breeding Seabirds of the Humboldt and Benguela Upwelling Regions," in *Seabirds: Feeding Ecology and Role in Marine Ecosystems*, ed. J. P. Croxall (Cambridge: Cambridge University Press, 1987), 327–46.

6. Ibid., 328.

7. D. H. Davies, as cited in A. P. Bowmaker, "Cormorant Predation on Two Central African Lakes," *Ostrich* 34, no. 1 (March 1963): 3.

8. Duffy and Siegfried, "Historical Variations in Food Consumption," 339.

9. H. H. Berry, "Mass Mortality of Cape Cormorants, Caused by Fish Oil, in the Walvis Bay Region of South West Africa," *Madoqua* 9, no. 4 (1976): 61; R. J. M. Crawford and J. Cooper, "Cape Cormorant *Phalacrocorax capensis*," in *Roberts' Birds of Southern Africa*, 7th ed., ed. P. A. R. Hockey, W. R. J. Dean, and P. G. Ryan (Cape Town: Trustees of the John Voelcker Bird Book Fund, 2005), 579.

10. Duffy and Siegfried, "Historical Variations in Food Consumption," 330.

11. Charles John Andersson, *Notes on the Birds of Damara Land and the Adjacent Countries of South-West Africa*, ed. John Henry Gurney, trans. L. Lloyd (London: John Van Voorst, 1872), 368.

12. Ibid.

13. G. J. van der Linde and M. A. Pitse, "The South African Fertiliser Industry," paper presented at AFA Conference (Cairo, 2006), 1; Arthur C. Watson, "The Guano Islands of Southwestern Africa," *Geographical Review* 20, no. 4 (October 1930): 634.

14. Skaggs, *Great Guano Rush*, 8; Watson, "Guano Islands," 634–35.

15. Skaggs, *Great Guano Rush*, 7–8. The first captain to visit Ichaboe estimated the deposit at between 1.4 to 1.6 million pounds: see Watson, "Guano Islands," 634.

16. Watson, "Guano Islands," 633, 640.

17. Skaggs, *Great Guano Rush*, 155.

18. Les Underhill, "Bird Rock: The Walvis Bay Guano Platform," February 13, 2003, Avian Demography Unit, University of Cape Town, http://web.uct.ac.za/depts/stats/adu/walvisbayguanoplatform.htm; H. H. Berry, "The History of the Guano Platform on Bird Rock, Walvis Bay, South West Africa," *Bokmakierie* 27 (1975): 62–63.

19. Jessica Kemper, personal communication, June 9, 2012; deNapoli, *Great Penguin Rescue*, 44; Andreas Vogt, "White Gold: A Visit to Bird Rock Island," May 11, 2009, *Allgemeine Zeitung* (Namibia), www.az.com.na/tourismus; Duffy, "Guano Islands of Peru," 74.

20. J. Cooper, "Biology of the Bank Cormorant, Part 1: Distribution, Population Size, Movements and Conservation," *Ostrich* 52 (1980): 208, 212.

21. Johan August Wahlberg, *Travel Journals (and Some Letters) South Africa and Namibia/Botswana, 1838–1856*, ed. Adrian Craig and Chris Hummel, trans. Michael Rob-

erts (Cape Town: Van Riebeeck Society, 1994), 13. The bank cormorant today is known sometimes as "Wahlberg's cormorant." Wahlberg wrote in this first description of the bank cormorant: "Iris ochre-yellow in adult birds, but green on the lower moiety [half], in younger specimens entirely a cinereous brown." As cited in Andersson, *Notes on the Birds of Damara Land*, 369–70.

22. Wahlberg, *Travel Journals*, 150. The two new ones here are the bank and the crowned.

23. As cited in Andersson, 370.

24. Ibid., xiii.

25. It was named as "neglected" and "ignored," meaning previously so. In other words, only until now described. R. J. M. Crawford and J. Cooper, "Bank Cormorant *Phalacrocorax neglectus*," in Hockey, Dean, and Ryan, *Roberts' Birds of Southern Africa*, 7th ed., 577–78.

26. Nelson, *Pelicans, Cormorants and Their Relatives*, 433; Michelle du Toit, "Bank Cormorant, *Phalacrocorax neglectus*," March 30, 2004, Avian Demography Unit, University of Cape Town, http://web.uct.ac.za/depts/stats/adu/species/bankcormorant.htm; BirdLife International, "*Phalacrocorax neglectus*," 2011, IUCN Red List of Threatened Species, version 2011.2, www.iucnredlist.org.

27. A. E. Burger, "Functional Anatomy of the Feeding Apparatus of Four South African Cormorants," *Zoologica Africana* 13, no. 1 (1978): 93, 97.

28. BirdLife International, "*Phalacrocorax neglectus*"; Timothée Cook, personal communication, September 28, 2012.

29. BirdLife International, "*Phalacrocorax neglectus*"; Timothée Cook, personal communications, July 22 and September 28, 2012; Crawford and Cooper, "Bank Cormorant," 577.

30. Timothée Cook, personal communication, July 22, 2012.

31. Timothée Cook, personal communication, September 28, 2012; R. J. M. Crawford, S. Davis, R. Harding, et al., "Initial Impact of the *Treasure* Oil Spill on Seabirds off Western South Africa," *South African Journal of Marine Science* 22 (2000): 163.

32. Michelle du Toit, http://web.uct.ac.za/depts/stats/adu/species/bankcormorant .htm; BirdLife International, "*Phalacrocorax neglectus*"; Timothée Cook, personal communication, July 22, 2012; R. J. M. Crawford, A. C. Cockcroft, B. M. Dyer, and L. Upfold, "Divergent Trends in Bank Cormorants *Phalacrocorax neglectus* Breeding in South Africa's Western Cape Consistent with a Distributional Shift of Rock Lobsters *Jasus lalandii*," *African Journal of Marine Science* 31 (2008): 161–66. See also R. J. M. Crawford, P. A. Whittington, A. P. Martin, A. J. Tree, and Azwianewi B. Makhado, "Population Trends of Seabirds Breeding in South Africa's Eastern Cape and the Possible Influence of Anthropogenic and Environmental Change," *Marine Ornithology* 37 (2009): 159–74.

33. R. L. Johnson, A. Venter, M. N. Bester, and W. H. Oosthuizen, "Seabird Predation by White Shark, *Carcharodon carcharias*, and Cape Fur Seal, *Arctocephalus pusillus pusillus*, at Dyer Island," *South African Journal of Wildlife Research* 36, no. 1 (2006): 23; Timothée Cook, personal communication, July 22, 2012. See also Anne Voorbergen, Willem

F. De Boer, and Les G. Underhill, "Natural and Human-Induced Predation on Cape Cormorants at Dyer Island," *Bird Conservation International* 22 (2012): 82–93.

34. H. H. Berry, "Physiological and Behavioural Ecology of the Cape Cormorant, *Phalacrocorax capensis,*" *Madoqua* 9, no. 4 (1976): 42.

35. Ibid.; Nelson, *Pelicans, Cormorants and Their Relatives*, 160.

36. Kazuto Hino, personal communication, June 12, 2012.

37. Hanson, *Feathers*, 220.

38. A. M. Rijke, "The Water Repellency and Feather Structure of Cormorants, Phalacrocoracidae," *Journal of Experimental Biology* 48 (1968): 187–89.

39. Hanson, *Feathers*, 221.

40. P. J. Jones, "A Possible Function of the 'Wing-Drying' Posture in the Reed Cormorant *Phalacrocorax africanus,*" *Ibis* 120, no. 4 (1978): 542.

41. David Grémillet, "'Wing-Drying' in Cormorants," *Journal of Avian Biology* 26, no. 2 (1995): 176.

42. Craig W. White, Graham R. Martin, and Patrick J. Butler, "Wing-Spreading, Wing-Drying, and Food-Warming in Great Cormorants *Phalacrocorax carbo,*" *Journal of Avian Biology* 39, no. 5 (September 2008): 578.

43. Melville, *Moby-Dick*, 234.

44. John Blight, "Cormorants," *Selected Poems 1939–1990* (St. Lucia: University of Queensland Press, 1992), 42.

45. Bruce Coultas, Elizabeth Cridland, et al., "How SANCCOB Began: The Story of Our Founder," South African Foundation for the Conservation of Coastal Birds, www .SANCCOB.co.za.

46. DeNapoli, *Great Penguin Rescue*, 54–56; Les Underhill, "A Brief History of Penguin Oiling in South African Waters," March 18, 2001, Avian Demography Unit, University of Cape Town, http://web.uct.ac.za/depts/stats/adu/oilspill/oilhist.htm.

47. Crawford, Davis, Harding, et al., "Initial Impact of the *Treasure*," 172, 175.

48. C. L. Griffiths, L. van Sittert, P. B. Best, et al., "Impacts of Human Activities on Marine Animal Life in the Benguela: A Historical Overview," *Oceanography and Marine Biology: An Annual Revue* 42 (2004): 322.

49. "Status of Restoration: Cormorants," *Exxon Valdez* Oil Spill Trustee Council, www.evostc.state.ak.us (accessed September 26, 2012).

50. BirdLife International, "*Phalacrocorax nigrogularis,*" 2011, IUCN Red List of Threatened Species, version 2011.2, www.iucnredlist.org; Nelson, *Pelicans, Cormorants and Their Relatives*, 430–31.

51. Mariam M. Al Serkal, "Release of Flamingos and Cormorants," November 28, 2012, GulfNews.com, gulfnews.com.

52. "Deepwater Horizon Bird Impact Data from the DOI-ERDC NRDA Database 12 May 2011," U.S. Fish and Wildlife Service, www.fws.gov/home/dhoilspill/pdfs/Bird %20Data%20Species%20Spreadsheet%2005122011.pdf; Alberto Velando, David Álvarez, Jorge Mouriño, Francisco Arcos, and Álvaro Barros, "Population Trends and Repro-

ductive Success of the European Shag *Phalacrocorax aristotelis* on the Iberian Peninsula following the *Prestige* Oil Spill," *Journal of Ornithology* 146 (2005): 116–17.

53. Ibid., 116.

54. Dan Gunderson, "BP Oil Spill Residue Found on Pelicans in Minnesota," May 16, 2012, Minnesota Public Radio, http://minnesota.publicradio.org.

55. Berry, "Mass Mortality of Cape Cormorants," 57, 59.

56. BirdLife International, "*Phalacrocorax neglectus.*"

57. Jessica Kemper, personal communication, June 9, 2012.

58. Elaine Hayes, personal communication, February 1, 2011.

59. Press release, "Rare Species Banking on Living Coasts," Living Coasts, July 11, 2009, www.livingcoasts.org.uk/news/293.

60. Staff at the Robben Island Museum, University of Cape Town, Department of Environmental Affairs, and SANCCOB were responsible for assisting Living Coasts with gathering these bank cormorant eggs.

61. Press release, "Heartbreak for Living Coasts," September 28, 2010, Living Coasts, www.livingcoasts.org.uk/news/359.

62. Ibid.

63. Elaine Hayes, personal communication, February 1, 2012.

12. GATES ISLAND, UNITED STATES

The epigraph is from Mendall's *The Home-Life and Economic Status of the Double-Crested Cormorant* (Orono: University of Maine, 1936), 136.

1. Army Corps of Engineers, Portland District, "Draft Environmental Assessment," 18; Associated Press, "OSU Students to Haze Columbia River Cormorants," *Seattle Times*, February 8, 2012, seattletimes.com.

2. Jeff Barnard, "Oregon Asks to Kill Salmon-Eating Birds," *East Oregonian* and Associated Press, April 27, 2012, http://www.eastoregonian.com.

3. Larry Coonrod, "Cormorant Hazing," *South Lincoln County News* (Waldport, OR), May 9, 2012, www.southlincolncountynews.com.

4. LSU Ag Center, "Tenn. Company Licenses 'Scarebot' from LSU AgCenter," *Catfish Journal* 25, no. 11 (July 2011): 11.

5. "John Kline to Testify on His Bill to Allow States to Manage Menacing Cormorant Overpopulation," March 28, 2012, Congressman John Kline, kline.house.gov.

6. Rena Sarigianopoulus, "Calls for Cormorant Control on Lake Waconia," Kare 11 / NBC, October 19, 2011, www.kare11.com; Barfsoup, "Huge Flock of Cormorants," October 9, 2006, YouTube, www.youtube.com.

7. Linda Wires, personal communication, January 12, 2009.

8. Terry Doyle, USFWS, personal communication, April 12 and June 1, 2012.

9. Ontario Parks, "Annual Report on the Management of Double-Crested Cormorants on Presqu'ile Provincial Park," Queens Printer for Ontario (2006), 18; Chip Weseloh,

personal communication, April 20, 2012; Terry Doyle, personal communication, June 1, 2012.

10. USFWS, "Final Environmental Impact Statement: Double-Crested Cormorant Management in the United States," 23; Tyson, Belant, Cuthbert, and Weseloh, "Nesting Populations," 20; USFWS, "Final Environmental Assessment: Extended Management of Double-Crested Cormorants," 5. This rough one to two million was before the federal management plan, but also does not include the natural growth of the species since 1999. Terry Doyle, USFWS wildlife biologist, wrote to me about my writing this percentage range: "This is a misleading statement. On average, less than 3 percent of the continental population of Double-crested Cormorants is taken each year. Summing the take from 2004–2011 does not make sense, unless you compare that to the sum of the continental population over that same period" (personal communication, June 1, 2012). Acquiring, or even modeling, the total population each year, however, is very difficult.

11. W. M. W. Fowler, *Countryman's Cooking* (Ludlow: Excellent Press, 2006), 30–31.

12. People around the world still eat cormorants, and the birds have a place in many historic cuisines. Apparently early Irish and Spanish coastal communities ate shag. There's a recipe for cormorant soup in Gaelic from the Hebrides held by the Scottish Council on Archives. Cormorants can apparently still be found on the odd menu in Iceland. See, for example: Hull, *Scottish Birds*, 98; Emma Cowing, "'Edible Archive' Needs Strong Stomach," *Scotland on Sunday*, August 21, 2011, www.scotsman.com/news/edible-archive-needs-strong-stomach-1-1801741; Jane Victoria Appleton and Lisa Gail Shannen, *Frommer's Iceland* (Chichester: John Wiley & Sons, 2010): Google eBook.

13. Jim Andrews and AccuWeather, "Do Peru's Marine Die-Offs Herald the Return of El Nino?" *Scientific American*, May 8, 2012, www.scientificamerican.com.

14. "New Zealand Oil Spill Ship Owners Charged," *Guardian*, April 5, 2012, www.guardian.co.uk.

15. "Kiso Cormorant Fishermen Add Woman to Fold," *Japan Times*, May 18, 2012, www.japantimes.co.jp.

16. "Project: Identifying the Causes for the Decline of the Endangered Bank Cormorant in Southern Africa," Animal Demography Unit and the Percy FitzPatrick Institute, University of Cape Town, February 16, 2012, https://sites.google.com/site/timotheecook/-phd-opportunity.

17. Audubon, *Ornithological Biography*, 3:193.

18. Wires and Cuthbert, "Historic Populations," 29–30.

19. Jeremy J. Hatch, Kevin M. Brown, Geoffrey G. Hogan, and Ralph D. Morris, "Great Cormorant (*Phalacrocorax carbo*)," *Birds of North America Online*, ed. A. Poole (Ithaca, NY: Cornell Lab of Ornithology, 2000), http://bna.birds.cornell.edu/bna/species/553; Elphick, Dunning, and Sibley, *Sibley Guide*, 161.

20. Fred J. Alsop III, *Birds of North America* (New York, DK Publishing / Smithsonian, 2001), 112; Hatch, Brown, Hogan, and Morris, "Great Cormorant."

21. Samuel de Champlain, *The Works of Samuel de Champlain*, vol. 1 (Toronto:

Champlain Society, 1922), 241, 243. Looking at accounts by Audubon and Lewis, Wires and Cuthbert suggest these were double-crested. See Wires and Cuthbert, "Historic Populations," 9.

22. Lewis, *Natural History*, 4.

23. William Wood, *New England's Prospect*, ed. Alden T. Vaughan (Amherst: University of Massachusetts Press, 1993), 51.

24. Ibid.

25. Mendall, "Home-Life," 6.

26. Paul J. Lindholdt, ed., *John Josselyn, Colonial Traveler: A Critical Edition of Two Voyages to New-England* (Hanover, NH: University Press of New England, 1988), 73–74.

27. John Brickell, *The Natural History of North-Carolina* (Dublin: James Carson, 1737), 211–12.

28. Mendall, "Home-Life," 7–8.

29. Audubon, *Ornithological Biography*, 3:387. Audubon reported this true for both double-crested and great cormorants.

30. Lewis, *Natural History*, 5–9; Wires and Cuthbert, "Historic Populations," 28.

31. Jane McCamant, "Double-Crested Cormorants in Fishers Island Sound: Our Role in the National Debate on Cormorant Control," undergraduate paper (Mystic, CT, April 2004), 11.

32. Peg Van Patten, "Aqua Kids Discover That South Dumpling Island Is for the Birds," *Wrack Lines* 11, no. 1 (Spring/Summer 2011): 12.

33. "Birds of Dumpling Island," Aqua Kids, November 14, 2011, www.youtube.com.

34. American sand lance, sculpin, cusk, and tautog are the most common components of cormorant diets in this area. See Svati Narula and Richard King, "Who's Afraid of the Big, Bad Cormorant?" *Wrack Lines* 12, no. 2 (Fall/Winter 2012/13): 12–15.

35. Brae Rafferty, personal communication, June 17, 2008.

36. Daniel J. Decker and Ken G. Purdy, "Toward a Concept of Wildlife Acceptance Capacity in Wildlife Management," *Wildlife Society Bulletin* 16, no. 1 (Spring 1988): 53. Discussed specifically in relation to cormorants in Wires and Cuthbert, "Historic Populations" (2006).

37. J. R. Sauer, J. E. Hines, J. E. Fallon, K. L. Pardieck, D. J. Ziolkowski Jr., and W. A. Link, "Double-Crested Cormorant *Phalacrocorax auritus*: BBS Trend Map," North American Breeding Bird Survey Results and Analysis 1966–2010, version 12.07.2011, USGS-Patuxet Wildlife Research Center (Laurel, MD: rev. April 16, 2012), www.mbr-pwrc.usgs.gov/bbs/tr2010/tr01200.htm.

38. Hatch, "Changing Populations," 11.

39. Remy Tumin, "Threat to Fish, Cormorants May Soon Be Driven from Area," *Vineyard Gazette*, February 20, 2012, www.mvgazette.com/article.php?34055.

40. Ibid.

41. Charley Soares, "Black Death — the Birds from Hell," *On the Water* 16, no. 3 (July 2012): 106.

42. Bill Sweetman, "The Navy's Swimming Spy Plane," *POPSCI* (February 21, 2006),

www.popsci.com/node/3747; Lockheed Martin Corp., "Lockheed Martin MPUAV Cormorant," 2008, www.youtube.com/watch?v=r-J8LNhCr8I.

43. Richard Grimmett, Carol Inskipp, and Tim Inskipp, *A Guide to the Birds of India, Pakistan, Nepal, Bangladesh, Bhutan, Sri Lanka, and the Maldives* (Princeton, NJ: Princeton University Press, 1999), 558.

44. Egremont and Rothschild, "Calculating Cormorants," 181.

45. Tim Birkhead, "Do Birds Have Emotions?" *Chronicle Review*, May 11, 2012, B14–B15.

46. Mendall, "Home-Life," 171.

47. For example, see Jackson and Jackson, "Double-Crested Cormorant," 126–27.

48. As in the note from chapter 2, this for double-crested: Weseloh and Casselman, "Calculated Fish Consumption," 64. Diamond, Aprahamian, and North summarize studies of daily food intake in great cormorants (*P. carbo sinensis* and *P. carbo carbo*) as between 0.26 to 1.9 pounds of fish per day (0.12–0.88 kg); they used 0.9 pounds (0.4 kg) for their larger estimates. See M. Diamond, M. W. Aprahamian, and R. North, "A Theoretical Assessment of Cormorant Impact on Fish Stocks in Great Britain," in *Interactions between Fish and Birds: Implications for Management*, ed. I. G. Cowx (Oxford: Blackwell Science, 2003), 48.

49. Bowmaker, "Cormorant Predation on Two Central African Lakes," 18.

50. See, for example, Callum Roberts, *The Unnatural History of the Sea* (Washington, DC: Island Press, 2007), 320–21. See a careful, technical discussion of the challenges of trying to estimate actual bird predation effects on fish stocks in W. Dekker and J. J. De Leeuw, "Bird-Fisheries Interactions: The Complexity of Managing a System of Predators and Preys," in Cowx, *Interactions between Fish and Birds*, 3–13. See also I. G. Cowx, "Interactions between Fisheries and Fish-Eating Birds: Optimising the Use of Shared Resources," 364–65, in the same volume.

51. Lewis, *Natural History*, 84.

52. Weseloh and Collier, "Rise of the Double-Crested Cormorant," 10.

53. See Sievert Rohwer, Christopher E. Filardi, Kimberly S. Bostwick, and A. Townsend Peterson, "A Critical Evaluation of Kenyon's Shag (*Phalacrocrax [Stictocarbo] kenyoni*)," *Auk* 117, no. 2 (April 2000): 308–20.

54. James T. Carlton, personal communication, November 28, 2011.

55. Paul Heikkila, personal communication, March 14, 2002.

56. See, for example, Brown and Connelly, "Lake Ontario Sportfishing," 19.

57. Jim Steeby, personal communication, April 5, 2012.

58. Sálim Ali and S. Dillon Ripley, *Handbook of the Birds of India and Pakistan*, vol. 1 (Oxford: Oxford University Press, 1968), 35. Ali and Ripley wrote: "Cormorants are notorious for their prodigious appetites, and their depredations on local fish populations can be potentially devastating. Nevertheless, the wholesale indiscriminate persecution of the birds on this account, without a proper scientific inquiry, is unjustified. Investigations on the food and feeding habits of these and other piscivorous birds elsewhere have shown that the majority of fishes taken are of low economic worth, or which themselves

often constitute a far greater menace to the spawn and fry of valuable food fishes than the birds."

59. "@Cormy Cormorant," April 18, 2012, Twitter, http://twitter.com/#!/Cormy Cormorant.

60. Jane Weinberger, personal communication, February 12, 2002. See Jane Weinberger, *Cory the Cormorant* (Mount Desert, ME: Windswept, 1992).

61. Celia Thaxter, *Drift-Weed* (Boston: Houghton Mifflin, 1890), 123–24.

Selected Bibliography

I am and ever have been, a great reader, and have read almost
everything—a library cormorant. I am deep in all out of the way
books, whether of the monkish times, or of the puritanical era.
I have read and digested most of the historical writers.

SAMUEL TAYLOR COLERIDGE, 1796

Please see the notes section for a complete list of all sources used and all interviews referenced. Included below are sources devoted primarily to cormorants, or that provide especially important source or background material. This bibliography represents an essential foundation for the sure-to-be-soon-blossoming field of Cormorant Studies, both for the general reader and the specialist.

The epigraph is from a letter written on November 19, 1796, and published in James Dykes Campbell, *Samuel Taylor Coleridge: A Narrative of the Events of His Life* (London: Macmillan and Co., 1896), 58.

NONFICTION BOOKS

Audubon, John James. *Ornithological Biography, or an Account of the Habits of the Birds of the United States of America.* 5 vols. Philadelphia/Edinburgh: Judah Dobson / Adam and Charles Black, 1831–39.

Cowx, I. G. *Interactions between Fish and Birds: Implications for Management.* Oxford: Blackwell Science, 2003.

Dawson, William Leon. *Dawson's Avian Kingdom.* Edited by Anna Neher. Berkeley, CA: Heyday Books, 2007.

Elphick, Chris, John B. Dunning Jr., and David Allen Sibley, eds. *The Sibley Guide to Bird Life and Behavior.* New York: Alfred A. Knopf, 2009.

Gill, Frank B. *Ornithology.* 3rd ed. New York: W. H. Freeman, 2007.

Harrison, Peter. *Seabirds: An Identification Guide.* Revised ed. Boston: Houghton Mifflin, 1985.

Hollett, David. *More Precious Than Gold: The Story of the Peruvian Guano Trade.* Madison, NJ: Farleigh Dickinson University Press, 2008.

Le Couteulx de Canteleu, Jean Baptiste Emmanuel Hector. *La Pêche au Cormoran.* Paris: Bureaux de la Revue Britannique, 1870.

Lewis, Harrison Flint. *The Natural History of the Double-Crested Cormorant.* Ottawa: H. C. Miller, 1929.

Mendall, Howard L. *The Home-Life and Economic Status of the Double-Crested Cormorant.* Orono, ME: University of Maine, 1936.

Murphy, Robert Cushman. *Bird Islands of Peru.* New York: G. P. Putnam, 1925.

———. *Oceanic Birds of South America.* 2 vols. New York: Macmillan, 1936.

Nelson, J. Bryan. *Pelicans, Cormorants and Their Relatives* (Oxford: Oxford University Press, 2005).

Nettleship, David N., and David C. Duffy, eds. *The Double-Crested Cormorant: Biology, Conservation and Management.* Special Publication 1, *Colonial Waterbirds* 18 (1995). [Several articles within were especially useful to this project; they are listed individually in the notes.]

Peck, George W. *Melbourne, and the Chincha Islands; with Sketches of Lima and a Voyage Round the World.* New York: Scribner, 1854.

Rankin, Niall. *Antarctic Isle: Wild Life in South Georgia.* London: Collins, 1951.

Skaggs, Jimmy M. *The Great Guano Rush: Entrepreneurs and American Overseas Expansion.* New York: St. Martin's Griffin, 1995.

Steller, Georg Wilhelm. *Journal of a Voyage with Bering, 1741–1742.* Translated and edited by O. W. Frost, translated by Margritt A. Engel. Stanford, CA: Stanford University Press, 1988.

Wild, Dennis. *The Double-Crested Cormorant: Symbol of Ecological Conflict.* Ann Arbor: University of Michigan Press, 2012.

Yalden, D. W., and U. Albarella. *The History of British Birds.* Oxford: Oxford University Press, 2009.

FICTION AND PLAYS

Gregory, Stephen. *The Cormorant.* Clarkston, GA: White Wolf, 1986.

Hill, Elizabeth Starr. *Bird Boy.* Illustrated by Lesley Liu. New York: Farrar, Straus and Giroux, 1999.

O'Dell, Scott. *Island of the Blue Dolphins.* New York: Dell, 1971.

O'Flaherty, Liam. *The Black Soul.* London: Jonathan Cape, 1936.

Sayemon, Enami no. "Ukai." In *The Nō Plays of Japan,* edited by Arthur Waley. London: George Allen & Unwin, 1921.

Vonnegut, Kurt. *Galápagos.* New York: Laurel, 1988.

Weinberger, Jane. *Cory the Cormorant.* Mount Desert, ME: Windswept, 1992.

POEMS AND SHORT STORIES

Bashō. "Ukai." In David Landis Barnhill, *Bashō's Journey: The Literary Prose of Matsuo Bashō.* Albany: SUNY Press, 2005. [See also the translation in Yamashita, 1987.]

Blight, John. "Cormorants." In *Selected Poems, 1939–1990.* St. Lucia: University of Queensland Press, 1992.

Coleridge, Samuel Taylor. "The Devil's Thoughts." In *The Complete Poetical Works* of *Samuel Taylor Coleridge*, edited by Ernest Hartley Coleridge. Vol. 1. Oxford: Clarendon Press, 1975.

Golte, Jürgen. *Los dioses de Sipán*. Lima: IEP Ediciones, 1993.

Jeffers, Robinson. "Birds and Fishes." In *The Wild God of the World: An Anthology of Robinson Jeffers*, edited by Albert Gelpi. Stanford, CA: Stanford University Press, 2003.

Lawless, Emily. "The Cormorant." *All Ireland Review* 3, no. 25 (August 23, 1902): 389. [Find this also in her 1902 collection *With the Wild Geese* or in anthologies such as *Sea Poems* (London: Muller, 1944).]

Longfellow, Henry Wadsworth. "The Skeleton in Armor." In *The Complete Poetical Works of Longfellow*. Boston: Houghton Mifflin, 1922.

MacDairmid, Hugh. "Shags' Nests." In *Stony Limits and Other Poems*. London: V. Gollancz, 1934.

Milton, John. *Paradise Lost*. Edited by Alastair Fowler. 2nd ed. London: Longman, 1998.

Minty, Judith. "Destroying the Cormorant Eggs." In *Walking with the Bear*. East Lansing: Michigan State University Press, 2000.

Neruda, Pablo. "Guanay Cormorant." In *The Poetry of Pablo Neruda*, edited by Ilan Stavans. New York: Farrar, Straus and Giroux, 2003.

O'Flaherty, Liam. "The Wounded Cormorant." In *Liam O'Flaherty's Short Stories*. 2 vols. London: New English Library, 1981.

Wilbert, Johannes, ed. "The Selfish Cormorant" and "The Revenge of the Tufted Cormorants." In *Folk Literature of the Yamana Indians*. Berkeley: University of California Press, 1977.

SCHOLARLY AND POPULAR ARTICLES, REPORTS, AND SIGNIFICANT ENTRIES OR CHAPTERS IN NONFICTION BOOKS

Adkins, Jessica Y., and Daniel D. Roby. "A Status Assessment of the Double-Crested Cormorant (*Phalacrocorax auritus*) in Western North America: 1998–2009." Final report submitted to U.S. Army Corps of Engineers, USGS–Oregon Cooperative Fish and Wildlife Research Unit (March 2010), 1–69.

Ainley, David G., Daniel W. Anderson, and Paul R. Kelly. "Feeding Ecology of Marine Cormorants in Southwestern North America." *Condor* 83, no. 2 (1981): 120–31.

Anderson, Cynthia D., Daniel D. Roby, and Ken Collis. "Foraging Patterns of Male and Female Double-Crested Cormorants Nesting in the Columbia River Estuary." *Canadian Journal of Zoology* 82 (2004): 541–54.

Associated Press. "OSU Students to Haze Columbia River Cormorants." *Seattle Times*, February 8, 2012. http://seattletimes.com/html/localnews/2017454602_apwahazingcormorants1stldwritethru.html.

Associated Press and Special to the Sentinel. "Untouched Cormorant Eggs Found." *Milwaukee Sentinel*, June 19, 1987, pt. 2, 10.

Audubon, John James. "Mississippi River Journal." In *John James Audubon: Writings and Drawings*, edited by Christoph Irmscher, 3–155. New York: Library of America, 1999.

Austin, Oliver Luther, Jr. "*Phalacrocorax carbo carbo*" and "*Phalacrocorax auritus auritus*." In *The Birds of Newfoundland Labrador*, Memoirs of the Nuttall Ornithological Club 7, 32–33. Cambridge, MA: Nuttall Ornithological Club, 1932.

Bayer, Range D. "Cormorant Harassment to Protect Juvenile Salmonids in Tillamook County, Oregon: Studies in Oregon Ornithology, No. 9." Gahmken Press (2000), 1–66.

Behn, F., J. D. Goodall, A. W. Johnson, and R. A. Phillippi. "The Geographic Distribution of the Blue-Eyed Shags, *Phalacrocorax albiventer* and *Phalacrocorax atriceps*." *Auk* 72 (January 1955): 6–13.

Behrens, Vivien, Felix Rauschmayer, and Heidi Wittmer. "Managing International 'Problem' Species: Why Pan-European Cormorant Management Is So Difficult." *Environmental Conservation* 35, no. 1 (2008): 55–63.

Bernstein, Neil P., and Stephen J. Maxson. "Absence of Wing-Spreading Behavior in the Antarctic Blue-Eyed Shag (*Phalacrocorax atriceps brandsfieldensis*)." *Auk* 99, no. 3 (July 1982): 588–89.

———. "Sexually Distinct Daily Activity Patterns of Blue-Eyed Shags in Antarctica." *Condor* 86 (1984): 151–56.

Berry, H. H. "The History of the Guano Platform on Bird Rock, Walvis Bay, South West Africa." *Bokmakierie* 27 (1975): 60–64.

———. "Physiological and Behavioural Ecology of the Cape Cormorant, *Phalacrocorax capensis*." *Madoqua* 9, no. 4 (1976): 5–55.

Bertrand, Sophie, Elisa Goya, and Jaime Silva. "Fishers and Seabirds Competing for the Same Fish: Foraging Strategies, Interactions and Consequences." Abstract. *IMARPE Boletin* 25, nos. 1, 2 (2010): 1–5.

Bevan, R. M., I. I. Boyd, P. Butler, K. Reid, A. Woakes, and J. Croxall. "Heart Rates and Abdominal Temperatures of Free-Ranging South Georgian Shags, *Phalacrocorax georgianus*." *Journal of Experimental Biology* 200 (1997): 661–75.

Boston Traveler. "Interesting from the Chincha Islands." *American Farmer* 6, no. 4 (June 1, 1854): 85–86.

Bowmaker, A. P. "Cormorant Predation on Two Central African Lakes." *Ostrich* 34, no. 1 (March 1963): 2–26.

Bregenballe, Thomas, Stephano Volponi, Mennobart R. van Eerden, et al. "Status of the Breeding Population of Great Cormorants *Phalacrocorax carbo* in the Western Palearctic in 2006." In *Proceedings 7th International Conference on Cormorants 2005*, edited by M. R. van Eerden, Stef van Rihn, and V. Keller (2011), 8–20.

Brown, Paul. "Cull 'Will Wipe Out Cormorants': RSPB Will Go to Court against Minister's Move to Help Anglers." *Guardian*, February 17, 2004. www.guardian.co.uk.

Brown, Tommy L., and Nancy A. Connelly. "Lake Ontario Sportfishing: Trends,

Analysis, and Outlook." Human Dimensions Research Unit, Cornell University, Series No. 09–3 (June 2009), 1–21.

Burger, A. E. "Functional Anatomy of the Feeding Apparatus of Four South African Cormorants." *Zoologica Africana* 13, no. 1 (1978): 81–102.

Carss, David N., ed. "Reducing the Conflict between Cormorants and Fisheries on a Pan-European Scale, Final Report: Summary." REDCAFE. Arberdeenshire: Natural Environment Research Council Centre for Ecology & Hydrology, 2003, 1–12.

Carss, David N., and Diet Assessment and Food Intake Working Group. "Techniques for Assessing Cormorant Diet and Food Intake: Towards a Consensus View." Wetlands International Cormorant Research Group (2007; posted online 2001). http://web.tiscalinet.it/sv2001/WI%20-%20CRSG/diet_foodintake.htm.

Casaux, R., M. Favero, P. Silva, and A. Baroni. "Sex Differences in Diving Depths and Diet of Antarctic Shags at the South Shetland Islands." *Journal of Field Ornithology* 72, no. 1 (Winter 2001): 22–29.

CBC News, Prince Edward Island. "Fishermen Want Cormorant Cull." CBC News, January 30, 2012. www.cbc.ca.

Clayton, Lawrence. "Chinese Indentured Labour in Peru." *History Today* 30 (June 1980): 19–23.

Cline, David R., and Eric Dornfeld. "The Agassiz Refuge Cormorant Colony." *Loon* 40, no. 3 (September 1978): 68–72.

Cocker, Mark, and Richard Mabey. "Cormorant Family." In *Birds Britannica*, 34–39. London: Chatto & Windus, 2005.

Coker, Robert E. "Habits and Economic Relations of the Guano Birds of Peru." *Proceedings of the United States National Museum* 56, no. 2298 (1920): 449–511.

———. "Peru's Wealth-Producing Birds." *National Geographic*, June 1920, 537–66.

———. "Regarding the Future of the Guano Industry and the Guano-Producing Birds of Peru." *Science* 28, no. 706 (July 10, 1908): 58–64.

———. "The Protection of Birds Made Profitable." *Science* 82, no. 2114 (July 5, 1935): 10–12.

Cook, Timothée R., Yves Cherel, and Yann Tremblay. "Foraging Tactics of Chick-Rearing Crozet Shags: Individuals Display Repetitive Activity and Diving Patterns over Time." *Polar Biology* 29, no. 7 (2006): 562–69.

Cook, Timothée R., and Guillaume LeBlanc. "Why Is Wing-Spreading Behaviour Absent in Blue-Eyed Shags?" *Animal Behaviour* 74 (2007): 649–52.

Cook, Timothée R., Amelie Lescroël, Yann Tremblay, and Charles-Andre Bost. "To Breathe or Not to Breathe? Optimal Breathing, Aerobic Dive Limit and Oxygen Stores in Deep-Diving Blue-Eyed Shags." *Animal Behavior* 76 (2008): 565–76.

Coonrod, Larry. "Cormorant Hazing." *South Lincoln County News*, May 9, 2012. www.southlincolncountynews.com.

Cooper, J. "Biology of the Bank Cormorant, Part 1: Distribution, Population Size, Movements and Conservation." *Ostrich* 52 (1980): 208–15.

Uncorrected Page Proof

Courtot, Karen N., Daniel D. Roby, Jessica Y. Adkins, Donald E. Lyons, D. Tommy King, and R. Scott Larsen. "Colony Connectivity of Pacific Coast Double-Crested Cormorants Based on Post-Breeding Dispersal from the Region's Largest Colony." *Journal of Wildlife Management* 76, no. 7 (September 2012): 1462–71.

Crawford, R. J. M., A. C. Cockcroft, B. M. Dyer, and L. Upfold. "Divergent Trends in Bank Cormorants *Phalacrocorax neglectus* Breeding in South Africa's Western Cape Consistent with a Distributional Shift of Rock Lobsters *Jasus lalandii*." *African Journal of Marine Science* 31 (2008): 161–66.

Crawford, R. J. M., and J. Cooper. "Bank Cormorant *Phalacrocorax neglectus*." In *Roberts' Birds of Southern Africa*. 7th ed. Edited by P. A. R. Hockey, W. R. J. Dean, and P. G. Ryan, 577–78. Cape Town: Trustees of the John Voelcker Bird Book Fund, 2005.

Crawford, R. J. M., S. Davis, R. Harding, L. Jackson, T. Leshoro, M. Meyer, R. Randall, et al. "Initial Impact of the *Treasure* Oil Spill on Seabirds off Western South Africa." *South African Journal of Marine Science* 22 (2000): 157–76.

Cushman, Gregory T. "'The Most Valuable Birds in the Word': International Conservation Science and the Revival of Peru's Guano Industry, 1909–1965." *Environmental History* 10, no. 3 (July 2005): 477–509.

Davis, Malcolm, and Herbert Friedmann. "The Courtship of the Flightless Cormorant." *Scientific Monthly* 42, no. 6 (June 1936): 560–63.

Decker, Daniel J., and Ken G. Purdy. "Toward a Concept of Wildlife Acceptance Capacity in Wildlife Management." *Wildlife Society Bulletin* 16, no. 1 (Spring 1988): 53–57.

De Ornellas, Kevin. "'Fowle Fowles?' The Sacred Pelican and the Profane Cormorant in Early Modern Culture." In *A Cultural History of Animals*. Vol. 3, *In the Renaissance*, edited by Bruce Boehrer, 27–50. New York: Berg, 2007.

Dorr, Brian S., Loren W. Burger, Scott C. Barras, and Kristina C. Godwin. "Double-Crested Cormorant Distribution on Catfish Aquaculture in the Yazoo River Basin of Mississippi." *Wildlife Society Bulletin* 36 (2012): 70–77.

Dorr, Brian S., Loren W. Burger, Scott C. Barras, and Kristina C. Godwin. "Economic Impact of Double-Crested Cormorant, *Phalacrocorax auritus*, Depredation on Channel Catfish, *Ictalurus punctatus*, Aquaculture in Mississippi, USA." *Journal of the World Aquaculture Society* 43, no. 4 (2012): 502–13.

Doucette, Jennifer L., Björn Wissel, and Christopher M. Somers. "Cormorant-Fisheries Conflicts: Stable Isotopes Reveal a Consistent Niche for Avian Piscivores in Diverse Food Webs." *Ecological Applications* 21, no. 8 (2011): 2987–3001.

Duffy, David Cameron, and W. Roy Siegfried. "Historical Variations in Food Consumption by Breeding Seabirds of the Humboldt and Benguela Upwelling Regions." In *Seabirds: Feeding Ecology and Role in Marine Ecosystems*, edited by J. P. Croxall, 327–46. Cambridge: Cambridge University Press, 1987.

Duffy, David C. "The Guano Islands of Peru: The Once and Future Management of a Renewable Resource." *Birdlife Conservation Series* 1 (1994): 68–76.

Dunn, Erica H. "Intake of Nestling Double-Crested Cormorants." *Auk* 92, no. 3 (July 1975): 553–65.

Egremont, Pamela, and Miriam Rothschild. "The Calculating Cormorants." *Biological Journal of the Linnean Society* 12 (September 1979): 181–86.

Estevez, Jordi, Ernesto Piana, Adrian Schiavini, and Nuria Juan-Muns. "Archaeological Analysis of Shell Middens in the Beagle Channel, Tierra del Fuego Island." *International Journal of Osteoarchaeology* 11 (2001): 24–33.

"Flightless Birds in Cargo for Zoo." *New York Times*, May 17, 1923, 14.

Fowler, W. M. W. "The Preparation and Cooking of a Cormorant . . ." In *Countryman's Cooking*. Ludlow, UK: Excellent Press, 2006, 30–31.

Glahn, James F., and D. Tommy King. "Bird Depredation." In *Biology and Culture of Channel Catfish*, edited by C. S. Tucker and J. A. Hargreaves, 503–29. Amsterdam: Elsevier B.V., 2004.

Glahn, J. F., D. S. Reinhold, and C. A. Sloan. "Recent Population Trends of Double-Crested Cormorants Wintering in the Delta Region of Mississippi: Responses to Roost Dispersal and Removal under a Recent Depredation Order." *Waterbirds* 23, no. 1 (2000): 38–44.

Green, Rhys E., Will J. Peach, and Norman Ratcliffe. "RSPB Comments on CSL's Revised Cormorant Cull Analysis." Royal Society for the Protection of Birds, March 31, 2005, 1–6.

Grémillet, David. "'Wing-Drying' in Cormorants." *Journal of Avian Biology* 26, no. 2 (1995): 176.

Griffiths, C. L., L. van Sittert, P. B. Best, A. C. Brown, B. M. Clark, P. A. Cook, R. J. M. Crawford, et al. "Impacts of Human Activities on Marine Animal Life in the Benguela: A Historical Overview." *Oceanography and Marine Biology: An Annual Revue* 42 (2004): 303–92.

Gudger, E. W. "Fishing with the Cormorant in Japan." *Scientific Monthly* 29, no. 1 (July 1929): 5–38.

Guillaumet, Alban, Brian Dorr, Guiming Wang, Jimmy D. Taylor 2nd, Richard B. Chipman, Heidi Scherr, Jeff Bowman, Kenneth F. Abraham, Terry J. Doyle, and Elizabeth Cranker. "Determinants of Local and Migratory Movements of Great Lakes Double-Crested Cormorants." *Behavioral Ecology* 22 (2011): 1096–1103.

Hanisch, Shauna L., and Paul R. Schmidt. "Resolving Double-Crested Cormorant *Phalacrocorax auritus* Conflicts in the United States: Past, Present, and Future." In *Waterbirds around the World*, eds. G. C. Boere, C. A. Galbraith, and D. A. Stroud (Edinburgh: Stationery Office, 2006): 826–28.

Hanson, Terrill R. "Catfish Farming in Mississippi." *Mississippi History Now*, Mississippi Historical Society (April 2006). mshistory.k12.ms.us/articles/217/catfishfarming-in-mississippi.

Harris, M. P. "Population Dynamics of the Flightless Cormorant *Nannopterum harrisi*." *Ibis* 121, no. 2 (1979): 135–46.

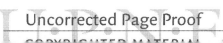

Harrop, Trevor (contact). "Press Release: Cormorant Campaign Goes to Parliament." Angling Trust, February 22, 2012. www.anglingtrust.net.

Harting, James Edmund. "Fishing with Cormorants." In *Essays on Sport and Natural History*, 423–40. London: Horace Cox, 1883.

Hatch, Jeremy J., and D. V. Weseloh. "Double-Crested Cormorant (*Phalacrocorax auritus*)." In *The Birds of North America*, edited by A. Poole and F. Gill, no. 441, 1–35. Philadelphia: Birds of North America, 1999.

Heydon, Matthew. "Licensing the Shooting of Cormorants to Prevent Serious Damage to Fisheries and Inland Waters." Natural England–Wildlife Management and Licensing (January 14, 2008), 1–5.

Ho, Erling. "Flying Fishes of Wucheng." *Natural History* 107, no. 8 (October 1988): 66–71.

Hobson, Keith A., Richard W. Knapton, and Walter Lysack. "Population, Diet and Reproductive Success of Double-Crested Cormorants Breeding on Lake Winnipegosis, Manitoba, in 1987." *Colonial Waterbirds* 12, no. 2 (1989): 191–97.

Hull, Robin. "Shag" and "Cormorant." *Scottish Birds: Culture and Tradition*, 96–98. Edinburgh: Mercat Press, 2001.

Jackson, Christine E. "Fishing with Cormorants." *Archives of Natural History* 24, no. 2 (1997): 189–211.

Jobaneck, George A., and David B. Marshall. "John K. Townsend's 1836 Report of the Birds of the Lower Columbia River Region, Oregon and Washington." *Northwestern Naturalist* 73, no. 1 (Spring 1992): 1–14.

Joseph, Lawrence E. "Man and Cormorant." *Audubon* 88 (May 1986): 38–41.

Kasuya, Shiro, Soichi Murase, and Yuichi Miyano. "Destructive Effects of the Estuary Dam on the Nagara River's Environment, and the Program for Its Regeneration." *Bulletin of the Faculty of Regional Studies, Gifu University* 20 (2007): 1–22.

Kato, A., J. P. Croxall, Y. Watanuki, and Y. Naito. "Diving Patterns and Performance in Male and Female Blue-Eyed Cormorants *Phalacrocorax atriceps* at South Georgia." *Marine Ornithology* 19 (1992): 117–29.

Kennedy, Martyn, Carlos A. Valle, and Hamish G. Spencer. "The Phylogenetic Position of the Galápagos Cormorant." *Molecular Phylogenetics and Evolution* 53 (2009): 94–98.

King, D. Tommy, Bradley F. Blackwell, Brian S. Dorr, and Jerrold L. Belant. "Effects of Aquaculture on Migration and Movement Patterns of Double-Crested Cormorants." *Human-Wildlife Conflicts* 4, no. 1 (Spring 2010): 77–86.

Kirk-Smith, Anna. "Easel Words: The Unfortunate Repercussions of Discovery and Survival.'" *Jackdaw* 101 (January/February 2012): 21.

Klimkiewicz, M. Kathleen, and Anthony G. Futcher. "Longevity Records of North American Birds, Supplement 1." *Journal of Field Ornithology* 60, no. 4 (Autumn 1989): 469–94.

Kline, John (office of). "John Kline to Testify on His Bill to Allow States to Manage Menacing Cormorant Overpopulation" (March 28, 2012). http://kline.house.gov.

Knauer, Elfriede R. "Fishing with Cormorants: A Note on Vittore Carpaccio's *Hunting on the Lagoon.*" *Apollo* 158, no. 499 (September 2003): 32–39.

Johnson, Cuthbert W. "On Guano." *Farmer's Magazine* 7, no. 3 (March 1843): 170–74.

Johnson, James H., Russell D. McCullough, and James Farquhar. "Double-Crested Cormorant Studies at Little Galloo Island, Lake Ontario in 2010: Diet Composition, Fish Consumption and the Efficacy of Management Activities in Reducing Fish Predation." NYSDEC Lake Ontario Annual Report (2010): section 14, 1–12.

Lambert, Robert A. "Seabird Control and Fishery Protection in Cornwall, 1900–1950." *British Birds* 96, no. 1 (January 2003): 30–34.

Laufer, Berthold. "The Domestication of the Cormorant in China and Japan." *Field Museum of Natural History No. 300: Anthropological Series* 18, no. 3 (1931): 201–62.

Lee, Richard. "These Birds Must Be Killed." *Angling Times*, no. 2266, December 4, 1996, 1, 8–9.

Léger, Claire, and Raymond McNeil. "Nest Attendance and Care of Young in Double-Crested Cormorants." *Colonial Waterbirds* 8, no. 2 (1985): 96–103.

Line, Les. "A Conflict of Cormorants." *Wildlife Conservation*, February 2002, 44–51.

Lyons, Donald. "Bioenergetics-Based Predator-Prey Relationships between Piscivorous Birds and Juvenile Salmonids in the Columbia River Estuary." PhD diss., Oregon State University, 2010.

Lyons, Donald, Daniel Roby, and Ken Collis. "Foraging Patterns of Caspian Terns and Double-Crested Cormorants in the Columbia River Estuary." *Northwest Science* 81, no. 2 (2007): 91–103.

MacKay, Barry Kent, and Liz White. "A Critical Analysis of Point Pelee National Park's Rationale for Killing the Middle Island Cormorants." Cormorant Defenders International (February 2008), 1–93.

Manzi, Maya, and Oliver T. Coomes. "Cormorant Fishing in Southwestern China: A Traditional Fishery under Siege." *American Geographical Society* 92, no. 4 (October 2002): 597–603.

Mathew, W. M. "A Primitive Export Sector: Guano Production in Mid-Nineteenth-Century Peru." *Journal of Latin American Studies* 9, no. 1 (May 1977): 35–57.

Matthews, L. H. "The Birds of South Georgia." In *The Discovery Reports*, vol. 1, 561–92. Cambridge: Cambridge University Press, 1929.

McCullough, Russell D., James F. Farquhar, and Irene M. Mazzocchi. "Cormorant Management Activities in Lake Ontario's Eastern Basin." NYSDEC Lake Ontario Annual Report (2010): section 13, 1–8.

McGrath, Susan. "Shoot-out at Little Galloo." *Smithsonian*, February 2003, 72–78.

McIlwraith, Thomas. "Family Phalacrocoracidæ: Cormorants." In *The Birds of Ontario*, 46–47. Hamilton, ON: A. Lawson, 1886.

McNab, Brian K. "Energy Conservation and the Evolution of Flightlessness in Birds." *American Naturalist* 144, no. 4 (October 1994): 628–42.

Mercer, Dacey. "Phylogeography and Population Genetic Structure of Double-Crested Cormorants (*Phalacrocorax auritus*)." PhD diss., Oregon State University, 2008.

Selected Bibliography

Moodie, Jim. "Cormorant Nests Dwindle on Huron Due to Natural Ebb and Vigilante Culls: MNR Survey Shows Nine Colonies 'Shot Up' on Georgian Bay, N. Channel." *Manitoulin Expositor*, March 19, 2008. www.manitoulin.ca.

Murphy, Robert Cushman. "Notes on American Subantarctic Cormorants." *Bulletin of the American Museum of Natural History* 35 (1916): 31–48.

Nemtzov, S. C. "Relocation of Pygmy Cormorants *Phalacrocorax pygmeus* Using Scare Tactics to Reduce Conflict with Fish Farmers in the Bet She-an Valley, Israel." *Conservation Evidence* 2 (2005): 3–5.

Newson, Stuart E., Graham R. Ekins, John H. Marchant, Mark M. Rehfisch, and Robin M. Sellers. "The Status of Inland and Coastal Breeding Great Cormorants *Phalacrocorax carbo* in England," BTO Research Report, no. 443 (April 2006), 1–36.

Newson, Stuart E., Baz Hughes, Ian C. Russell, Graham R. Ekins, and Robin M. Sellers. "Sub-specific Differentiation and Distribution of Great Cormorants *Phalacrocorax carbo* in Europe." *Ardea* 92, no. 1 (2005): 3–10.

New York State Department of Environmental Conservation Bureau of Fisheries and USGS, Biological Resources Division. "Final Report: To Assess the Impact of Double-Crested Cormorant Predation on the Smallmouth Bass and Other Fishes of the Eastern Basin of Lake Ontario" (February 1, 1999). 12 sections.

O'Connor, Helene. "Liam O'Flaherty: Literary Ecologist." *Éire-Ireland* 7, no. 2 (Summer 1972): 47–54.

Ogilvie-Grant, W. R. "Phalacrocoracidæ." In *Catalogue of the Birds of the British Museum*, vol. 26, 330–410. London: Order of the Trustees of the British Museum (Natural History), 1898.

Olinger, John Peter. "The Guano Age in Peru." *History Today*, June 1980, 13–18.

Ontario Parks. "Annual Report on the Management of Double-Crested Cormorants on Presqu'ile Provincial Park," 1–42. Queens Printer for Ontario (2006).

———. "Resource Management Implementation Plan High Bluff and Gull Islands," 1–65. Queen's Printer for Ontario (2011).

Palacios, Eduardo, and Eric Mellink. "Nesting Waterbirds on Islas San Martín and Todos Santos." *Western Birds* 31 (2000): 184–89.

Palmer, Henry Spencer. "Cormorant Fishing in Japan." In *Letters from the Land of the Rising Sun*, 167–75. Yokohama: Japan Mail, 1894.

Parsons, N. J., and L. G. Underhill. "Oiled and Injured African Penguins *Spheniscus demersus* and Other Seabirds Admitted for Rehabilitation in the Western Cape, South Africa, 2001 and 2002." *African Journal of Marine Science* 27, no. 1 (2005): 289–96.

G.W.P. [Peck, George Washington]. "From the Chincha Islands." *New York Times*, January 7, 1854, 2.

Pimm, Stuart, Peter Raven, Alan Peterson, and Paul R. Ehrlich. "Human Impacts on the Rates of Recent, Present, and Future Bird Extinctions." *National Academy of Sciences of the USA* 103 (July 18, 2006): 10941–46.

Quintana, Flavio, Rory P. Wilson, and Pablo Yorio. "Dive Depth and Plumage Air

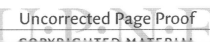

in Wettable Birds: The Extraordinary Case of the Imperial Cormorant." *Marine Ecology Progress Series* 334 (2007): 299–310.

Rasmussen, Pamela C., and Philip S. Humphrey. "Wing-Spreading in Chilean Blue-Eyed Shags (*Phalacrocorax atriceps*)." *Wilson Bulletin* 100, no. 1 (March 1988): 140–44.

Revkin, Andrew C. "A Slaughter of Cormorants in Angler Country." *New York Times,* August 1, 1998, A1, B5.

Ribak, Gal, Daniel Weihs, and Zeev Arad. "How Do Cormorants Counter Buoyancy during Submerged Swimming?" *Journal of Experimental Biology* 207 (2004): 2101–14.

Rijke, A. M. "The Water Repellency and Feather Structure of Cormorants, Phalacrocoracidae." *Journal of Experimental Biology* 48 (1968): 185–89.

Roberts, T. S. "Double-Crested Cormorant." In *Birds of Minnesota,* 1:166–72. Minneapolis: University of Minnesota Press, 1932.

Robbins, William G. "The World of Columbia River Salmon: Nature, Culture, and the Great River of the West." In *The Northwest Salmon Crisis: A Documentary History,* edited by Joseph Cone and Sandy Ridlington, 2–24. Corvallis: Oregon State University Press, 1996.

Roby, Daniel D., and Ken Collis (principal investigators). "Research, Monitoring, and Evaluation of Avian Predation on Salmonid Smolts in the Lower and Mid-Columbia River: Final 2011 Report," 1–91, appendices. U.S. Geological Survey–Oregon Cooperative Fish and Wildlife Research Unit and Real Time Research Inc. (August 2012).

Romero, Simon. "Peru Guards Its Guano as Demand Soars Again." *New York Times,* May 30, 2008, A6, A10.

Roselaar, C. S. "An Inventory of Major European Bird Collections." *Bulletin of the British Ornithologists' Club* 123A (2003): 253–337.

Rosenberg, Daniel K., Carlos A. Valle, Malcolm C. Coulter, and Sylvia A. Harcourt. "Monitoring Galapagos Penguins and Flightless Cormorants in the Galapagos Islands." *Wilson Bulletin* 102, no. 3 (September 1990): 525–32.

Ross, Robert M., and James H. Johnson. "Fish Losses to Double-Crested Cormorant Predation in Eastern Lake Ontario, 1992–97." Symposium on Double-Crested Cormorants: Population Status and Management Issues in the Midwest, Technical Bulletin 1879, edited by Mark E. Tobin (USDA-APHIS, 1999), 61–70.

Rothschild, Walter. "*Carbo perspicillatus.*" In *Extinct Birds,* 89. London: Hutchinson & Co., 1907.

Sauer, J. R., J. E. Hines, J. E. Fallon, K. L. Pardieck, D. J. Ziolkowski Jr., and W. A. Link. "Double-Crested Cormorant *Phalacrocorax auritus*: BBS Trend Map." North American Breeding Bird Survey Results and Analysis 1966–2010, version 12.07.2011. USGS–Patuxet Wildlife Research Center (Laurel, MD: rev. April 16, 2012). www .mbr-pwrc.usgs.gov/bbs/tr2010/tr01200.htm.

Schiavini, A. C. M., P. Yorio, and E. Frere. "Distribución reproductiva y abundancia de las aves marinas de la Isla Grande de Tierra del Fuego, Isla de los Estados e Islas de

Uncorrected Page Proof

Año Nuevo (Provincia de Tierra del Fuego, Antártida e Islas del Atlántico Sur)." In *Atlas de la distribucion reproductiva de aves marinas en el litoral Patagonico Argentino*, edited by Pablo Yorio et al., 179–213. Puerto Madryn: Fundación Patagonia Natural, 1998.

Sharpe, R. Bowdler, ed. "The Hon. Walter Rothschild Sent . . ." *Bulletin of the British Ornithologists' Club* 7, no. 54 (May 25, 1898): 51–54.

Shaw, P. "Factors Affecting the Breeding Performance of Antarctic Blue-Eyed Shags *Phalacrocorax atriceps*." *Ornis Scandinavica* 17, no. 2 (May 1986): 141–50.

Sibley, Charles G., and Jon E. Ahlquist. "The Totipalmate Swimmers: Traditional Order Pelecaniformes." In *Phylogeny and Classification of Birds: A Study in Molecular Evolution* (New Haven, CT: Yale University Press, 1990), 491–503.

Sibley, Charles G., and Burt L. Monroe. "Superfamily Phalacrocoracoidea." In *Distribution and Taxonomy of Birds of the World* (New Haven, CT: Yale University Press, 1990): 299–302.

Siegel-Causey, Douglas. "Phylogeny of the Phalacrocoracidae." *Condor* 90 (1988): 885–905.

Siegel-Causey, Douglas, Christine Lefevre, and Arkadii B. Savinetskii. "Historical Diversity of Cormorants and Shags from Amchitka Island, Alaska." *Condor* 93 (1991): 840–52.

Snow, Barbara K. "Observations on the Behaviour and Ecology of the Flightless Cormorant *Nannopterum harrisi*." *Ibis* 108, no. 2 (April 1966): 265–80.

Somers, C. M., V. A. Kjoss, F. A. Leighton, and D. Fransden. "American White Pelicans and Double-Crested Cormorants in Saskatchewan: Population Trends over Five Decades." *Blue Jay* 68, no. 2 (June 2010): 75–86.

"Spontaneous Combustion of Guano." *Chemist* 3, no. 26 (February 1845): 78.

Stejneger, Leonhard. "Family Phalacrocoracidae: Results of Ornithological Explorations in the Commander Islands and in Kamtschatka." *Bulletin No. 29 of the United States National Museum* 39:180–91. Washington, DC: Government Printing Office, 1885.

———. "Contributions to the History of the Commander Islands: No. 1, Notes on the Natural History, Including Descriptions of New Cetaceans." *Proceedings of the United States National Museum* 6, no. 5 (1883): 58–89.

Stejneger, Leonhard, and Frederic A. Lucas. "Contributions to the Natural History of the Commander Islands." *Proceedings of the United States National Museum* 12:83–94. Washington, DC: Government Printing Office, 1890.

Takahashi, Tetsumi, Kayoko Kameda, Megumi Kawamura, and Tsuneo Nakajima. "Food Habits of Great Cormorant *Phalacrocorax carbo hanedae* at Lake Biwa, Japan, with Special Reference to Ayu *Plecoglossus altivelis altivelis*." *Fisheries Science* 72, no. 3 (2006): 477–84.

Taverner, P. A. "The Double-Crested Cormorant (*Phalacrocorax auritus*) and Its Relation to the Salmon Industries on the Gulf of St. Lawrence." *Museum Bulletin* no. 13 / Biological Series, no. 5 (April 30, 1915): 1–24.

Townsend, Charles Haskins. "The Flightless Cormorant in Captivity." *Auk* 46, no. 2 (April 1929): 211–13.

Tremblay, Y., T. R. Cook, and Y. Cherel. "Time Budget and Diving Behaviour of Chick-Rearing Crozet Shags." *Canadian Journal of Zoology* 83 (2005): 971–82.

Tyson, Laura A., Jerrold L. Belant, Francesca J. Cuthbert, and D. V. (Chip) Weseloh. "Nesting Populations of Double-Crested Cormorants in the United States and Canada." Symposium on Double-Crested Cormorants: Population Status and Management Issues in the Midwest. Technical Bulletin 1879, edited by Mark E. Tobin, 17–25. USDA-APHIS (1999).

U.S. Army Corps of Engineers, Portland District. "Draft Environmental Assessment: Double-Crested Cormorant Dissuasion Research on East Sand Island in the Columbia River Estuary" (February 2, 2012), i–iii, 10–57.

U.S. Department of Agriculture, Wildlife Services. *Living with Wildlife: Cormorants* (pamphlet, n.d.), 1–4.

U.S. Fish and Wildlife Service. "Final Environmental Impact Statement: Double-Crested Cormorant Management in the United States," 1–165. Arlington, VA, 2003.

———. "Final Environmental Assessment: Extended Management of Double-Crested Cormorants," 1–50. Arlington, VA, 2009.

Valle, Carlos A. "The Flightless Cormorant: The Evolution of Female Rule." In *Galápagos: Preserving Darwin's Legacy*, edited by Tui De Roy, 162–69. Buffalo, NY: Firefly Books, 2009.

Van Patten, Peg. "Aqua Kids Discover That South Dumpling Island Is for the Birds." *Wrack Lines* 11, no. 1 (Spring/Summer 2011): 11–13.

Van Rijn, Stef, and Mennobart R. van Eerden. "Cormorant Research Group Bulletin," IUCN-Wetlands International, no. 7 (August 2011), 1–56.

Van Tets, G. F. "Australia and the Origin of Shags and Cormorants, Phalacrocoracidae." *Proceedings of the 16th International Ornithological Congress* (1974), 121–24.

Velando, Alberto, David Álvarez, Jorge Mouriño, Francisco Arcos, and Álvaro Barros. "Population Trends and Reproductive Success of the European Shag *Phalacrocorax aristotelis* on the Iberian Peninsula following the *Prestige* Oil Spill." *Journal of Ornithology* 146 (2005): 116–20.

Voorbergen, Anne, Willem F. De Boer, and Les G. Underhill. "Natural and Human-Induced Predation on Cape Cormorants at Dyer Island." *Bird Conservation International* 22 (2012): 82–93.

Wanless, S., T. Corfield, M. P. Harris, S. T. Buckland, and J. A. Morris. "Diving Behaviour of the Shag *Phalacrocorax aristotelis* (Aves: Pelecaniformes) in Relation to Water Depth and Prey Size." *Journal of Zoology* (London) 231 (1993): 11–25.

Wanless, S., and M. P. Harris. "Use of Mutually Exclusive Foraging Areas by Adjacent Colonies of Blue-Eyed Shags (*Phalacrocorax atriceps*) at South Georgia." *Colonial Waterbirds* 16, no. 2 (1993): 176–82.

Wanless, S., M. P. Harris, and J. A. Morris. "Diving Behaviour and Diet of the Blue-Eyed Shag at South Georgia." *Polar Biology* 12 (1992): 713–19.

Watanuki, Yutaka, Francis Daunt, Akinori Takhashi, Mark Newell, Sarah Wanless, Katsufumi Sato, and Nobuyuki Miyazaki. "Microhabitat Use and Prey Capture of a Bottom-Feeding Top Predator, the European Shag, Shown by Camera Loggers." *Marine Ecology Progress Series* 356 (2008): 283–93.

Watson, Arthur C. "The Guano Islands of Southwestern Africa." *Geographical Review* 20, no. 4 (October 1930): 631–41.

Weimerskirch, Henri, Sophie Bertrand, Jaime Silva, Jose Carlos Marques, and Elisa Goya. "Use of Social Information in Seabirds: Compass Rafts Indicate the Heading of Food Patches." *PLoS ONE* 5, no. 3 (2010): 1–8.

Weseloh, D. V., and B. Collier. "The Rise of the Double-Crested Cormorant on the Great Lakes: Winning the War against Contaminants." Great Lakes Fact Sheet. Environment Canada, Ontario Region. Catalogue No. EN 40–222/2–1995E (1995), 1–12.

Weseloh, D. V., and J. Casselman. "Calculated Fish Consumption by Double-Crested Cormorants in Eastern Lake Ontario." Abstract. *Colonial Waterbird Society Bulletin* 16, no. 2 (1992): 63–64.

Wetlands International Cormorants Research Group. "Cormorants in the Western Palearctic." Leaflet, 2008, 1–2.

Wetmore, A. "The Amount of Food Consumed by Cormorants." *Condor* 29 (November 1927): 273–74.

White, Craig R., Norman Day, Patrick J. Butler, and Graham R. Martin. "Vision and Foraging in Cormorants: More Like Herons Than Hawks." *PloS ONE* 7 (July 2007): 1–6.

White, Craig R., Graham R. Martin, and Patrick J. Butler. "Wing-Spreading, Wing-Drying, and Food-Warming in Great Cormorants *Phalacrocorax carbo.*" *Journal of Avian Biology* 39, no. 5 (September 2008): 576–78.

Whitehead, Chas. S. "Loon Lake." *Ornithologist and Oölogist* 12, no. 2 (February 1887): 21–22.

Wilson, Rory P., F. Hernán Vargas, Antje Steinfurth, Philip Riordan, Yan Ropert-Coudert, and David W. Macdonald. "What Grounds Some Birds for Life? Movement and Diving in the Sexually Dimorphic Galápagos Cormorant." *Ecological Monographs* 78, no. 4 (2008): 633–52.

Wires, Linda R., and Francesca J. Cuthbert. "Historic Populations of the Double-Crested Cormorant (*Phalacrocorax auritus*): Implications for Conservation and Management in the 21st Century." *Waterbirds* 29, no. 1 (2006): 9–37.

Wires, L. R., F. J. Cuthbert, D. R. Trexel, and A. R. Joshi. "Status of the Double-Crested Cormorant (*Phalacrocorax auritus*) in North America." Final Report to USFWS. St. Paul: University of Minnesota Department of Fish and Wildlife, 2001.

Yamashita, Zempei. "Cormorant Fishing on the Nagara River." *Oceanus* 30, no. 1 (Spring 1987): 83–85.

UNPUBLISHED JOURNALS

Crowell, Nelson. Journal, May 8, 1852, through July 18, 1853. Log 850. Mystic Seaport Museum, Mystic, CT.

———. Journal, March 16, 1855, through November 20, 1856. Log 893. Mystic Seaport Museum, Mystic, CT.

Fish, Simeon G. Journal, June 4, 1856, through October 30, 1857. Mystic Seaport Museum, Mystic, CT.

WEBSITES

"Bank Cormorant, *Phalacrocorax neglectus.*" Michelle Du Toit, Avian Demography Unit, University of Cape Town. March 30, 2004. http://web.uct.ac.za/depts/stats/adu/species/bankcormorant.htm.

BirdLife International. "Phalacrocoracidae." IUCN Red List of Threatened Species. www.iucnredlist.org/apps/redlist/search.

Bird Research Northwest. Real Time Research Inc. www.birdresearchnw.org/.

"The Birds of North America Online." Edited by A. Poole. Cornell Lab of Ornithology. 2012. http://bna.birds.cornell.edu/bna.

"Cormorant Defenders International." www.zoocheck.com/cormorant/.

"Cormorants." IUCN-Wetlands International Cormorant Research Group. Stefano Volponi (webmaster). Updated July 7, 2012. http://cormorants.freehostia.com/index.htm.

"EU Cormorant Platform." European Commission, Environment Directorate General and CorMan. February 22, 2012. http://ec.europa.eu/environment/nature/cormorants/home_en.htm.

"Spectacled Cormorant *Phalacrocorax perspicillatus* Pallas, 1811." 2005. National Museum of Natural History Naturalis, Leiden. http://nlbif.eti.uva.nl/naturalis/detail.php?lang=uk&id=61.

"Ukai: Cormorant Fishing on the Nagara River." 2012. Gifu Convention and Visitors Bureau, Gifu, Japan. www.gifucvb.or.jp/en/01_sightseeing/01_01.html.

FILMS AND VIDEOS

"Birds of Dumpling Island." Aqua Kids. Posted online November 14, 2011. www.youtube.com.

Collyns, Dan (reporter). "How a Peruvian Island Is Making Money from Bird Poo." BBC News, September 2, 2010. www.bbc.co.uk/news/business-11156842.

Conil, Chris. "Christian Funnell: Cormorant." August 16, 2012. www.youtube.com.

"Cormorant Watch." 2011. Angling Trust. www.cormorantwatch.org/video.html.

Dawkins, Richard. "Vestigial Organs: Wings of the Flightless Cormorant." Richard Dawkins Foundation. May 5, 2010. www.youtube.com.

"Deep Dive." Wildlife Conservation Society. July 31, 2012. www.wcs.org/news-and-features-main/cormorant-deep-sea-dive.aspx.

Fougea, Frédéric (director/cowriter) and Zhu Xiao Ling (cowriter). *He Dances for His Cormorants*. Boréales Production, 1993.

"Lockheed Martin MPUAV Cormorant." Lockheed Martin Corp., 2008. www.youtube.com.

Markham, Peter (director), and Peter Ransley (screenplay). *The Cormorant*. BBC television, 1993. www.youtube.com.

Postgate, Oliver, Peter Firmin, and Smallfilms. *Noggin and the Ice Dragon*. BBC television, c. 1982. www.youtube.com.

Sarigianopoulus, Rena (reporter). "Calls for Cormorant Control on Lake Waconia." KARE 11 (NBC), October 19, 2011. www.kare11.com.

Satchell, Graham (reporter). "Thousands Sign Petition Calling for Cormorant Cull." BBC News, February 22, 2012. www.bbc.co.uk/news/science-environment-17123858.

Weir, Peter (director and screenwriter). *Master and Commander: The Far Side of the World*. Twentieth Century Fox, 2003.

Permissions & Credits

A few portions of the book have been published previously in a different form in *The Log of Mystic Seaport* 55 (2004); *Maritime Life and Traditions* 25 and 31 (2004 and 2006); *Beyond the Anchoring Grounds: More Cross-Currents in Irish and Scottish Studies* (2005); *Natural History* 118 (2009); and *Wracklines* (Fall/Winter 2012).

I am grateful to the Berkshire Taconic Community Foundation for permission to publish Amy Clampitt's "The Cormorant in Its Element," and to Jack Schmitt to publish his translation of Pablo Neruda's "Guanay Cormorant." Robinson Jeffers's "Birds and Fishes" is copyright 1987, Jeffers Literary Properties; published with the permission of Stanford University Press, sup.org.

Kind permission and credit for the images are as follows. Figures 1 and 3: courtesy of Williams College; figure 2: courtesy of Kazuto Hino and the Gifu Convention and Visitors Bureau; figures 4, 5, 11, 14, 17, 18, and 21: author's photograph or personal collection; figures 6, 7, and 12: courtesy of Connecticut College; figure 8: Ralph Lee Hopkins, National Geographic Stock; figure 9: Daniel Roby, Bird Research Northwest; figure 10: Adam Peck-Richardson, Bird Research Northwest; figure 13: courtesy Angling Trust, UK; figure 15: Ingo Arndt, Minden Pictures, National Geographic Stock; figure 16: Heather Carr (heatherunderground.com); figure 19: courtesy Gebr. Mann Verlag; figure 20: courtesy Nola Parsons, SANCCOB; figure 22: Chris Ross, Aurora Collection, Getty Images. Susan Schnur created the world map of cormorant distribution. Thanks to Brian Andrews, Susan Schnur, and Linda Wires for their assistance with the North American map. Audubon's "Florida Cormorant" on the cover is courtesy Lilly Library, Indiana University, Bloomington, Indiana.